DEBURRING

CAPABILITIES AND LIMITATIONS

Written and Edited By:

LaRoux K. Gillespie

Library of Congress Catalog Card Number: 76-47179
International Standard Book Number: 0-87263-038-2

PREFACE

When Eli Whitney contracted to produce 10,000 muskets in 1798, he introduced two major manufacturing concepts familiar to each of us:

The concept of mass production,

The concept of an unending pile of parts waiting for hand deburring.

Through the persistent and innovative efforts of thousands since that time, mass production concepts have evolved to a level that even today's public cannot always grasp. Yet if Eli Whitney walked in the door of most manufacturing plants today there is one area he would instantly recognize--the "burr benches."

Those involved in manufacturing know one all too obvious fact--burrs presently cost us a large sum of money. Manufacturers in the United States annually spend over two billion dollars to remove burrs and radius edges on machined components. This represents roughly 2 1/2% of the cost of component parts. If detailed cost figures were available, the actual cost to consumers would probably be much higher.

Despite these high costs and the universality of the problem, many of the processes used for deburring are not well known. This publication is designed to fill that void by describing the capabilities of each of the available deburring processes.

While finding better deburring processes or improvements in existing processes can result in major reductions in deburring costs, there are four other approaches which can be equally effective. These include analyzing the real product needs, practicing burr prevention, minimizing burrs, and better utilizing machine operator's time. Since the minimization of deburring costs is generally a more desirable goal than just finding another, less expensive process, all these facets of burr related cost reductions are described.

This material has been abstracted from papers presented at 1974 and 1975 SME conferences on deburring, as well as from a review of over 1900 references on burr related subjects. Excerpts taken from several papers have been reassembled under the appropriate topical headings so specific credits cannot be referenced. The author wishes to express his gratitude for the material utilized by extending credit to all who have contributed. Each of the following individuals have contributed to this publication.

Ajiouny, Nad S. - Bendix Corporation

Bateman, James E. - Anderson Corporation
Brandt, Jerry N. - King-Seeley Thermos Company

Campbell, William D. - Ford Motor Company

Gillespie, LaRoux K. - Bendix Corporation
Golden, Richard L. - Bendix Corporation

Harbin, Raymond - Centri-Spray Corporation
Hasegawa, Yosnio - Osaka University (Japan)
Hignett, J. Bernard - Harper Buffing Machine Company

Kittredge, John B. - Roto-Finish Company

Matsunaga, M. - University of Tokyo (Japan)
Montag, A. C. - Surftran Company
Morris, Eric - Herbert Machine Tools (England)

Narins, Robert R. - Deburring Laboratories Incorporated

Perry, Winfield B. - Dynetics Corporation

Rhoades, Lawrence J. - Extrude Hone Corporation

Schafer, Friederich - University of Stuttgart (Germany)
Shimizu, Akihiko - IJR Incorporated (Japan)
Siegel, Bruno R. - Poligrat Corporation GMBH (Germany)
Siwert, David E. - Extrude Hone Corporation
Stackhouse, John - Dynetics Corporation

West, Richard F. - Osborn Manufacturing Company

Yuki, Akiyasu - Oita University (Japan)

Azima, Shigeo - Shizuoka University (Japan)

CONTENTS

CHAPTER 1

BASIC APPROACHES TO MINIMIZING DEBURRING COSTS

A value engineering study of burrs and deburring reveals five distinct approaches to minimizing deburring costs:

1. Improve product design
2. Prevent burrs
3. Minimize burr properties
4. Remove burrs during the machining cycle
5. Develop or obtain better deburring processes

Through subtle or functional design changes, it is possible to design many parts which do not require deburring. This is obviously one of the most desirable approaches to reducing costs since, if done correctly, it does not require any additional metalworking equipment or operations. In addition, it reflects the fact that the item was thoughtfully designed and engineered.

In some instances it is possible to produce parts without producing burrs. While this approach eliminates deburring costs, it also may increase machining costs or cause other problems.

Minimizing the size of burrs produced in machining operations can greatly decrease deburring costs. As an example, a burr only 0.0001 inch thick by 0.0001 inch high (2.54 μm x 2.54 μm) can be easily removed by any process without affecting size or surface finishes. In contrast a burr 0.010 inch thick by 0.050 inch high (0.25 mm x 1.25 mm) can only be removed in most deburring processes by using long cycles and also sacrificing some workpiece dimensions.

A number of techniques are available to remove large burrs during the machining cycle. This operation can sometimes be performed while the primary cutter is also cutting or as a finishing step in the machining operation. In many cases, machine operators can remove the large part of the burr while he is waiting for the machine to finish its cycle.

Developing new processes, improving existing processes, or purchasing new types of deburring equipment should only be done after each of the previous four approaches has been fully explored. While it is easy to buy off-the-shelf deburring machines, their use requires additional floor space, capital investment, labor, and flow time which might not be necessary. In addition, each of the deburring processes and machines today have their own limitations which create secondary problems.

Each of these approaches is described more fully in subsequent chapters. In analyzing each of these approaches, it is essential to remember that the overriding goal generally is to minimize costs.

Although literature on deburring costs is practically nonexistent, many of the major elements are obvious. In addition to the direct costs, there are several more subtle cost elements which should also be considered. These include:

- .Costs associated with rejection from inspection
- .Costs due to unit failures caused by burrs
- .Costs of deburring during machine cycle
- .Costs resulting from cut hands due to burrs

Every time parts are rejected from inspection departments, manufacturing costs increase. If this rejection is because of the presence of unallowable burrs, the filling out of rejection paperwork is an added hidden cost of burrs. The rework required to remove the burrs and reinspect the parts is a real, although, hidden, cost. The time required by foremen or engineers to decide what to do with these burr related rejects is another real cost hidden in overhead. Warrenty costs because of damage caused by loose or sharp burrs is a very real cost in the automotive industry. Although machine operators can often remove burrs on their machine more economically than in a separate deburring operation, it is essential to remember that the time spent is still a real cost which should be considered in any calculations. Each time an employee has to leave his workpost because of a laceration from a burr, a cost is incurred. This cost includes the direct cost away from the workpost, the cost of medical attention, and the cost of reduced efficiency until the cut heals. While these costs may appear minute, over a year's time they can be measurable in a large organization. This is particularly true if in-house or governmental regulations require reports in writing of such safety-related events. Paperwork of any kind is expensive.

Summary

In summarizing this introductory section, it is important to point out that minimizing deburring costs does not necessarily mean obtaining different deburring equipment. While product designs may dictate burr-free edges, that may not require burr removal processes. Changing product design and preventing burrs are techniques that can be used. In many cases costs can be reduced by simply keeping burrs small. Removing burrs during the machining operations is also a widely used technique. The selection of better deburring processes is the most widely used approach, but new processes also demand additional expertise.

CHAPTER 2

PRODUCT DESIGN INFLUENCES ON DEBURRING

Deburring efforts can be minimized by two design approaches:

1. Design components with configurations which eliminate or minimize the need for deburring.

2. Specify the actual edge quality required.

Designing to Eliminate Deburring

Some components and assemblies can obviously operate adequately without deburring. The mechanisms in many children's moving toys for example need not be deburred. Sheet metal edges are often more aesthetic and trouble free if a rolled edge is produced. In this case deburring is not required on the hidden edge. Both of these examples are a direct result of design requirements. In the first case the designer somehow had to specify that edges could have burrs. In the second case the designer utilized the geometry of the part to reduce deburring. The majority of assemblies may not lend themselves to such obvious design changes. The point is, however, that if deburring can be eliminated from even one part in an assembly there is a consequent cost savings. Two common examples in which burr removal is not required are shown in Figures 2-1 and 2-2. Pins which are pressed into a hole often do not have to be entirely burr free. In Figure 2-2b the part was machined so that the burr was thrown into the shoulder relief. Since the burr does not interfere with part function and cannot escape from the relief, deburring is not required.

Recent research indicates that when a cutter passes over an edge having an angle much larger than 90 degrees little or no burr forms on the edge. Conversely when the edge angle is small (Figure 2-3) a large burr forms because there is no support for the metal being cut. Figure 2-4 illustrates the influence the workpiece edge angle has on burr size when a shell mill is used. As seen there no noticeable burr is produced when edge angles exceed 150 degrees. There is essentially no space left for material to bend into with these large angles. While the data presented in Figure 2-4 was taken from low carbon steel in an unhardened condition the basic phenomenon can be observed in all materials.

Figure 2-5 illustrates a practical example of how this phenomenon can be used in production. The left hand view illustrates the heavy burr (solid black) which occurred when a casting was faced with a mill to provide a locating surface. Because of the small angle ν and a 90 degree edge perpendicular to plane AA', a heavy web of material extended across the inner diameter of the part after milling. This could only be removed by hand or by adding a special trimming operation. By redesigning the

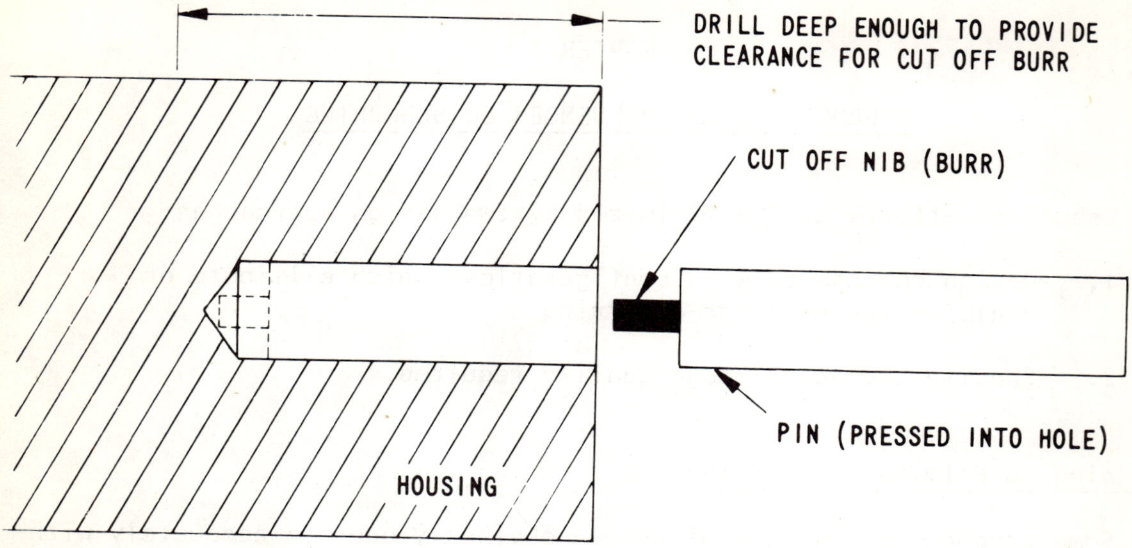

FIGURE 2-1: TAKE ADVANTAGE OF PART DESIGN TO MINIMIZE DEBURRING

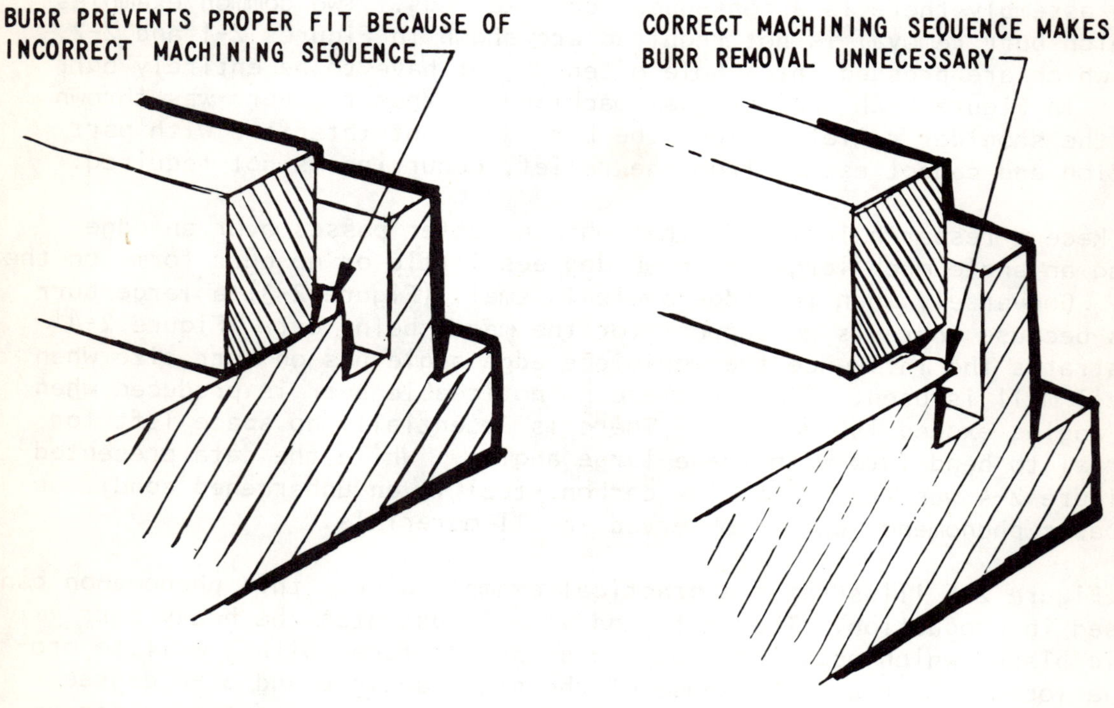

FIGURE 2-2: IMPACT OF BURR LOCATION

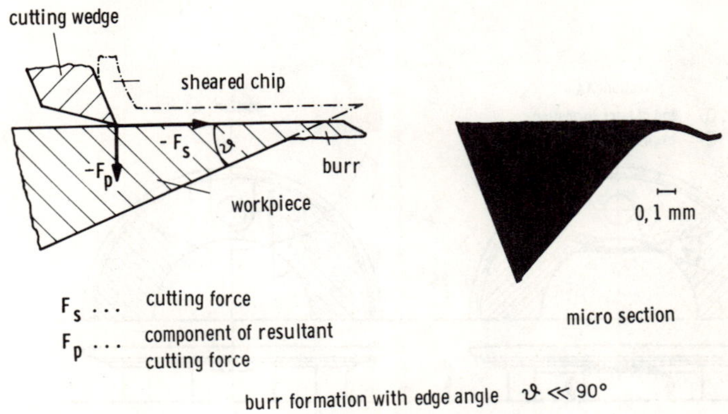

FIGURE 2-3: BURR FORMATION AT END OF CUT

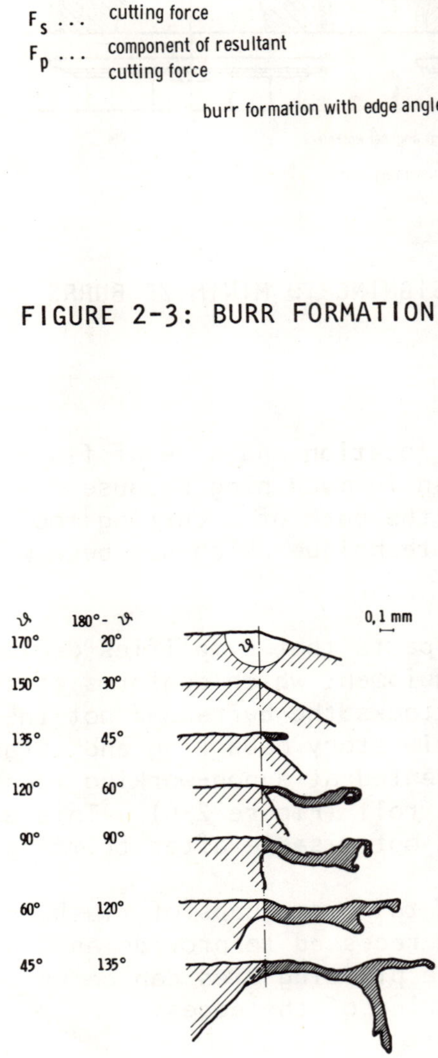

FIGURE 2-4: EFFECT OF EDGE ANGLE ON SIZE OF BURR AT END OF CUT

casting such that the cutter exited over a large angle ν a much smaller burr was produced which could be removed by abrasive jet deburring.

In reviewing Figure 2-5, it is obvious that by redesigning the casting such that a large draft angle is used on the boss surrounding the hole, the burr is further reduced in size. Thus in this particular example, burr size was minimized by using large angles in two planes and by eliminating some of the material which would have produced the burr.

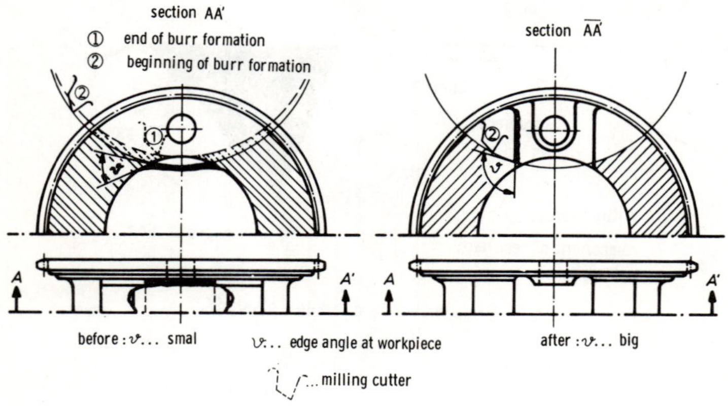

FIGURE 2-5: DESIGNING TO MINIMIZE BURRS

Designing to Minimize Deflashing

It is easier to predict the location and size of flash from die casting and molding processes than in machining because flash corresponds to die configuration rather than the path of a cutting tool. As a result designing to minimize flash is a technique which has been widely used for a number of years.

The second aspect of molded parts that simplifies deflashing is that generally trimming or clipping equipment which conforms to the part contour is used. This type of process attacks the burrs and not the entire workpiece surface as is the case in vibratory deburring and abrasive jet deburring. If flash cannot be prevented at a non-working (aesthetic) surface, it should be placed on a roll (Figure 2-6). This allows trimming to be used and any mold offset is not visable after trimming (Figure 2-7).

When surfaces must be ground to remove caps of flash around holes any projecting features should be recessed to provide an unobstructed grinding path (Figure 2-8). While piercing dies can be used to remove these caps they can result in tearing of the edges.

Die designs which result in rolls are shown in Figure 2-9. The labels of "poor design" and "good design" are only appropriate for the aesthetic conditions described above. In many functional applications, Figure 2-9a

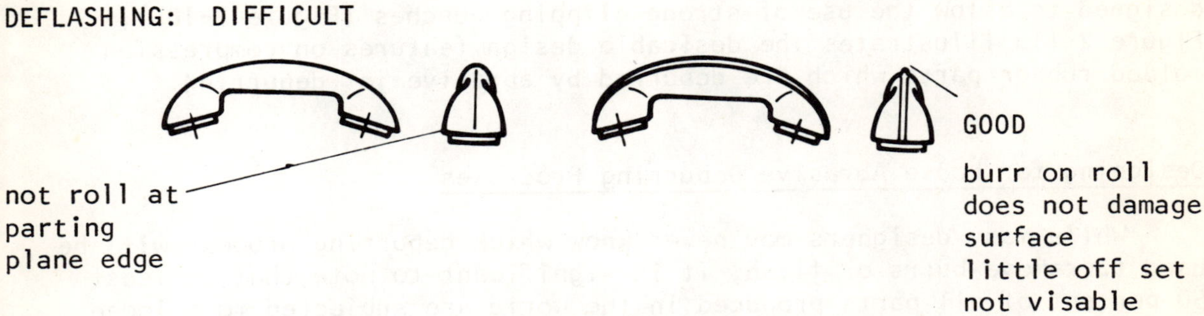

FIGURE 2-6: LOCATING FLASH FOR AESTHETICS AND EASY REMOVAL

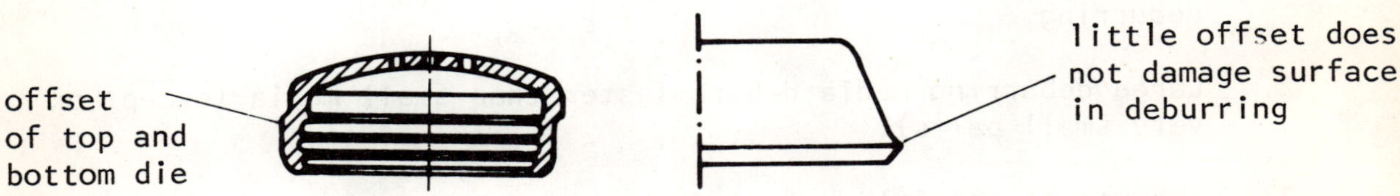

FIGURE 2-7: SLIGHT MOLD OFFSET IS MASKED IN DEFLASHING OPERATION

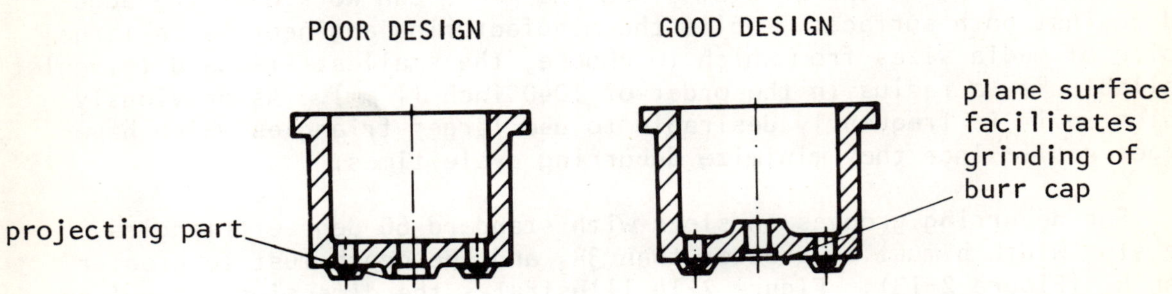

FIGURE 2-8: DESIGN FOR EASE OF MACHINING WHEN MACHINING OR GRINDING IS REQUIRED

is the most applicable design. Design considerations such as rolls and unobstructed surfaces are major factors in much of the aluminum, zinc, and plastic die casting industry since deflashing costs represent 20 to 35 percent of the total manufacturing cost.

The design of gating systems, ejector pins, and parting lines must all consider deflashing approaches. Overflow wells for example should be designed to allow the use of strong clipping punches (Figure 2-10b). Figure 2-11b illustrates the desirable design features on compression molded rubber parts which are deburred by abrasive jet deburring.

Designing for Loose Abrasive Deburring Processes

While many designers may never know which deburring process will be used to remove burrs or flash, it is significant to note that at least 50 percent of all parts produced in the world are subjected to a loose abrasive deburring process such as barrel tumbling, vibratory finishing, spindle finishing, or centrifugal barrel tumbling. Because these processes utilize the same basic deburring media and removal mechanism, a few considerations about design effects on them are in order.

The essential facts in these types of processes are:

1. The media must be able to slide over all surfaces which require deburring.

2. Large deburring media deburr faster than small media (except on very small parts).

3. The use of special preformed shapes such as triangles, stars, and cylinders have many advantages over gravel-like media.

As an example, consider Figure 2-12. In the left hand view the step height H_1 is smaller than the radius R_s on the deburring media which in this case is a triangle. As a result the media will often bend the burr over rather than remove it. Complete removal can be obtained only with long cycle times. The use of a larger step height as shown in the right hand view eliminates the problem since the media can work over the edge and against both surfaces. While the manufacturing engineer has a large choice of media sizes from which to choose, the smallest standard triangle available has a radius in the order of .040 inch (1 mm). As previously indicated it is frequently desirable to use larger triangles which have larger radii since they minimize deburring cycle times.

For deburring grooves on slots with standard 60 degree triangles, the slot width b must be greater than $3R_s$ and the depth must be greater than R_s (Figure 2-13). Figure 2-14 illustrates the dimensional requirements for holes when tetrahedrons are used for deburring. The smallest standard tetrahedron available has a 0.5 inch (12.5 mm) base and a similar height.

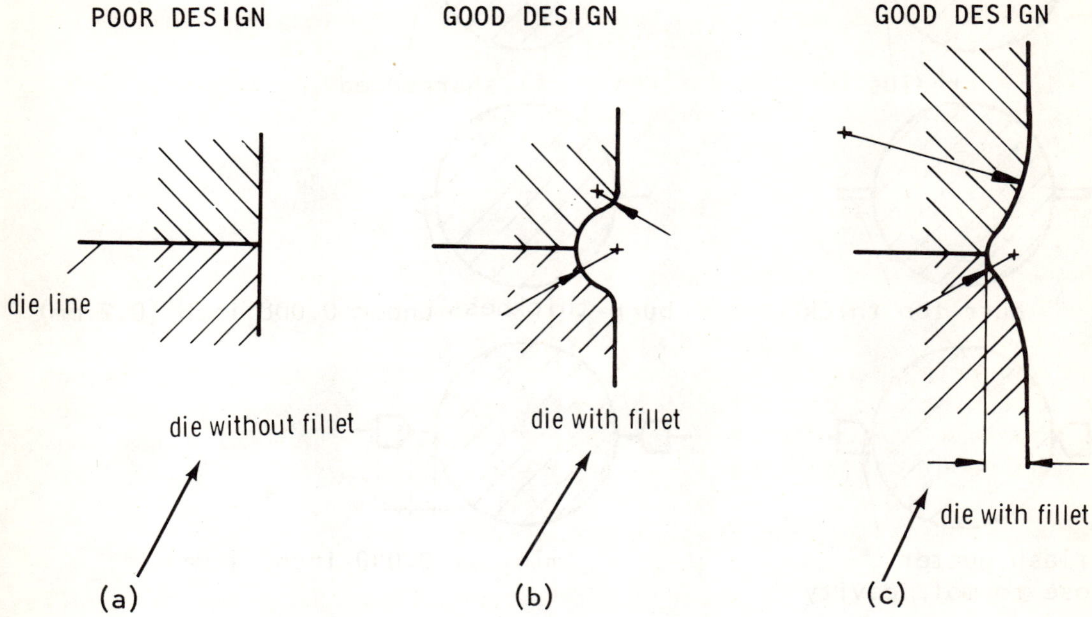

FIGURE 2-9: DESIGNING FOR AESTHETICS AND EASY FLASH REMOVAL

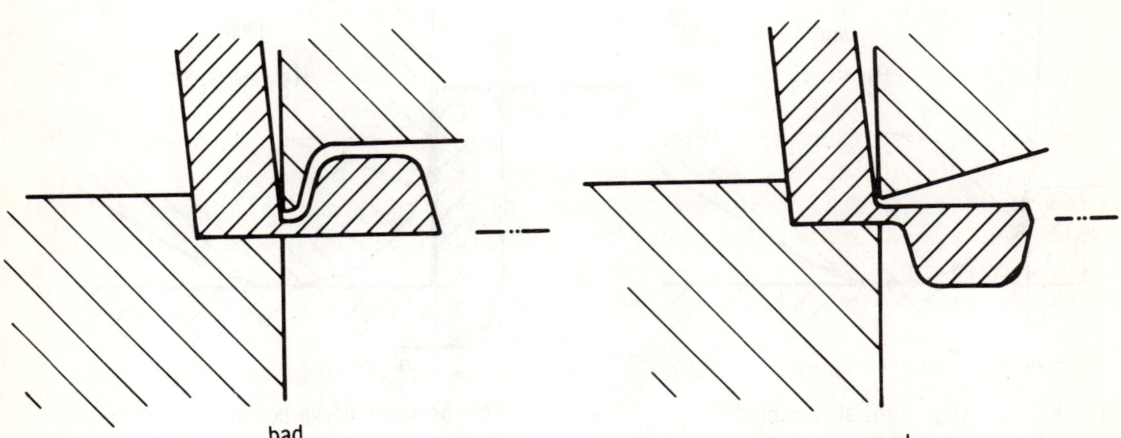

FIGURE 2-10: DESIGNING OVERFLOW WELLS FOR EASY TRIMMING AND LONG TOOL LIFE

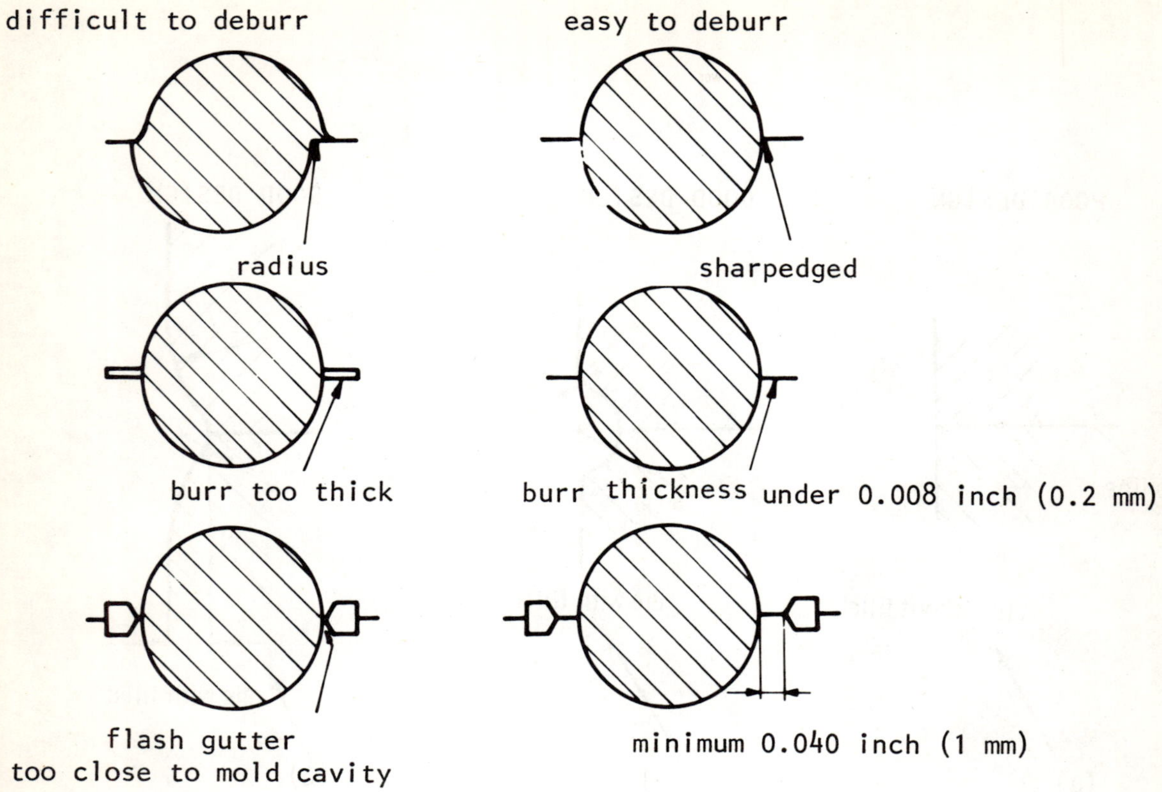

FIGURE 2-11: DESIGN CONSIDERATIONS FOR SUCCESSFUL DEFLASHING OF COMPRESSION MOLDED RUBBER COMPONENTS

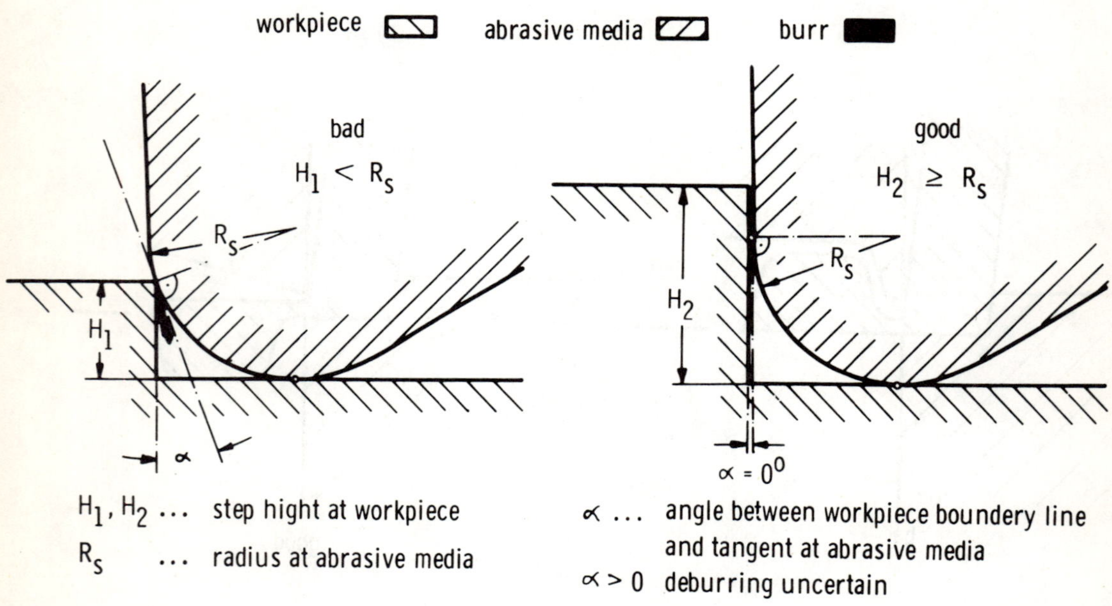

FIGURE 2-12: COMPLETE DEBURRING REQUIRES THAT THE ANGLE $\alpha = 0$

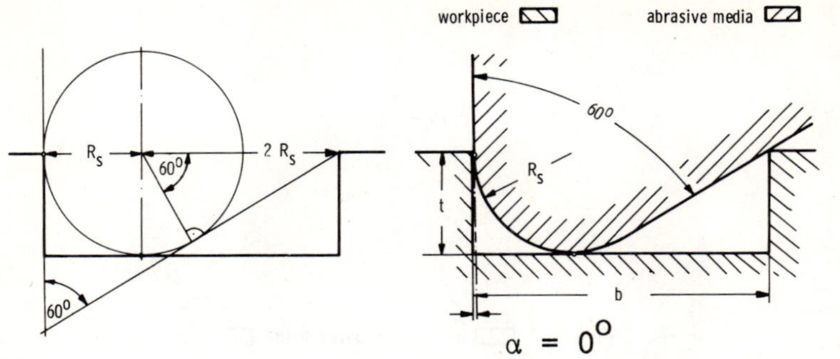

for $\alpha = 0^0$: $h \geq R_s$ $b \geq 3R_s$

h = depth of groove
b = width of groove

FIGURE 2-13: DEBURRING SLOT EDGES WITH 60 DEGREE TRIANGLES

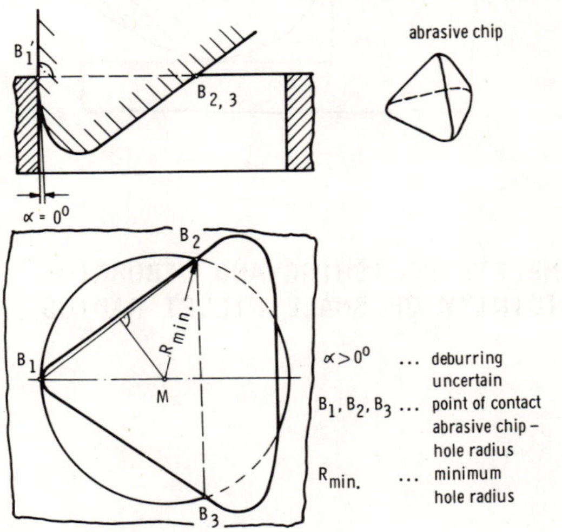

FIGURE 2-14: DEBURRING CONSTRAINTS ON HOLES

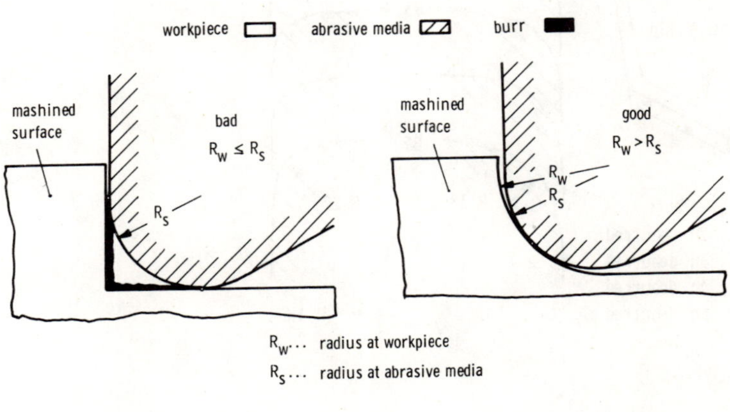

FIGURE 2-15: DEBURRING ACTION AT FILLETS

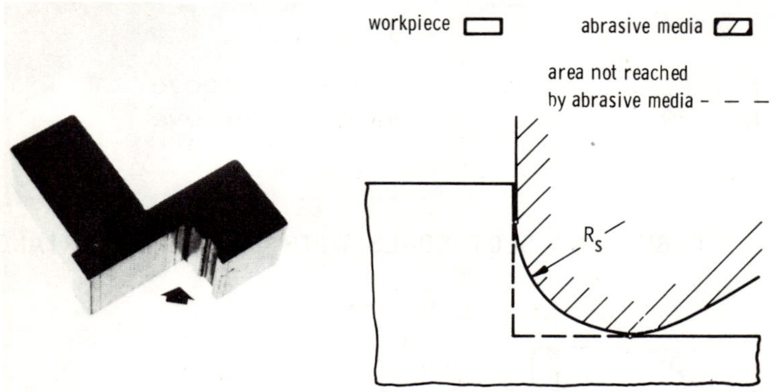

FIGURE 2-16: INCOMPLETE POLISHING AND DEBURRING IN VICINITY OF SMALL FILLET RADIUS

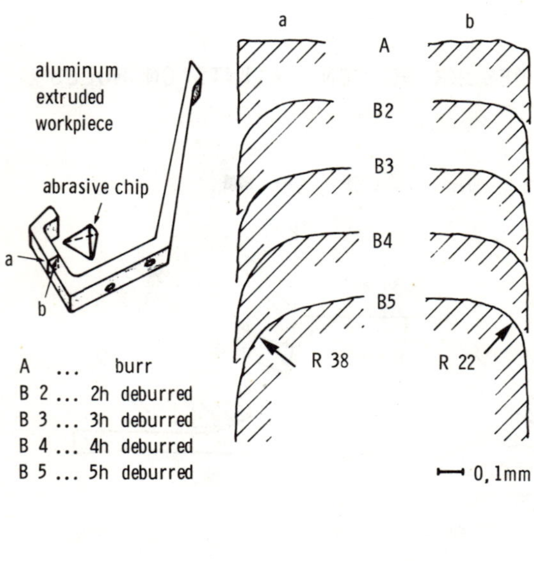

FIGURE 2-17: EFFECT OF WORKPIECE GEOMETRY ON EDGE RADIUSING

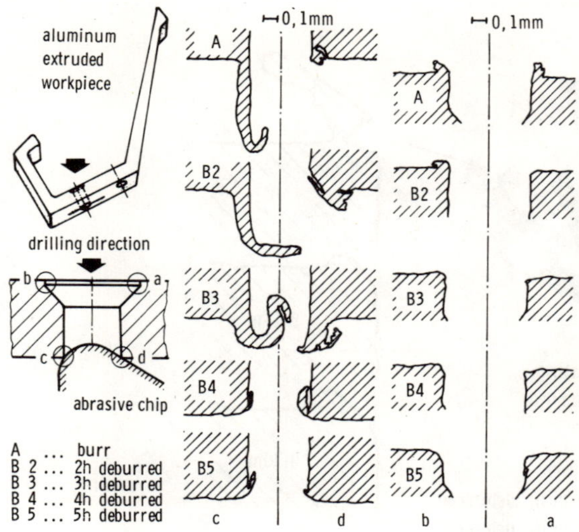

FIGURE 2-18: EFFECT OF PART GEOMETRY AND MEDIA SIZE ON HOLE DEBURRING

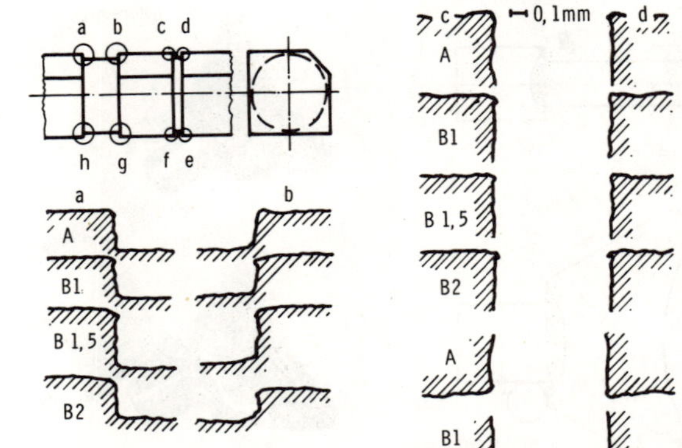

FIGURE 2-19: LARGE MEDIA CAN GENERATE BURRS IN NARROW SLOTS

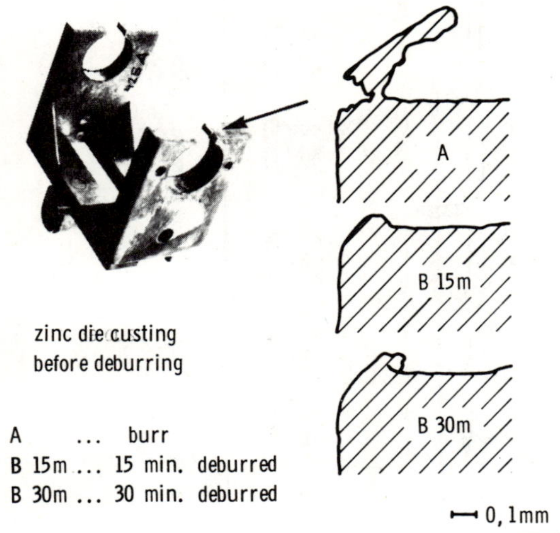

FIGURE 2-20: LARGE MEDIA AND SMALL HOLES CAN RESULT IN MORE BURRS

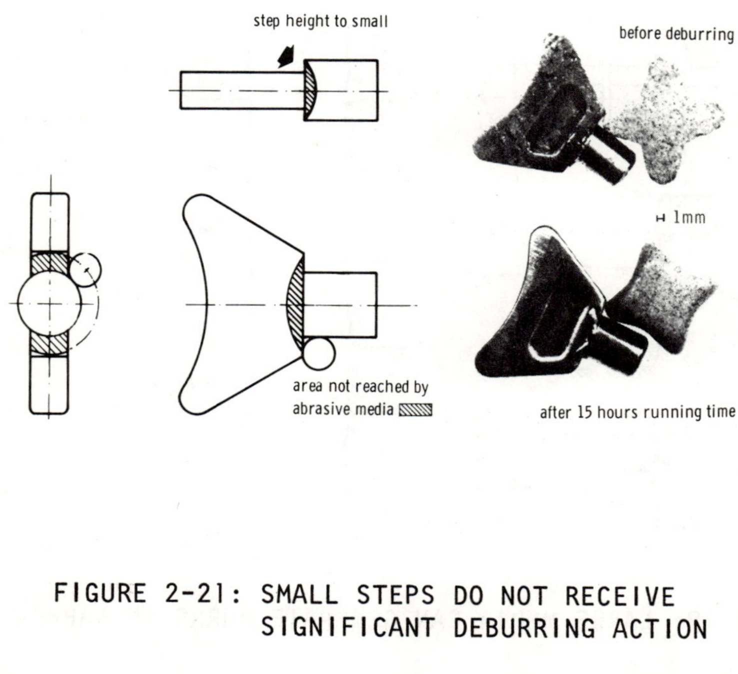

FIGURE 2-21: SMALL STEPS DO NOT RECEIVE SIGNIFICANT DEBURRING ACTION

When inner surfaces near fillet radii must be smoothed by loose abrasive processes it is essential that the fillet radii equal or exceed the radius on the media (Figure 2-15). This is a particularly significant consideration when the feature is an extruded item which is controlled by the die shape. Figure 2-16 shows a fillet radius which was not smoothed because the part radius was too small. While the lack of polishing is visable, the incomplete deburring on the edges of the part around the fillet radius is more difficult to distinguish.

When the same edge radius and deburring quality is required on all edges of a part, processes which attack the entire workpiece surface may not be suitable. As shown by Figure 2-17, radiusing action is largest where the media can most readily contact the part. The outside edge 'a' is accessible to all the media. Edge 'b' is partly shielded by the workpiece geometry. After a five hour run in tetrahedrons, edge 'a' had a radius of 0.015 inch (0.38mm) while edge 'b' had a radius of only 0.0087 inch (0.22mm). While such differences are not critical on many parts, they are on many others. The condition would be further aggrevated if the burr at 'b' were larger than at 'a'.

Figure 2-18 illustrates the incomplete deburring which occurs when the burr is totally accessable by the media. The chamfered side of the hole allows the media to work over the chamfer. Because the hole is smaller than the media, after five hours a burr is still left in the unchamfered side of the hole. It is significant to note that a burr occurs on both sides of the hole but the burr on the chamfered side is much smaller than the one on the drill breakthrough side.

If the media cannot work completely over an edge, it is possible that the media can itself create a burr (Figures 2-19 and 2-20). Small steps in the workpiece will not receive significant deburring action even with small media or balls because size and geometry prohibit contact (Figure 2-21).

Chamfered edges are nicely radiused on all workpieces after only a short deburring cycle time (Figure 2-22). The reason for this effect is that media can readily work over all edges. As just mentioned, burr sizes are also much smaller after chamfering than after other processes.

An impact action as well as an abrasion mechanism occurs in loose abrasive deburring processes. This can lead to noticeable edge deformation on soft parts (Figure 2-23), but it also reduces detrimental internal tensile stresses.

Each of these factors should be considered while parts are still in the design phase. While smaller deburring media or different shapes can be used, they create other problems. Whenever possible workpiece drawings should be reviewed by manufacturing engineers responsible for machining and deburring before the drawings are finalized.

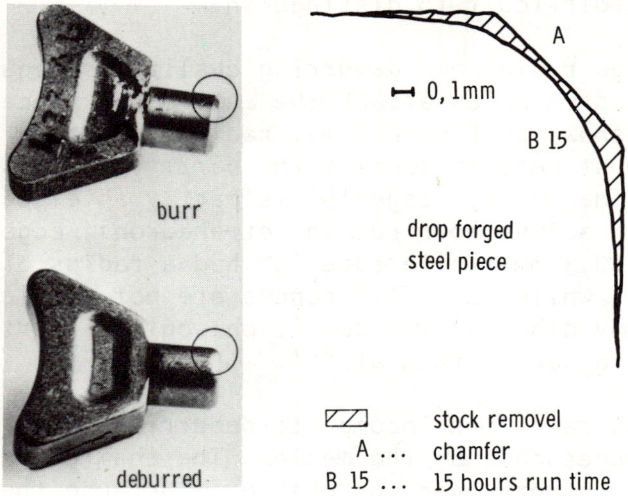

FIGURE 2-22: CHAMFERING AIDS DEBURRING

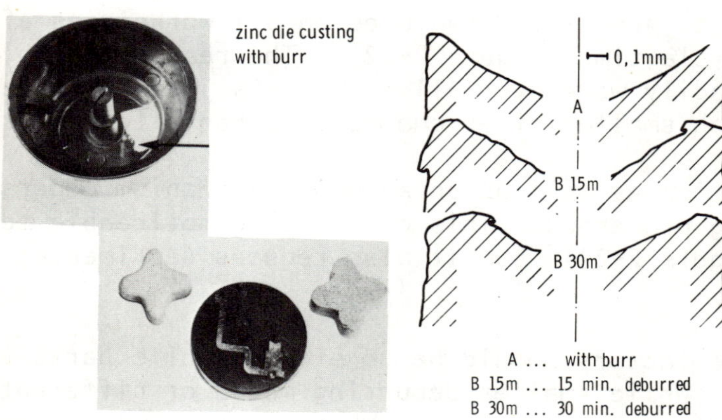

FIGURE 2-23: EDGE DEFORMATION FROM LOOSE ABRASIVE PROCESSES

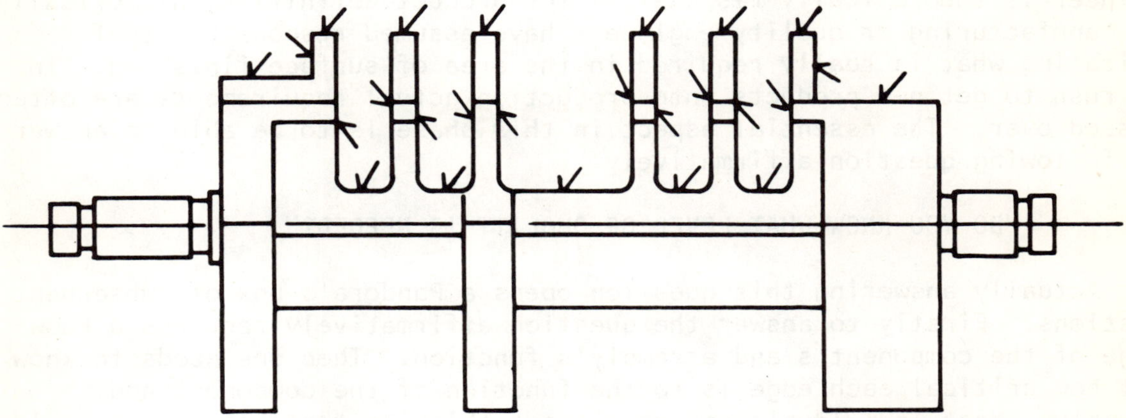

CHECK MARK INDICATES EDGES TO BE DEBURRED AT THIS OPERATION

FIGURE 2-24: IDENTIFY THE EDGES TO BE DEBURRED

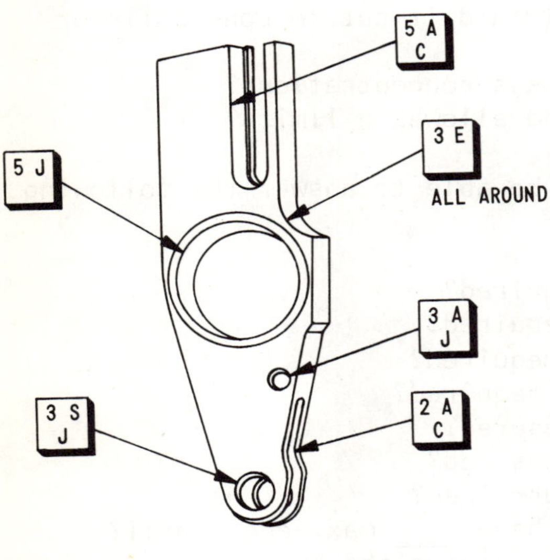

CODE	INSTRUCTION
1	DEBURR & BREAK EDGE 0.015-0.025" R.
2	DEBURR & BREAK EDGE 0.005-0.010" R.
3	DEBURR & BREAK EDGE 0.005" MAX. R.
4	DEBURR & BREAK EDGE 0.002"/0.003" R.
5	DEBURR & BREAK EDGE 0.002" MAX. R.
11	REMOVE HEAVY BURR ONLY
12	REMOVE FEATHER EDGE
13	CHAMFER FIRST AND LAST THREAD
A	BURR KNIFE
B	WIRE BRUSH
C	NYLON BRUSH WITH AL_2O_3
D	240 GRIT PAPER
E	STRING BRUSH WITH AL_2O_3
G	FILE
J	CRATEX "BULLET"
S	BURR BALL

FIGURE 2-25: INSTRUCTIONS ON EDGE REQUIREMENTS AND DEBURRING APPROACH

Specifying Edge Requirements

Probably the one area capable of the largest cost savings is that of understanding what edge requirements actually are. Although the product engineer is theoretically responsible for product definition, historically the manufacturing or quality engineers have assumed responsibility for indicating what is really required in the area of surface finishing. In the rush to get new products into production actual requirements are often glossed over. The essential aspect in this phase is to be able to answer the following question affirmatively.

"DO YOU KNOW WHAT LEVEL OF QUALITY IS NEEDED?"

Actually answering this question opens a Pandora's box of subsequent questions. Firstly to answer the question affirmatively requires a knowledge of the component's and assembly's function. Then one needs to know just how critical each edge is to the function of the component and assembly. Most individuals assume without thinking that all edges should have the same edge radius or burr free condition. In most situations this is not true.

While in-house workmanship standards are a necessary requirement for most companies, they do not absolve designers from the responsibility of specifying actual requirements. In many cases deburring costs could be lowered if designers asked themselves (and found the answers to) the following questions:

IS A BURR ALLOWABLE?

- Would it cause an electrical short circuit?
- Would it jam a mechanism?
- Would it cause interference fits?
- Would it cause misalignment?
- Would it be a safety hazard? (Would it cut someone's finger during assembly?)
- Would it cause unallowable stress concentrations?
- Would it accelerate wear beyond allowable limits?

The manufacturing engineer must also be able to answer the following questions.

- Why is burr free condition required?
- Why is ____ max. edge radii required?
- Where is burr free condition required?
- Where is ____ max. edge radii required?
- How is burr free condition measured?
- How is edge break condition measured?
- What happens if part is not burr free?
- What happens if part does not have ____ max. edge radii?
- How can part be redesigned to minimize the burr?

For many designers, manufacturing engineers, and inspectors the task of defining what is allowable has been difficult. The following approaches however have been used successfully by many large and small companies.

METHODS OF DEFINING AN ALLOWABLE BURR OR EDGE CONDITION

- Define it on the print
- Define it in a Process Engineering Specification (Manufacturing Specification)
- Define it on the production traveler (routing sheets)
- Define it by interpretive memo. Such a memo could include sketches, photos, measuring techniques, etc.
- Define it on the inspection traveler
- Define it with photos of acceptable and unacceptable conditions
- Define it by the use of comparative masters (the master is given a tool or gage number, or a visual aid or visual standard number).
- Define it by go-no go. If it fits the gage, the burr is acceptable.
- Define it by taking specific exception to general workmanship specifications
- Define it by such phrases as "Firmly adhered burrs or raised metal is allowable in this area provided a micro tool 90° hook will not dislodge them."

By some means the production operators as well as the engineers must be informed of the actual quality required on all edges. In some cases this can be controlled by the choice of deburring technique. On many parts, however, the engineer must designate not only the quality required but also the edges which must be deburred at a given operation. In some cases an illustration such as Figure 2-24 is adequate. Each edge to be deburred is marked by a check mark. In other cases a code such as shown in Figure 2-25 is required.

Schafer[1] recently described a technique which illustrates graphically what the allowable edge quality can be. His classification system is shown in Figure 2-26. As seen in the right hand view, this system has nine classes of quality. Class 1 allows projections up to 0.01 mm (0.0004 inch). Class 9 allows projections or radii up to 2.50 mm (0.10 inch). The symbols r_k, h_l, and b_k are defined in the left hand view. As seen there, r_k indicates the size of radius allowed, b_k indicates either a radius or the size of chamfer allowed, and h_l indicates the height of a projection which can be allowed. The left hand view also shows the many types of conditions that can be found at an edge. If one extended the planes of the workpiece surface, four quadrants are produced. It is possible for a burr to exist in any of these. It is also possible that as a result of edge deformation material can exist in more than one quadrant (Figure 2-27).

To illustrate the system, consider a 0.3 mm (0.012 inch) radius on an edge (Figure 2-27 upper left). From Figure 2-26 this falls within the range 0.16 to 0.315 mm and is thus a class 6 edge. Since it is a radius occuring in the fourth quadrant (the quadrant which the workpiece itself is in) the edge specification is

0	0
0	6

The upper right hand view shows a 0.17 mm (0.007 inch) high burr. This too falls within the range of a class 6 edge. In this case the burr projects into the first quadrant. Since the workpiece is always in the fourth quadrant for the position shown, anything in another quadrant must be a burr. Thus

0	6
0	0

indicates a burr which is between 0.16 and 0.315 mm high extending into the first quadrant.

The upper middle illustration portrays the case where a burr is in the third quadrant but some radiusing exists on the part. In this case edge quality number exists in two of the four quadrants.

This system can be used on edges with included angles less than 90 degrees (Figure 2-28). In this case however b_k is not equal to the radius r_k as is the case with 90 degree edges. As shown in Figure 2-27 when edge angles are different than 90 degrees, there will be two big and two little quadrants.

Summary

In summary, designers can minimize deburring costs by specifying only what is actually required at an edge and by recognizing that not all edges require the same quality. The geometry of the workpiece represents a major role in determining the size of burrs produced and the time required to remove them.

Designing to minimize deburring is a relatively new approach to minimizing deburring costs. Despite its newness, it has been shown to be a significant major approach to minimizing problems and costs.

Sources of Additional Information

[1] F. Schafer, "Deburring Processes in Perspective," SME Paper MR75-482, 1975.

[2] F. Schafer, Entgraten, Krausskopf-Verlag, West Germany, 1975 (in German).

[3] L. K. Gillespie, "Designing for Economical Deburring," Handbook of Production Design, McGraw-Hill, New York, to be published.

[4] L. K. Gillespie, "The Effect of Product Geometry on Burr Size," SME paper to be published.

[5] L. K. Gillespie, The Effect of Edge Angle on the Radius Produced by Vibratory Finishing, Bendix Corporation (Kansas City Division), Report BDX-613-1279, February, 1975 (available from NTIS).

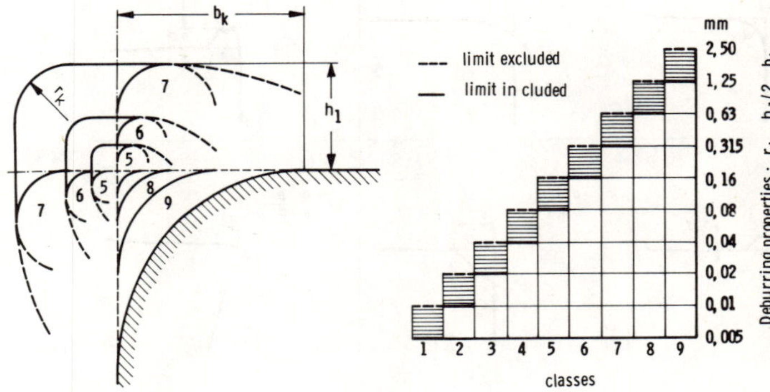

FIGURE 2-26: CLASSIFICATION SYSTEM FOR DEFINING ALLOWABLE EDGE QUALITY

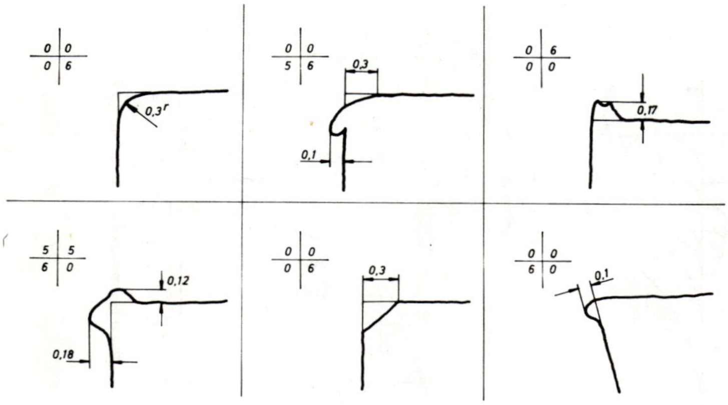

FIGURE 2-27: EDGE CONDITIONS AND THEIR DESIGNATION IN SCHAFER'S SYSTEM

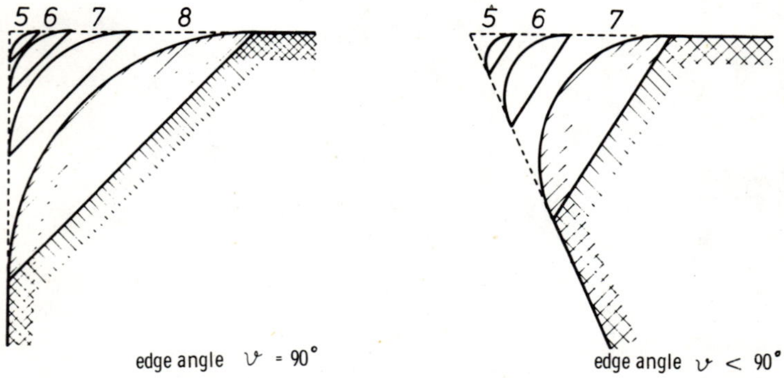

FIGURE 2-28: TWO EDGE ANGLE POSSIBILITIES

CHAPTER 3

BURR PREVENTION

If burrs could be prevented from forming then deburring would not be required. Unfortunately altering speeds, feeds, and tool geometries will not prevent burrs. Both analytical and empirical studies have shown that while tool sharpness and cutting conditions can minimize burr size and control burr repeatability, they cannot prevent burrs. Conventional machining techniques always produce burrs. Theoretically, supporting the workpiece with a piece of backup material should prevent burrs. From a practical standpoint however backup material only helps to minimize burrs.

This can be seen by looking closely at a workpiece. Most operations produce burrs at more than one location. In drilling for example one obtains a burr at both hole entrance and hole exit. In a milling operation burrs can be produced on up to ten edges. Thus at best one minimizes only burrs on one side of the workpiece. While minimizing burr size is a distinct advantage, it is not as desirable as burr prevention.

Theoretically it would be possible to completely cover a part with "backup" material and prevent all burrs. From a practical standpoint this is not very realistic because the "backup" material must have the same properties as the workpiece to prevent burr formation.

Burrs can be prevented by employing some of the nontraditional processes. As seen in Table 3-1 most of the nontraditional processes do not produce burrs. Despite many statements to the contrary EDM, EBM, and Laser Machining (LBM) do produce burrs or burr-like projections of recast material. While little test data is available on edge quality produced, two sources provide some insight. McBride's study[1] documents the effects of EDM parameters on EDM burr size, and the book Non-Traditional Machining Processes[2] discusses the edge conditions produced by EBM and LBM. Recent research on LBM however indicates that when a high velocity air blast is synchronized with the laser the majority of the burr is blown out before it can solidify on the workpiece. Thus in the future, LBM may fall in the category of a non-burr producing process.

Whenever possible processes such as CHM, ECG, ECM, ECH, ELP, and ESM should be utilized. They not only eliminate deburring costs but they also provide excellent surface finishes and minimize welding, brazing, and plating problems caused by media impregnation or improper cleaning. In addition the elimination of unnecessary operations reduces paperwork costs and shortens production flow time.

The disadvantages of using the nontraditional processes include high equipment costs, limitations to certain geometries and workpiece materials, and workpiece tolerance and surface integrity problems. These factors are discussed in detail elsewhere.[2]

TABLE 3-1
NONTRADITIONAL MACHINING CAPABILITIES

Process	Typically Makes Burr?*	Typical Edge Radius Produced (inch)	Typical Machining Tolerance (inch)
AJM Abrasive Jet Machining	No	0.003	
CHM Chemical Machining	No	Unknown	± .002
EBM Electron Beam Machining	Yes	--	± .001
ECDM Electro Chemical Discharge	Unknown	Unknown	Unknown
ECG Electro Chemical Grinding	No	0.003	± 0.002
ECM Electro Chemical Machining	No	0.001	± 0.002
ECH Electro Chemical Honing	No	0.0005	± 0.0002
EDM Electrical Discharge Machining	Yes	--	± 0.0006
ELP Electropolishing	No	0.001	± 0.0005
ESM Electro Stream Machining	No	0.002	± 0.001
HCG Hot Chlorine Gas	No	0.002	± 0.003
IBM Ion Beam Machining	No	0.00005	± 0.0001
LBM Laser Beam Machining	Yes	--	± 0.001
PAM Plasma Arc Machining	Yes	--	± 0.003
USM Ultrasonic Machining	No	0.001	± 0.001
WJM Water Jet Machining	No	Unknown	± 0.003

*Where burr is visible under 30X magnification.

As seen in Chapter 2, some burrs can be prevented by the correct choice of edge angles. While this approach to burr prevention should be pursued, it does not solve the total deburring problem as many cutting tools produce burrs on many edges with a single pass. It is not feasible to prevent these burrs on all edges.

Summary

In summary, the nontraditional machining processes are the only machining approach that can be used to prevent burrs, but their use is contingent upon such factors as workpiece tolerance, geometry, surface integrity, and availability of such equipment.

Sources of Additional Information

[1] James N. McBride, The Magnitude of Burrs Caused by Electrical-Discharge Machining, M. S. Thesis, Utah State University, Logan, Utah, 1969.

[2] R. K. Springborn (ed.), Non-Traditional Machining Processes, SME Dearborn, Michigan, 1967.

CHAPTER 4

BURR FORMATION AND PROPERTIES

Any analysis of minimizing deburring costs would not be complete without at least a rudimentary understanding of how burrs form. In addition, it is essential to have an order of magnitude appreciation for the size of burrs which exist on parts. The following pages provide a primer on such aspects as

- The theories of burr formation
- Predicting burr size
- Burr measurement techniques
- Typical burr sizes
- Minimizing burr sizes produced in drilling

BURR FORMATION

Basic Mechanisms

The formation of burrs, flash, and related protrusions occurs by six physical principles. Of the six, three represent the underlying basic mechanisms which form what is most generally defined as a burr. The remaining three are responsible for burr-like protrusions which are also described as burrs by many individuals. For completeness all six are discussed in this chapter.

Physical Principle of Formation	Name of Protrusion
Lateral flow of material	Poisson Burr
Bending of material (such as chip roll-over)	Roll-Over Burr
Tearing of chip from workpiece*	Tear Burr
Redeposition of material	Recast Burr
Incomplete cutoff	Cutoff Burr
Flow of material into cracks	Flash

*A Tear Burr also forms in stamping operations when the punch tears the part from the stock.

True burrs form by one of the first three principles listed. These three mechanisms of formation involve plastic deformation of the workpiece material. Solidification of material on the working edges, the fourth principle of formation, forms a burr-like projection. The fifth type of burr occurs when the workpiece is allowed to fall from the part before the cut is completed. Flash forms whenever the pressure on molten molding material is sufficient to force it between the two halves of a die.

Poisson Burr

The Poisson Burr occurs whenever the cutting edge extends past an edge of the workpiece (Figure 4-1). In essence, it is the lateral deformation that occurs whenever any solid is compressed. The term Poisson Burr is derived from the fact that this burr is directly related to Poisson's ratio. The extent of this deformation is a function of the workpiece material, the size and shape of the contacting body, and the applied load. In the case of a cutting tool in which no rubbing occurs on clearance surfaces, the effective shape of the indentor is a cylinder. Because cutting tools have cutting edge radii of 0.0002-0.0050 inch (5.08-127.0 μm) the extent of the lateral deformation (burr size) is also relatively small. The actual size of the burr is proportional to cutting edge radius and the applied pressure. It is also a complex function of several material variables (Equations 1 and 2).

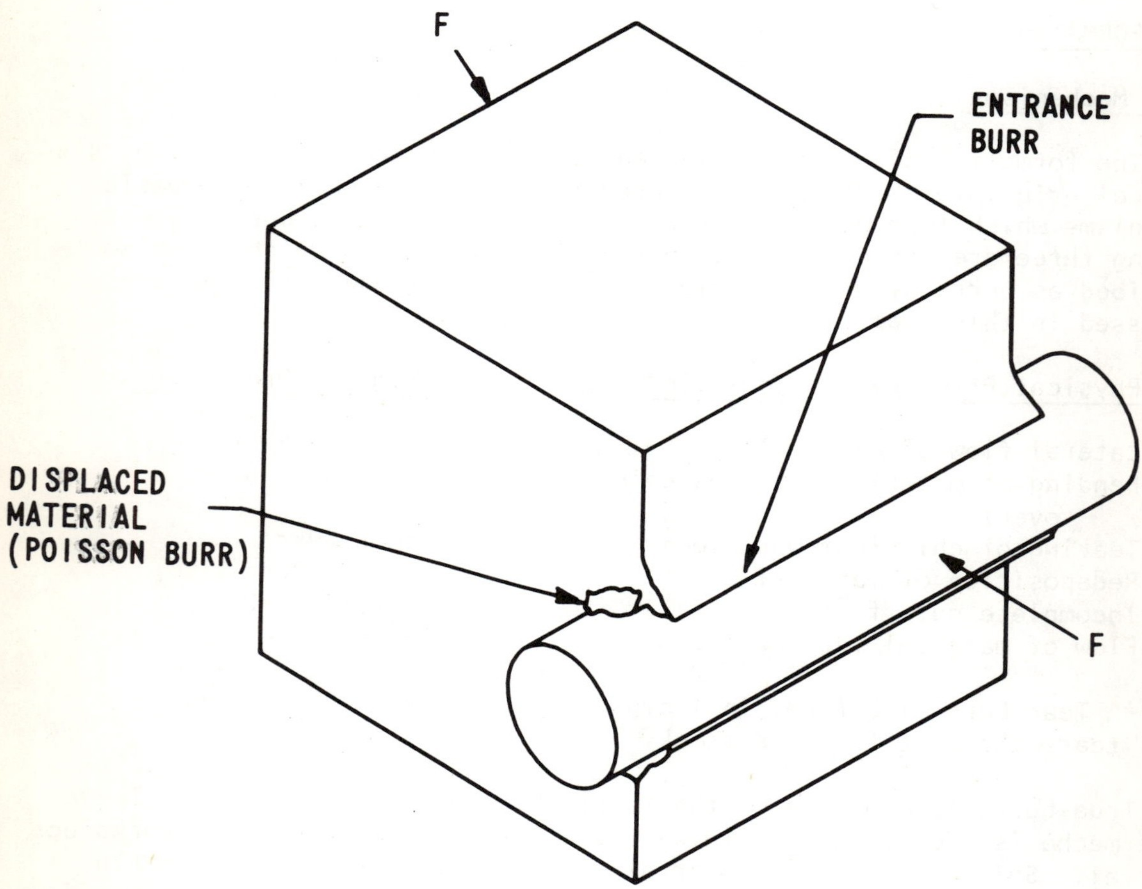

FIGURE 4-1: FORMATION OF A POISSON BURR

$$b_L = \left[\frac{h(1+\nu)\sigma_e e^{-\sqrt{3}\,\phi_a}}{\sqrt{3}\;\;E}\right]\left[-\frac{\sin\phi}{2(\sqrt{3}\,\cos\phi + \sin\phi)}\right], \tag{1}$$

$$b_t = r\left[e^{-\sqrt{3}\,\phi_a}\cos\phi_a - 1\right], \text{ and} \tag{2}$$

$$\phi_a = -\sin^{-1}\left[\frac{\sqrt{3}\,P}{2\sigma_e}\right] + \pi/6 \tag{3}$$

where

b_L = burr length
b_t = burr thickness
h = depth of cut
ν = Poisson's ratio for elastic stresses
σ_e = yield stress of a perfectly plastic material
E = modulus of elasticity
r = cutting edge radius
p = pressure on cutting edge
ϕ = angle defining state of plastic flow

Typical values predicted by these equations are shown in Figures 4-2 and 4-3. While measured burr lengths range within the values shown, measured burr thicknesses typically are 20 times larger than shown.

Entrance Burr

When the cutting edge first indents the workpiece, it is possible for another type of burr to form (Figure 4-4). This Entrance Burr is material which has flowed opposite the direction of the tool. It is similar to the ridge which forms around the indentation made by a Brinnell Hardness Tester.

Whether or not a burr forms at this point depends on the workpiece properties and, probably, the actual shape of the cutting edge. Assuming that a cylinder indenting metal produces effects somewhat similar to those of a ball indenting metal, then strain hardening plays an important role. Based on Brinnell Hardness results it has been shown that a lip of material forms when the material has a low strain hardening exponent (Figures 4-5 and 4-6). Materials having high strain hardening exponents cause a bulge but not a sharp burr. Assuming constancy of volume, this bulge will be wide but short. In the previous case, the burr will be long but narrow.

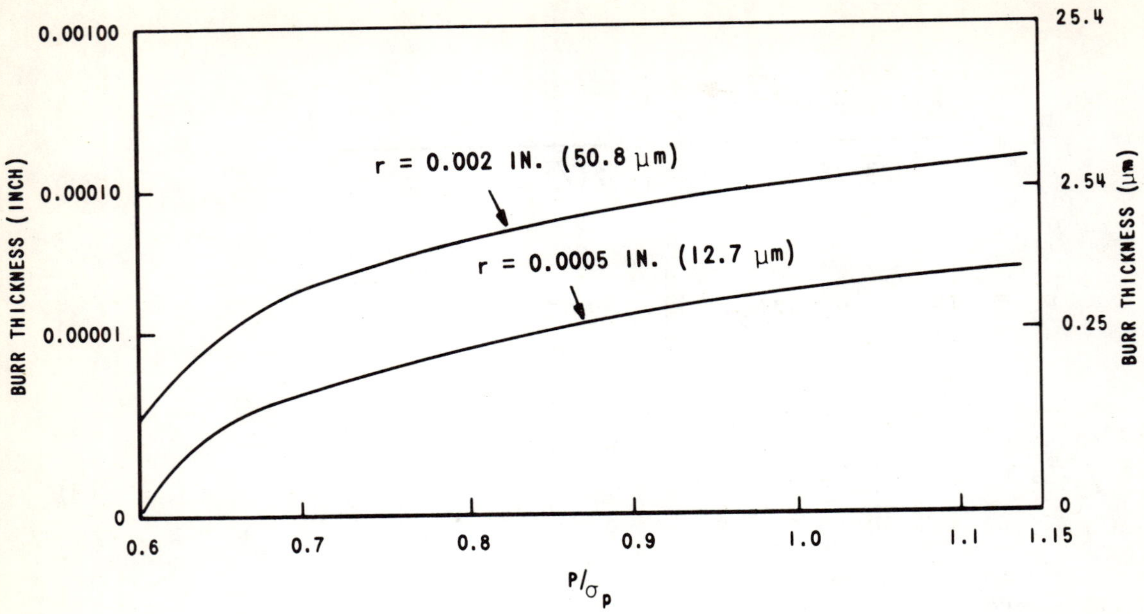

FIGURE 4-2: EFFECT OF CUTTING PRESSURES ON BURR THICKNESS FOR PERFECTLY PLASTIC MATERIAL

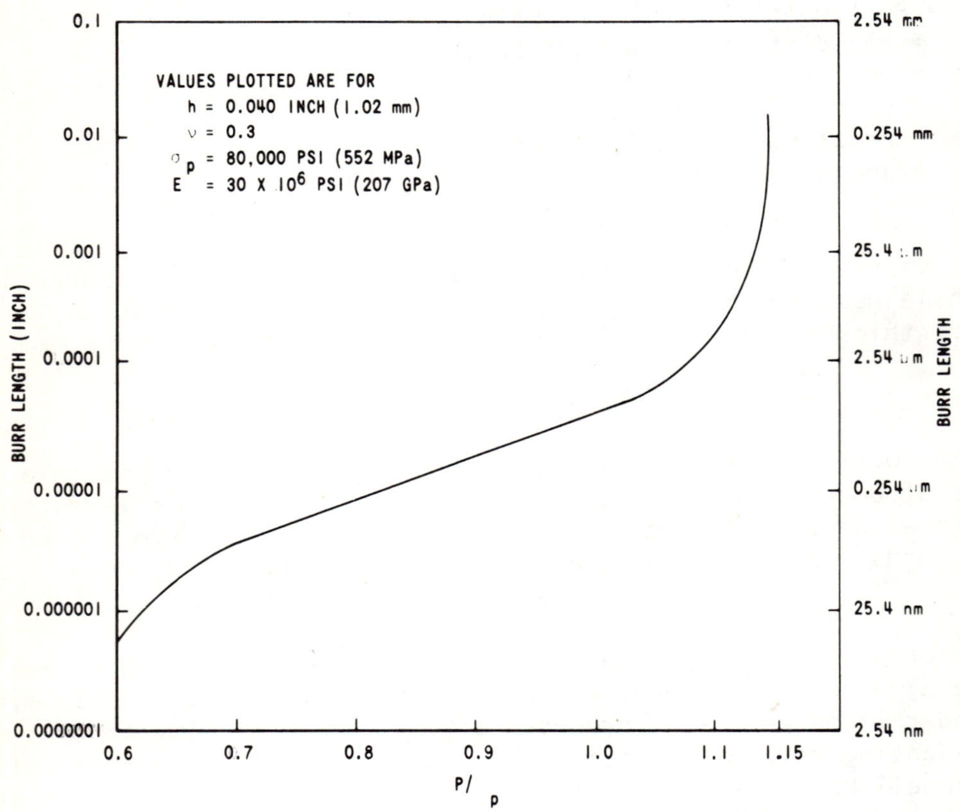

FIGURE 4-3: EFFECT OF CUTTING PRESSURES ON BURR LENGTH FOR PERFECTLY PLASTIC MATERIAL

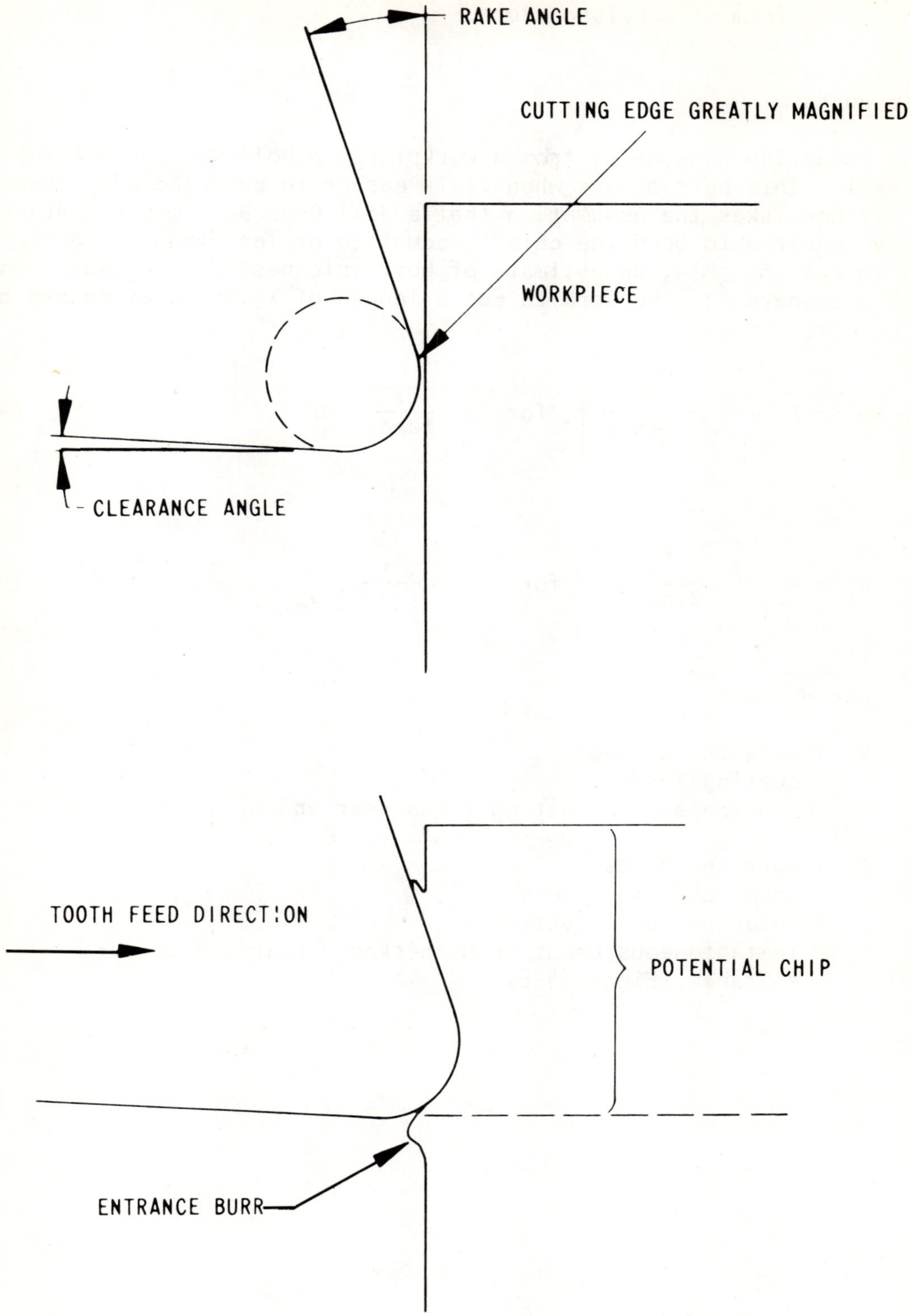

FIGURE 4-4: CUTTING EDGE PRODUCES INDENTATION BURR AS IT ENTERS WORKPIECE

In the case of high strain hardening exponents, the bulge will probably be short enough that it is difficult to detect. This Entrance Burr is essentially one form of a Poisson Burr.

Roll-Over Burrs

When a cutting edge exits from a workpiece, a Roll-Over Burr forms (Figure 4-7). This burr occurs when it is easier to bend the chip than to cut it. If one makes the assumption that a Roll-Over Burr occurs whenever the energy required to bend the chip is equal to or less than the energy required to cut the chip, an estimate of burr thickness can be made. As indicated elsewhere [2,3] the work to cut a length of λ can be expressed by

$$W_c = F_c \left[\lambda - \frac{t}{2\tan\phi} \right], \text{ for } \lambda \geq \frac{t}{\tan\phi}, \text{ or} \tag{4}$$

$$W_c = F_{c_i} \left[\frac{t_i}{2\tan\phi} \right], \text{ for } \lambda < \frac{t}{\tan\phi}, \tag{5}$$

where

W_c = work in cutting
F_c = cutting force
F_{c_i} = instantaneous cutting force near end of cut = $\frac{F_c t_i}{t}$
λ = burr thickness
t = uncut chip thickness
ϕ = shear angle in cutting
t_i = instantaneous uncut chip thickness near end of cut
= $\lambda\tan\phi$ (Figure 4-8)

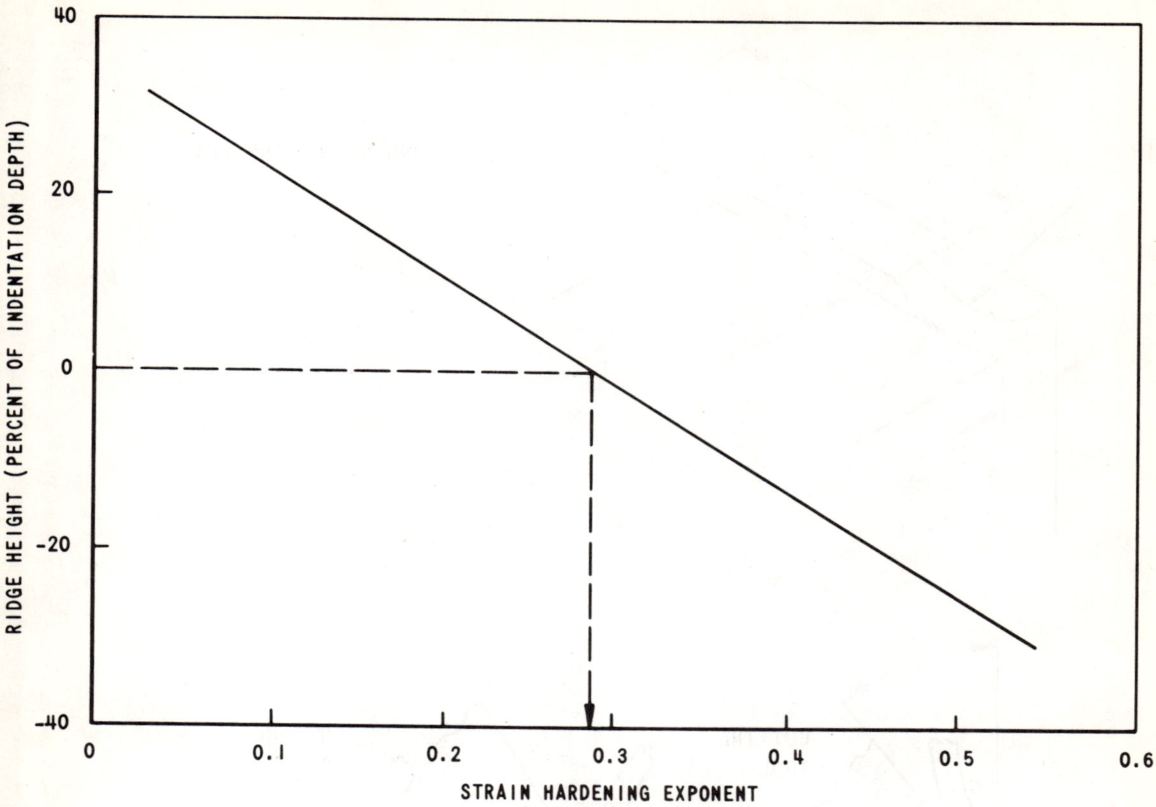

FIGURE 4-5: EFFECT OF STRAIN HARDENING EXPONENT OF RIDGING BURR FORMATION[1]

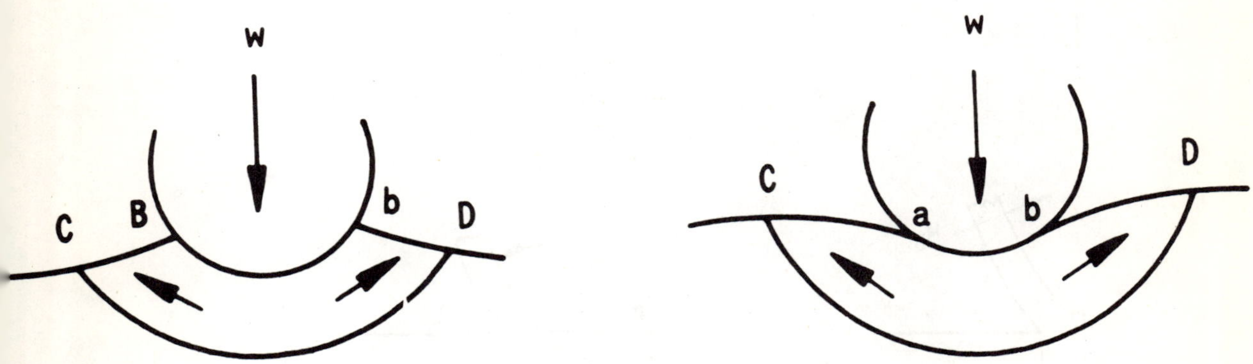

FIGURE 4-6: MATERIAL DISPLACEMENT USING SPHERICAL INDENTORS (For highly worked metals (non strain hardenable), the flow of metal around the indentor produces "piling up" (left). For annealed metals the displacement of metal occurs at regions at a small distance from the indentor so that "sinking in" occurs.[1]

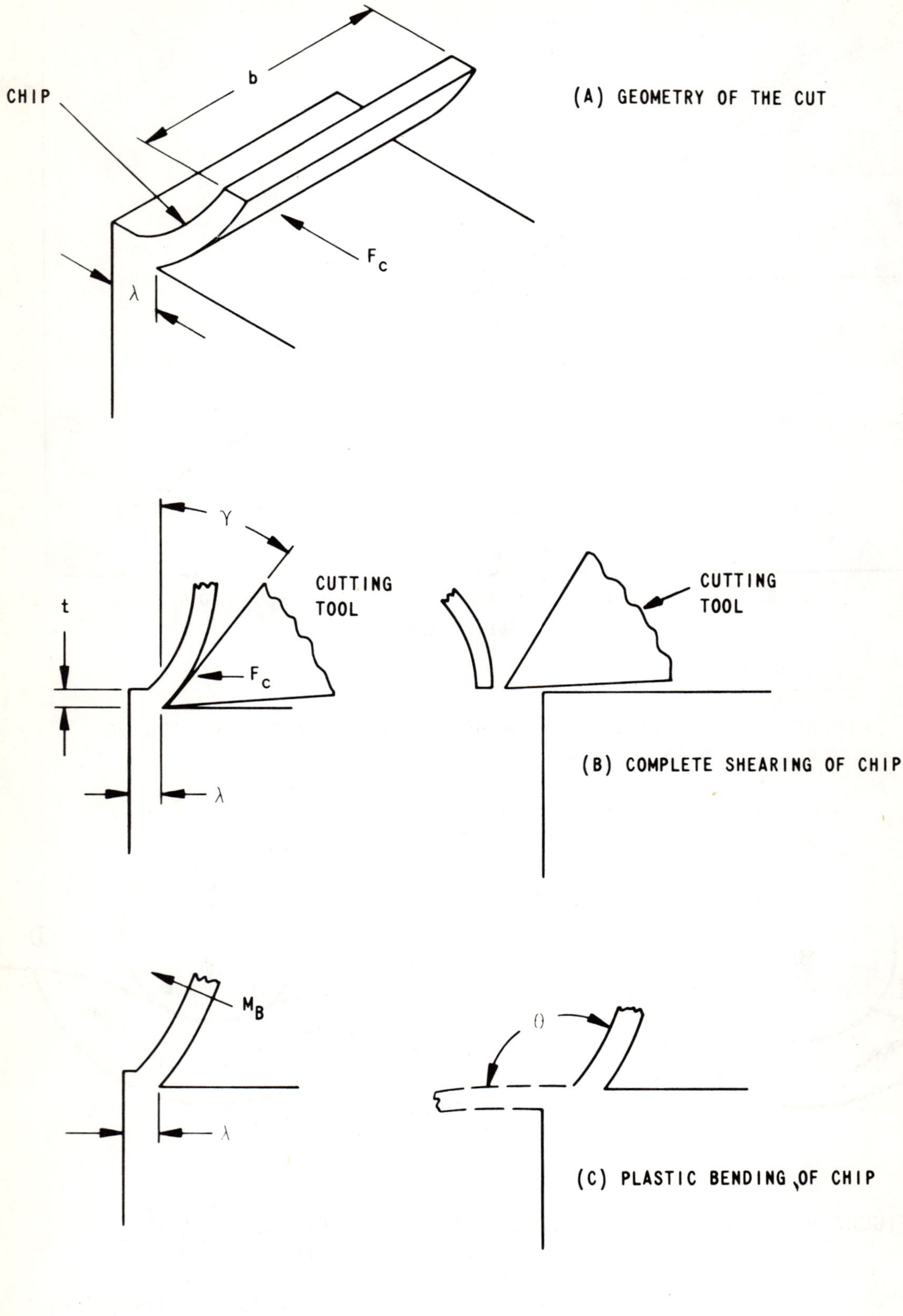

FIGURE 4-7: CHIP DEFORMATION MODES

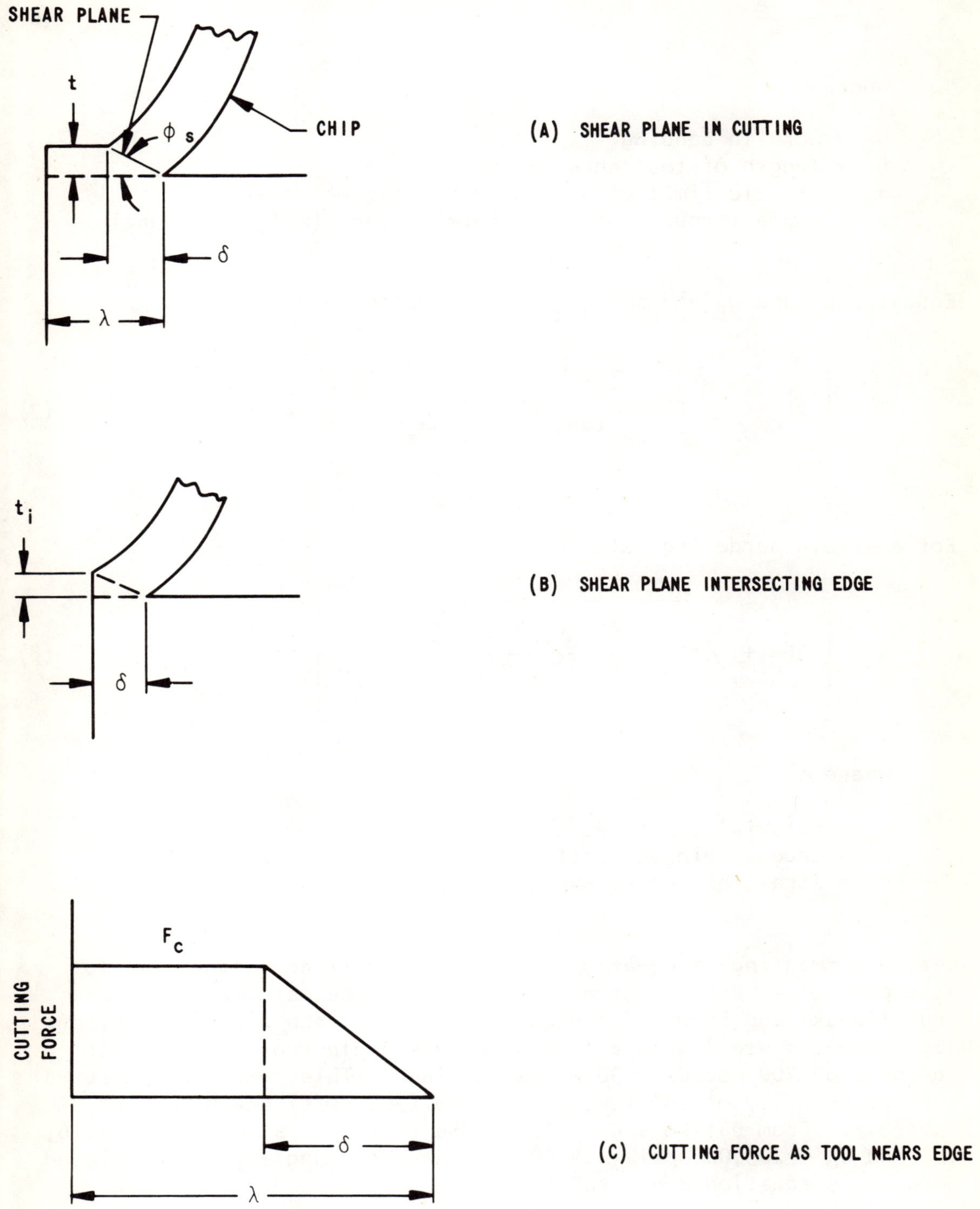

FIGURE 4-8: CUTTING ACTION NEAR EDGE

The work in bending a burr on a perfectly plastic material is given by

$$W_B = \frac{b\lambda^2 \sigma_e \theta}{4} \qquad (6)$$

where

W_B = work in bending
b = length of tool engaged in cut
σ_e = elastic limit of perfectly plastic material
θ = angle through which the burr is bent ($\pi/2$ + rake angle)

Equating W_c and W_B and solving for λ results in

$$\lambda = \left[2F_c + \sqrt{4F_c^2 - \frac{2b\sigma_e \theta t F_c}{\tan\phi}} \right] \left[\frac{1}{b\sigma_e \theta} \right] \text{ for } \lambda \geq \frac{t}{\tan\phi} \qquad (7)$$

For a strain hardening material,

$$\lambda = \left[2F_c + \sqrt{4F_c^2 - \frac{2b\sigma_o \varepsilon_f^n t\theta F_c}{(n+1)\tan\phi}} \right] \left[\frac{n+1}{b\sigma_o \theta \varepsilon_f^n} \right] \text{ for } \lambda \geq \frac{t}{\tan\phi} \qquad (8)$$

where

σ_o = material stress at a true strain of 1.0
ε_f = true strain at fracture
n = strain hardening exponent

Before commenting on equations 7 and 8, it will be instructive to look at a plot of W_c and W_B for a specific workpiece material. Extrapolating Vilenski and Shaw's force data[4] for a 3/8 inch wide (9.5 mm) cut in 303Se stainless steel with a 10 degree rake angle tool, a principle cutting force of 200 pounds (890 N) is obtained. This force was measured at a surface velocity of 100 feet per minute (508 mm/s) and 0.002 ipr (50.8 μm/rev). From Datsko's data for 303Se stainless steel[5], $\varepsilon_f = 1.16$, $n = 0.5$, and σ_o = 205,000 psi (1.4 GPa). The shear angle ϕ is calculated from Merchant's equation and data[6] as

$$\phi = \frac{C - \tau + \alpha}{2} \qquad (9)$$

where

C is a material constant
τ = friction angle = arc tan μ
μ = coefficient of friction between chip and tool
α = tool rake angle

Merchant indicates that for 303Se stainless steel (hot rolled) μ = 0.80 and C = 92.0 degrees. If these properties are also valid for the annealed condition then for α = 10 degrees, ϕ = 31.8 degrees. From the values already given, θ = 90 degrees + α = 100 degrees (1.745 radians)*, b = 3/8 inch, and t = 0.002 inch (50.8 μm).

To obtain W_B, one must define a value of σ_e which can be used to treat 303Se as a pseudo perfectly plastic material. As noted by Kalpakjian[7], the following expression can be used:

$$\sigma_e = \frac{\sigma_o \varepsilon_f^n}{n+1} \quad , \tag{10}$$

which, for the defined properties, results in σ_e = 147,200 psi (1.01 GPa). Note that substituting Equation 10 into Equation 7 produces Equation 8.

As seen in Figure 4-9 for these values, the work required to cut the chip at the end of the cut is larger than the work required to bend it provided λ is less than 0.0062 inch (157.5 μm). Thus by the assumptions made, burr thickness in this case would be 0.0062 inch. Actual measurements under roughly equivalent conditions indicate that the burr tends to be on the order of 0.001 to 0.002 inch (25.4 to 50.8 μm) thick.

The difference between actual and measured results is most likely the result of redundant work in bending. The assumptions made in the bending formulas are highly idealized and do not account for any of the redundant work which accompanies all plastic deformation processes. The cutting formulas and data include redundant work. As seen in Figure 4-9, only a small increase in bending work will produce a significant reduction in the calculated thickness of a burr.

The length of a Roll-Over Burr is a function of the cutting conditions and the plasticity of the workpiece. The length cannot be longer than the total depth of cut. If in bending, the strain exceeds the strain required to fracture, then the majority of the burr will break off leaving only a short burr.

Since hardness is a direct function of strength (and therefore of strain also), it is obvious that the highly strained burrs are harder than the basic workpiece. For all practical purposes the average burr hardness at the base of the burr approaches one half the sum of the parent material hardness and the hardness of the same material in the fully work hardened condition

---

*θ must be expressed in radians for use in Equations 7 and 8.

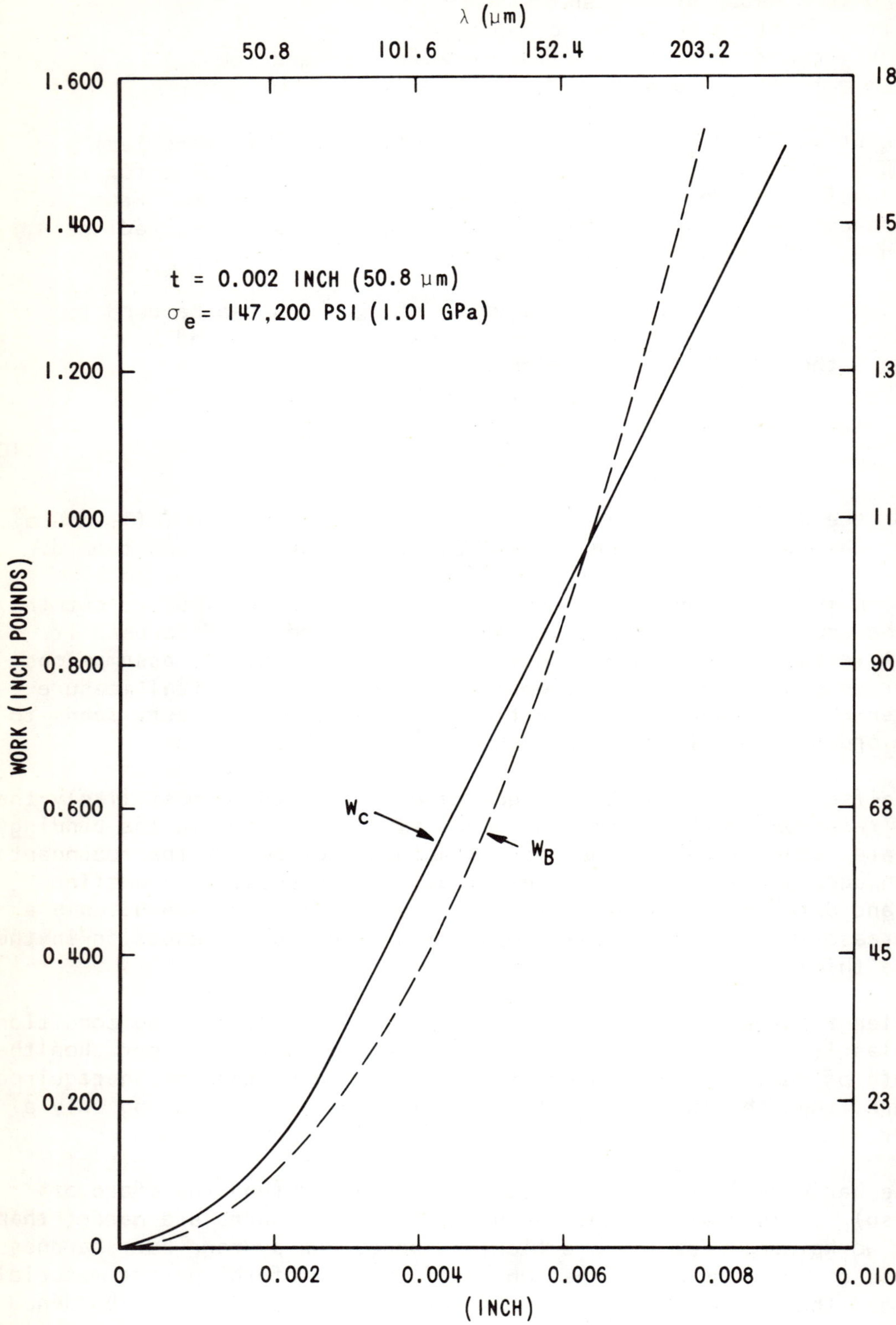

FIGURE 4-9: CUTTING AND BENDING WORK AS A FUNCTION OF DISTANCE λ

To determine the effect that depth of cut (t) has on λ, it is essential to note that F_c is directly proportional to t. Thus for 303Se and the conditions already described, λ can be expressed by

$$\lambda = \left[\frac{2(200)t}{0.002} + \sqrt{4\left(\frac{200t}{0.002}\right)^2 - \frac{2b\sigma_o\varepsilon_f^n t\theta(200)t}{(0.002)\tan\phi(n+1)}} \right] \left[\frac{(n+1)}{4b\sigma_o\theta\varepsilon_f^n} \right] \quad (11)$$

$$= 3.087t \quad \text{for } \lambda \geq \frac{t}{\tan\theta}$$

This burr thickness is a linear function of depth of cut (t).

Tear Burrs

Tear Burrs form when the chip is torn rather than sheared from the workpiece. Although this burr can form in most of the basic cutting processes, it is easiest to conceive for side milling operations. As seen in Figure 4-10, the cutter tooth forces the chip up and forward. As it does so, the sides of the chip are torn from the workpiece. The torn portion remaining on the workpiece is the Tear Burr. This burr can actually form by two mechanisms (Figure 4-11). The tooth can stretch the material in such a way that the burr is formed by a bending mechanism, or it can shear the metal. The mechanism which predominates appears to be a function of cutting velocity and workpiece properties. At this time, however, no quantitative theory exists for producing Tear Burr properties.

The Poisson Burr, the Roll-Over Burr, and the Tear Burr all have one property which has not been discussed. A radius occurs on the back side of these burrs (Figure 4-12). The total shape of the burr cross section then can be expressed by length b_l, thickness b_t, and radius b_r. While some individuals consider b_{min} as burr thickness, it is obvious that when all the burr must be removed it is b_t which best defines thickness. Perhaps more significantly b_{min} cannot be easily defined for some burrs (Figure 4-13).

For Poisson Burrs this radius is actually an exponential function (Equation 2). Although a theory has not been developed for b_r for a Roll-Over Burr, it would appear that some of the theory developed for sheet metal bending is applicable.

Since all three of the plastic deformation burrs are the result of plastic strains, and since most materials are strain hardening, the burrs are harder than the parent material. Using the Meyer Hardness Number (MHN) for example,

$$\text{MHN} = 2.8\ \sigma_o\varepsilon^n \quad (12)$$

In practice, Equation 12 is difficult to use because the strain varies across the cross section and length of the burr.

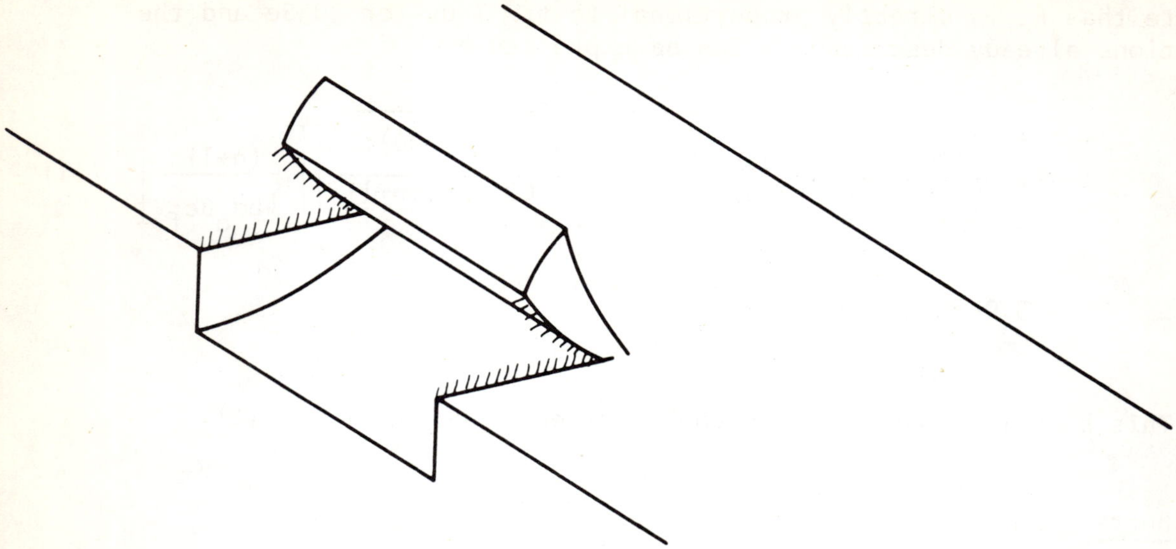

FIGURE 4-10: SEPARATION OF INITIAL CHIP IN SIDE MILLING

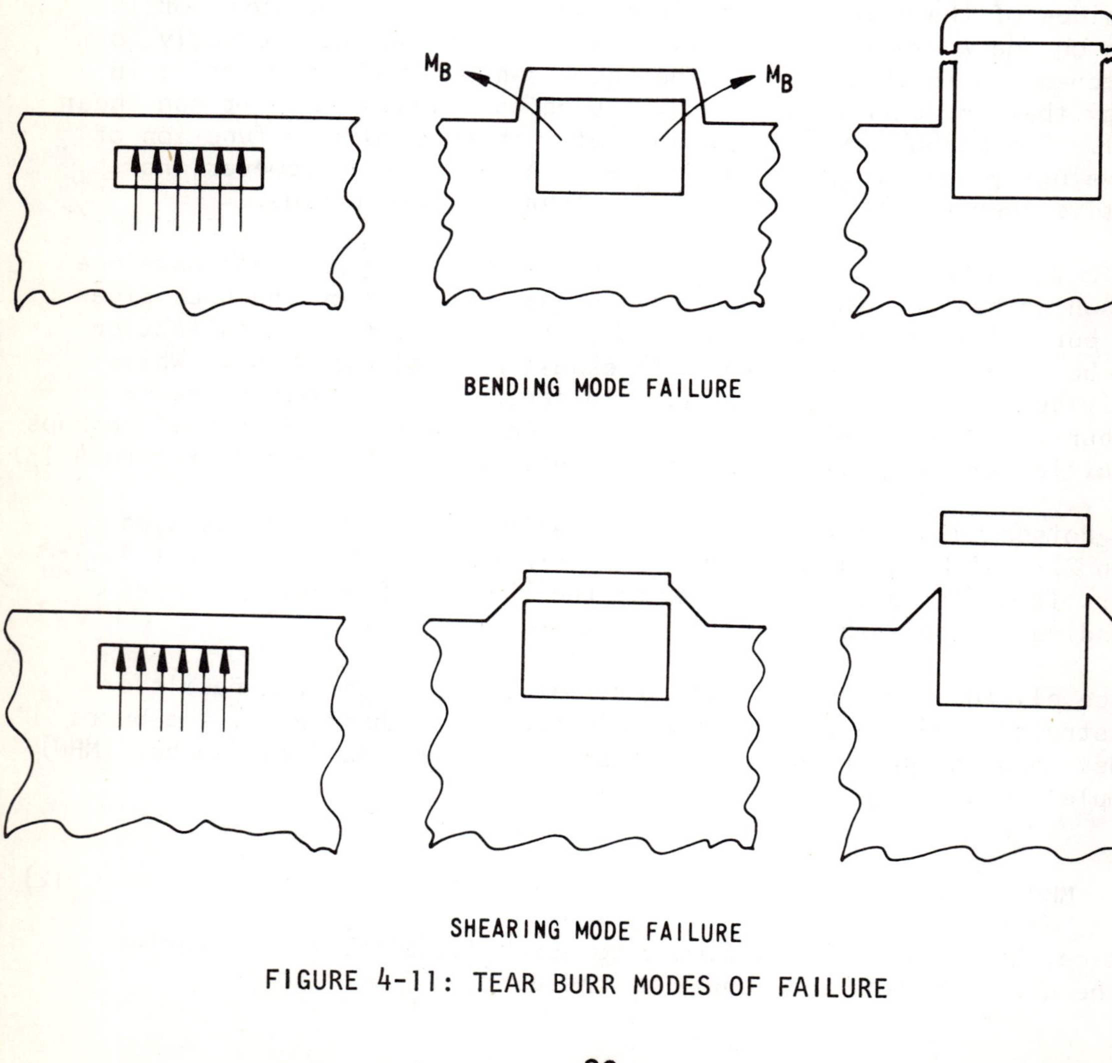

FIGURE 4-11: TEAR BURR MODES OF FAILURE

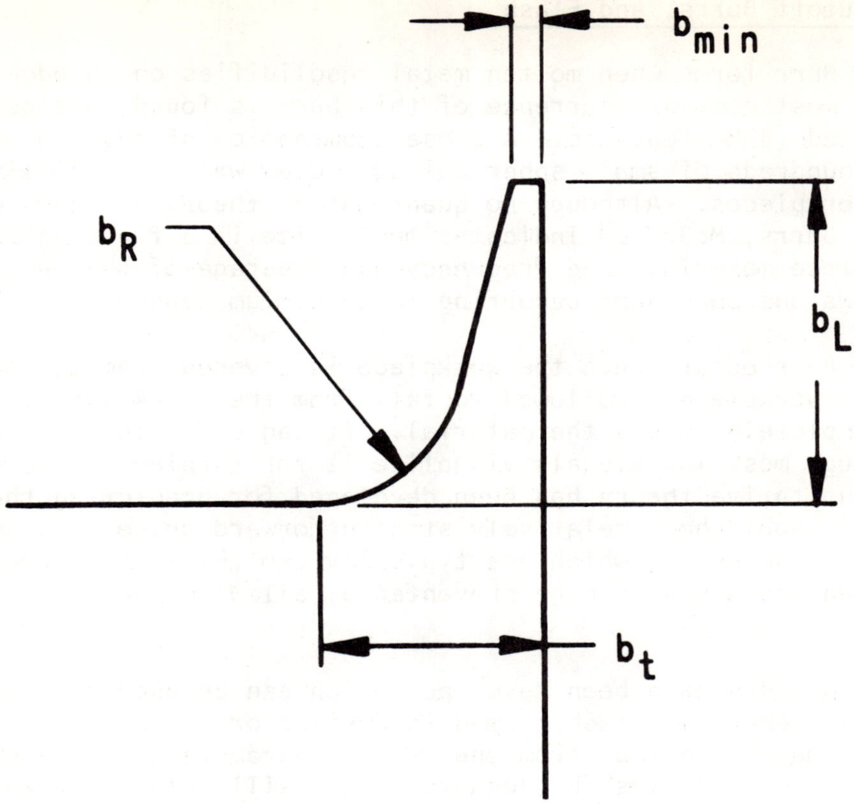

FIGURE 4-12: CROSS SECTION OF TYPICAL MACHINING BURR

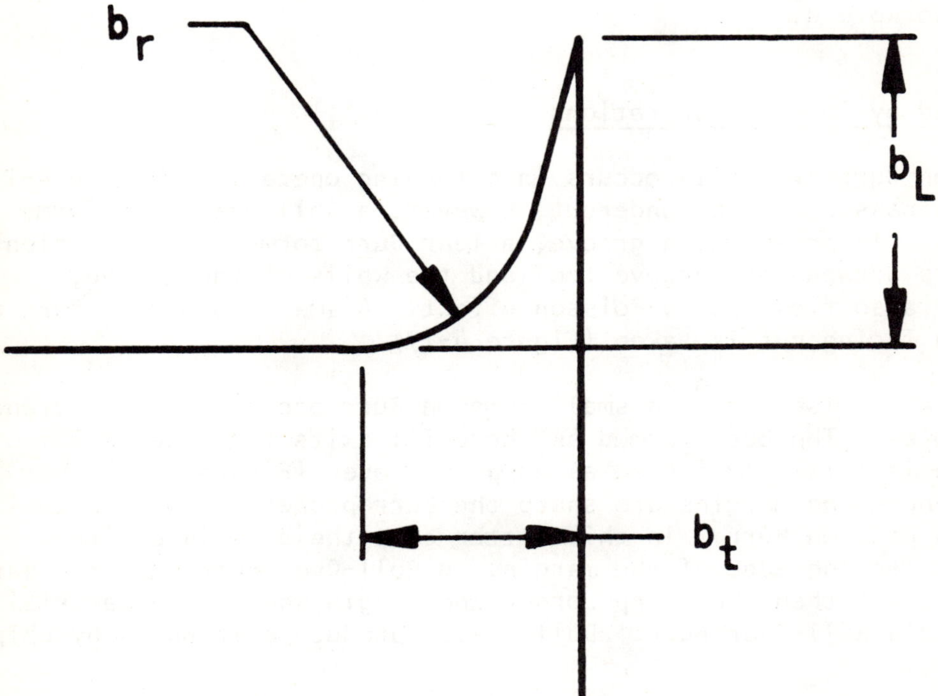

FIGURE 4-13: CROSS SECTION OF BLANKING AND SOME MACHINING BURRS

Recast Burrs, Cutoff Burrs, and Flash

The Recast Burr forms when molten metal resolidifies on an edge of the workpiece. The most common occurrence of this burr is found on electrical discharge machined (EDM) features. A close examination of edges produced by EDM reveals hundreds of small spherical particles which are firmly bonded to the workpieces. Although no quantitative theory has been developed for these burrs, McBride[8] indicates their size is a function of polarity, workpiece material, and frequency and amperage of the current. Figure 4-14 shows one such burr occurring in beryllium copper.

The Cutoff Burr occurs when the workpiece is severed from the barstock. In this case the workpiece is allowed to fall from the stock before the parting tool completely severs the material. It can occur in any parting operation although most individuals visualize it for turning operations. Although no quantitative theory has been developed for predicting the size of this burr, it should be a relatively straightforward process to develop such a theory. These burrs, which are typically too thick to be removed by most deburring processes, can be prevented by allowing the cut to be completed.

Although equations have been developed which can be used to predict the thickness and length of flash formed in casting or forging processes, these properties may be obvious from the process parameters. Provided the clamping pressure between dies is adequate, flash will only occur when die halves are not perfectly matched. Burr thickness then is equal to the thickness of the voids between mold halves, and burr height is equal to or shorter than the length of these voids. The hardness of these burrs is the same as the workpiece.

Burrs Produced by Specific Operations

A Poisson Burr typically occurs in a turning operation (Figure 4-15). When the tool passes over an undercut, however, a Roll-Over Burr forms (Figure 4-16). In producing a groove, a Tear Burr forms. If sufficient rubbing occurs between the groove tool and the walls of the groove, material will also flow from a Poisson effect. A small Roll-Over Burr also occurs when a facing cut is taken (Figure 4-17).

As indicated elsewhere[9] a small Poisson Burr occurs at the entrance of drilled holes. The burr formed as the drill exits from the bottom of the workpiece is initially formed as chip roll over (Figure 4-18), but if the drill corners and margins are sharp the burr properties will actually be those of a Poisson Burr. In this latter case the burr is entirely formed by the leading edge of the margin. A Roll-Over Burr is first partially formed, and then the sharp corner and margin shears the material at the base of this Roll-Over Burr. Dull drills produce exit burrs by chip roll over only.

In a side milling slotting operation, burrs are produced on eight edges (Figure 4-19). A Roll-Over Burr occurs at edges when the cutter

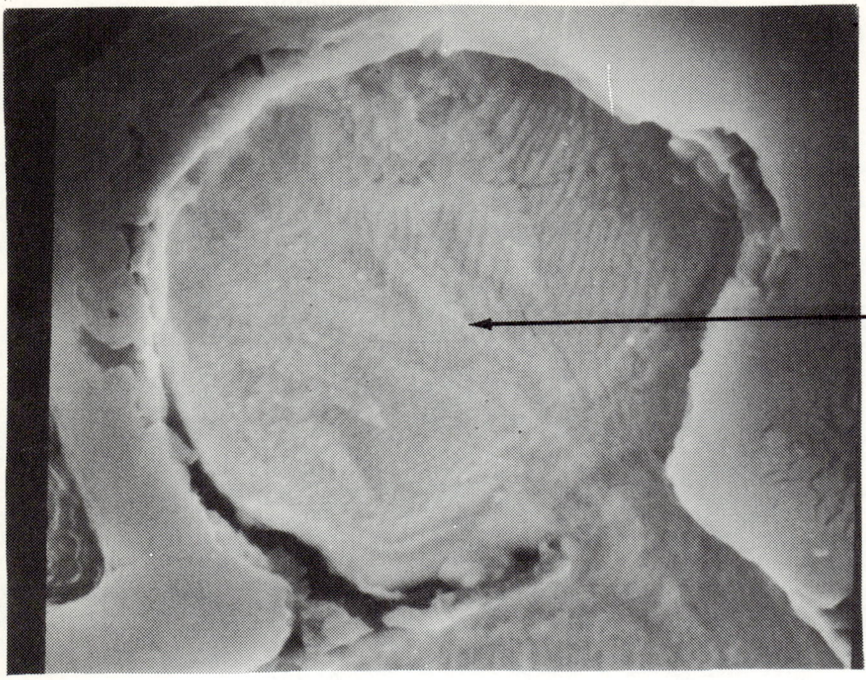

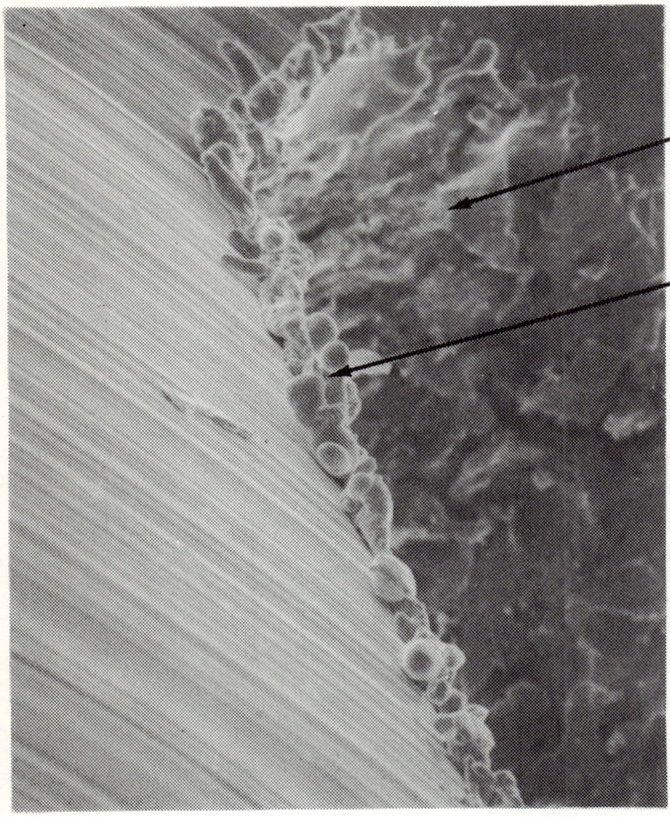

ABOVE: 6000X SCANNING ELECTRON MICROSCOPE PHOTO OF EDM BURR CROSS SECTION ILLUSTRATING FINE GRAIN DETAIL AND JUNCTURE TO PARENT MATERIAL.

BELOW: EDM BURRS ON BERYLLIUM COPPER SLOT (300X).

FIGURE 4-14: EDM BURRS

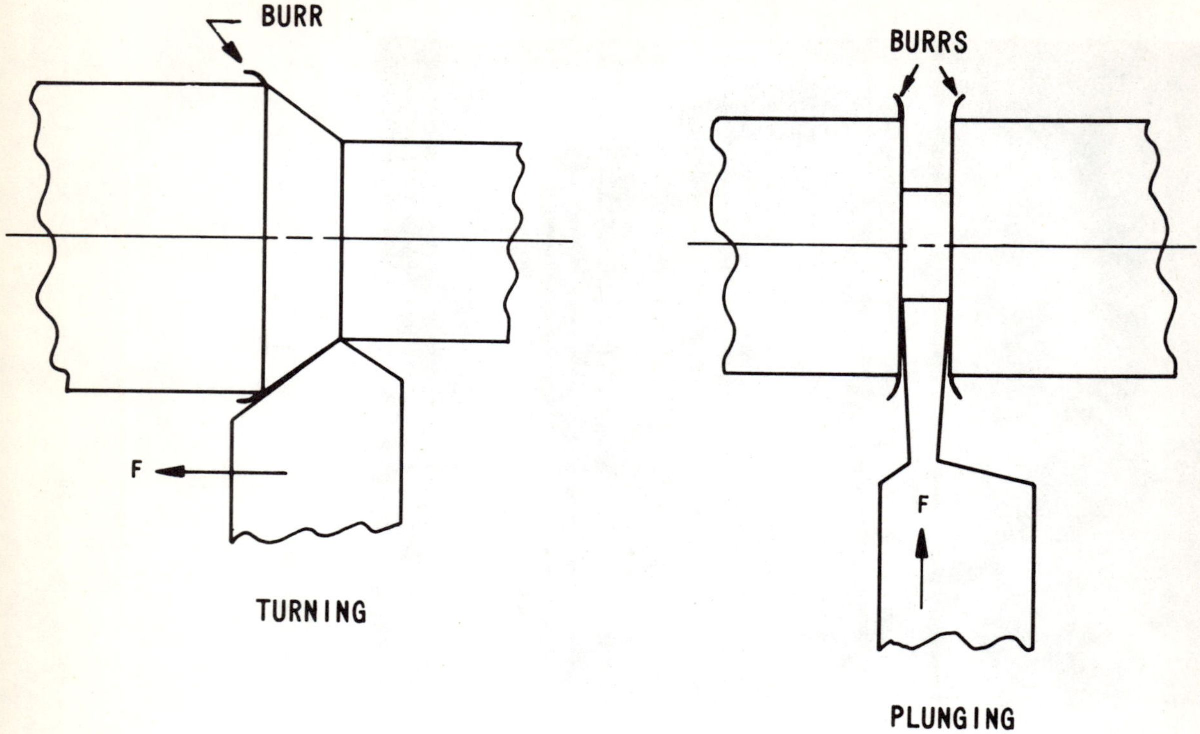

FIGURE 4-15: BURRS FORMED IN THE BASIC TURNING OPERATIONS

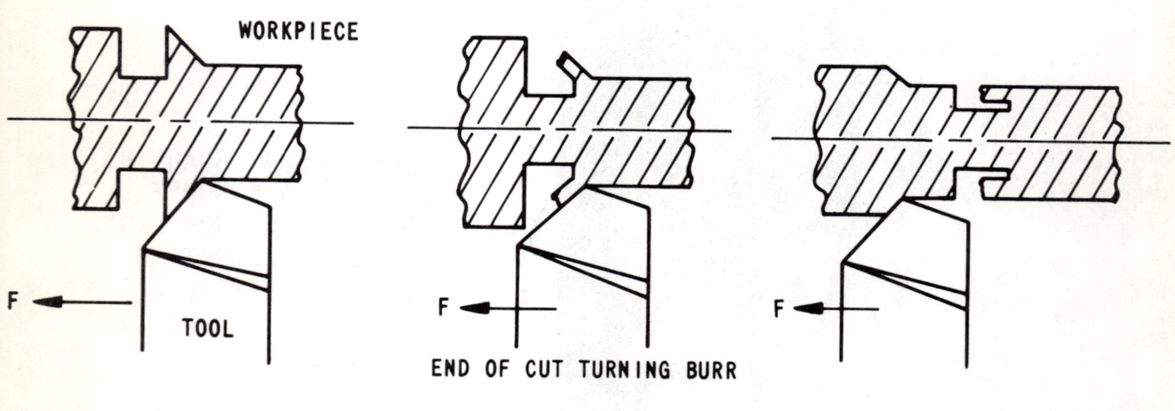

FIGURE 4-16: BURRS PRODUCED IN TURNING

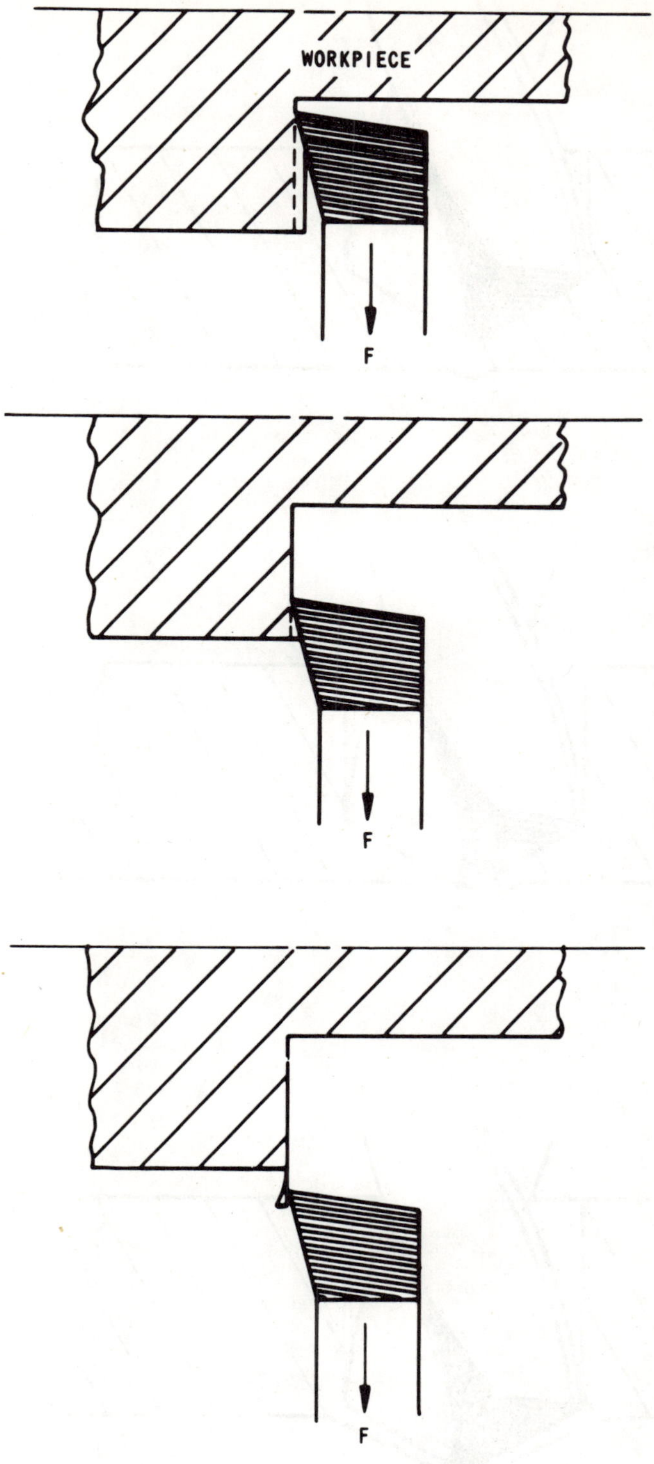

ROLL-OVER BURR PRODUCED IN FACING

FIGURE 4-17: BURRS PRODUCED IN FACING

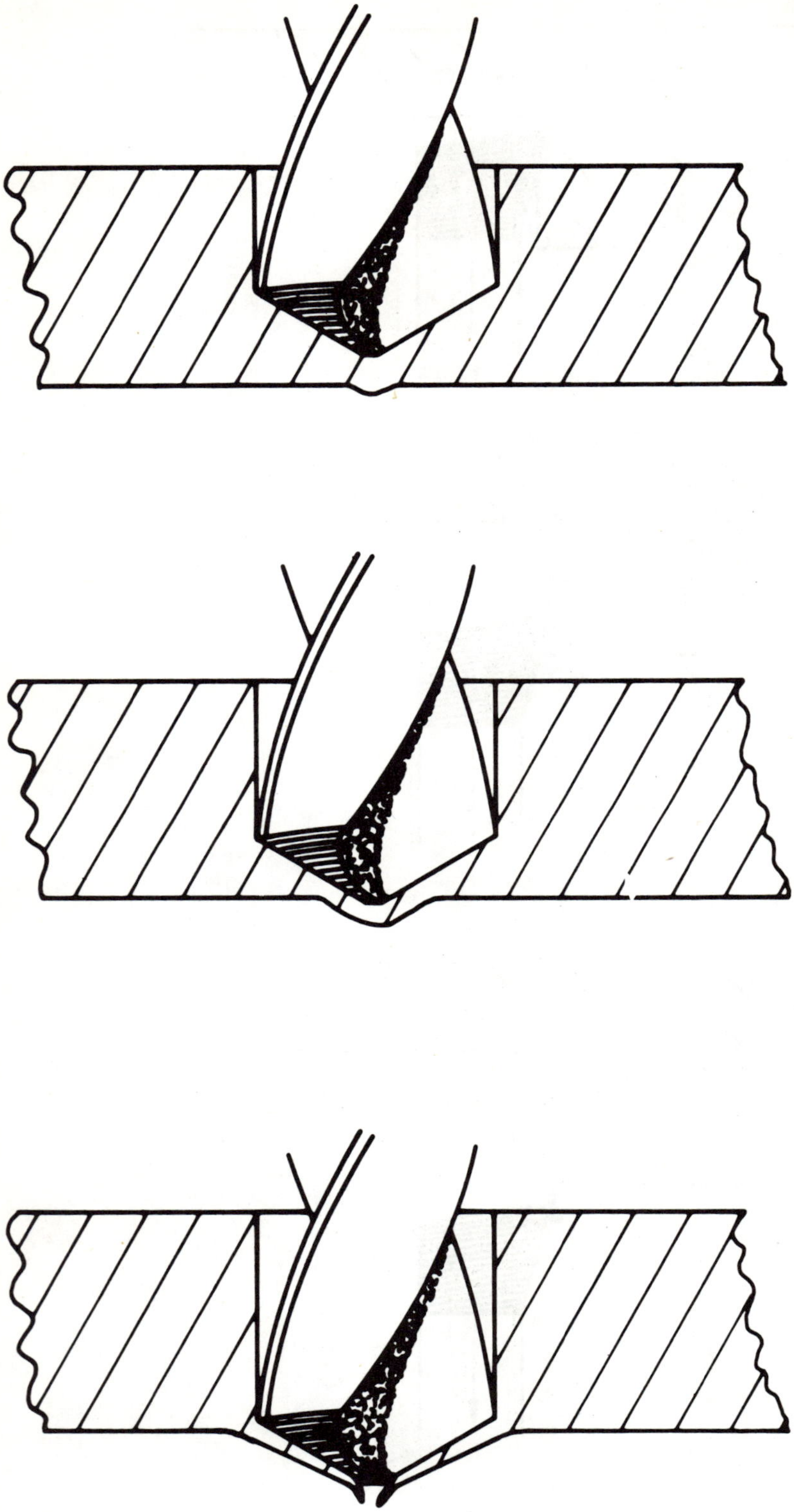

FIGURE 4-18: FORMATION OF DRILL EXIT BURR

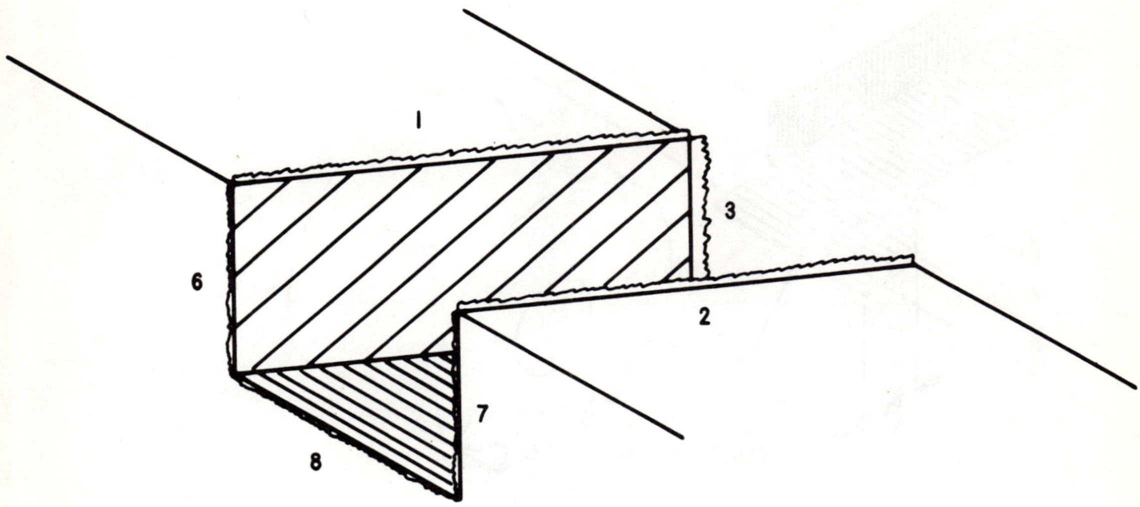

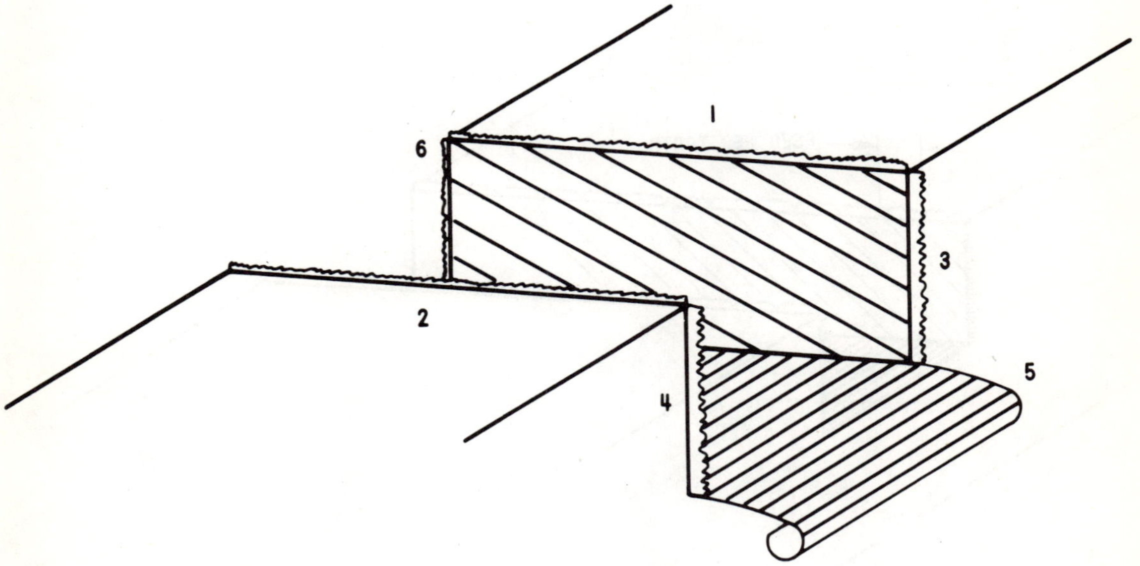

FIGURE 4-19: BURRS PRODUCED IN SIDE MILLING (CONVENTIONAL CUT)

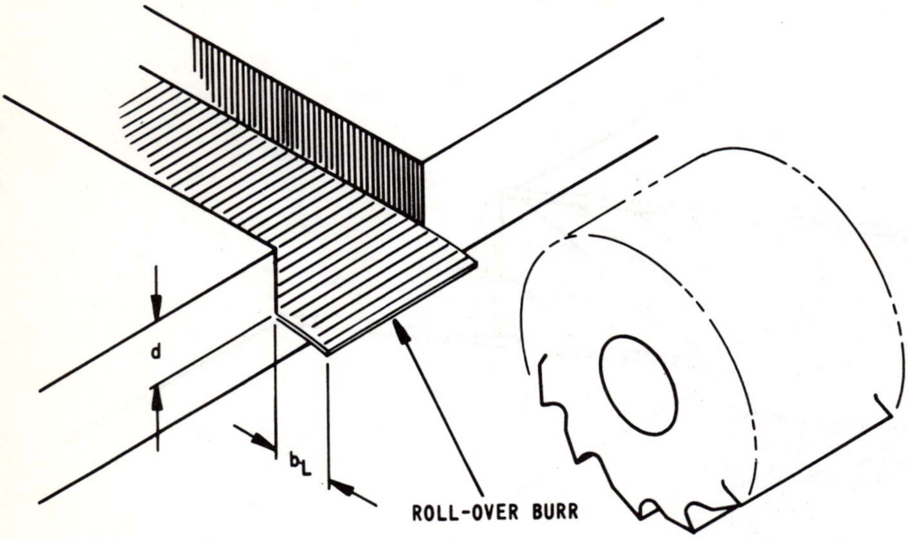

FIGURE 4-20: LOCATION OF ROLL-OVER BURR

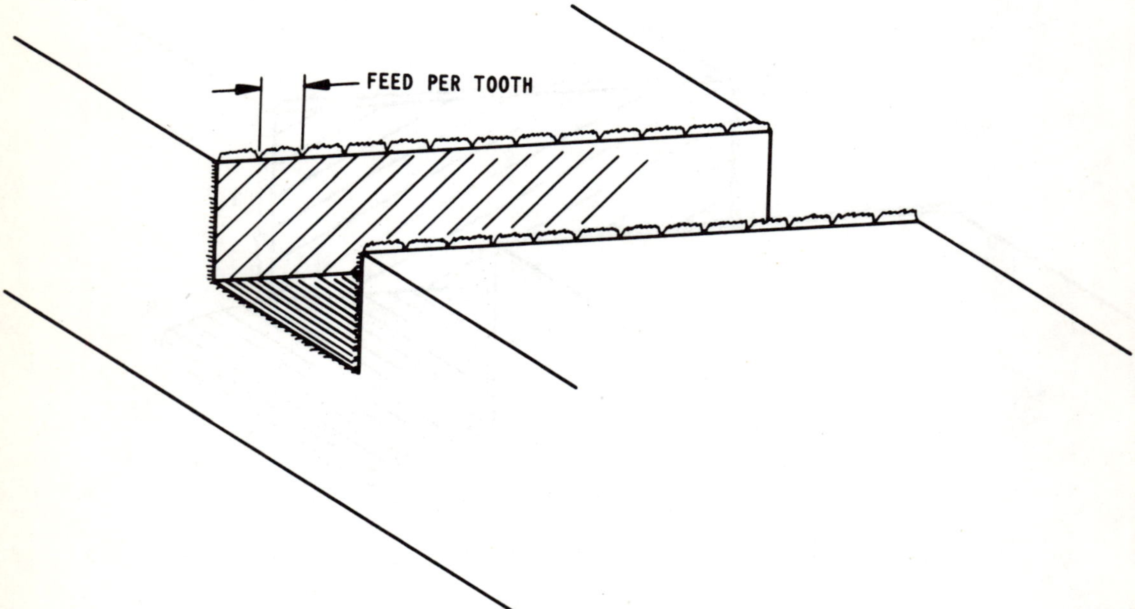

FIGURE 4-21: PERIODIC NATURE OF SIDE MILLING BURR

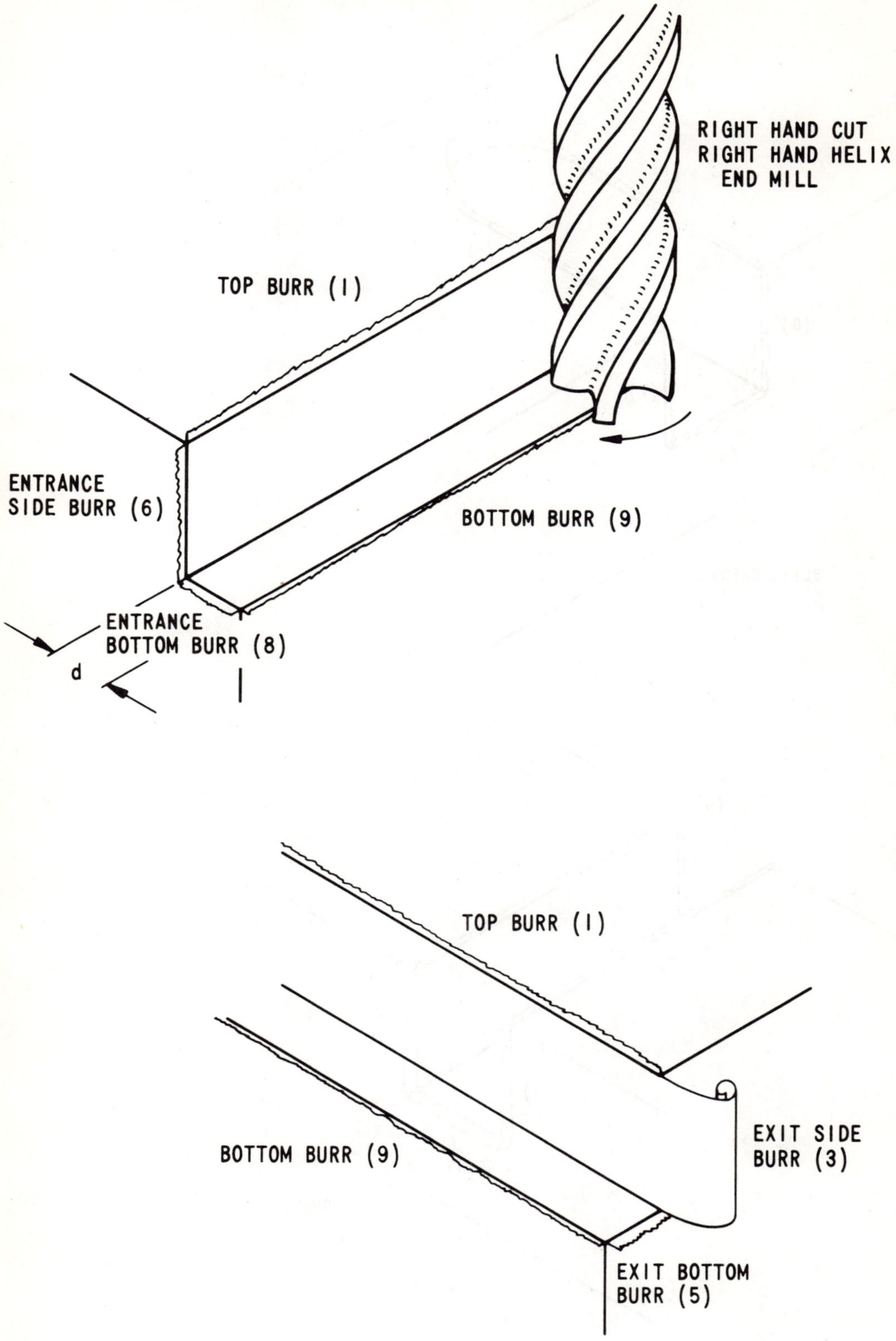

FIGURE 4-22: IDENTIFICATION OF BURR LOCATIONS IN END MILLING

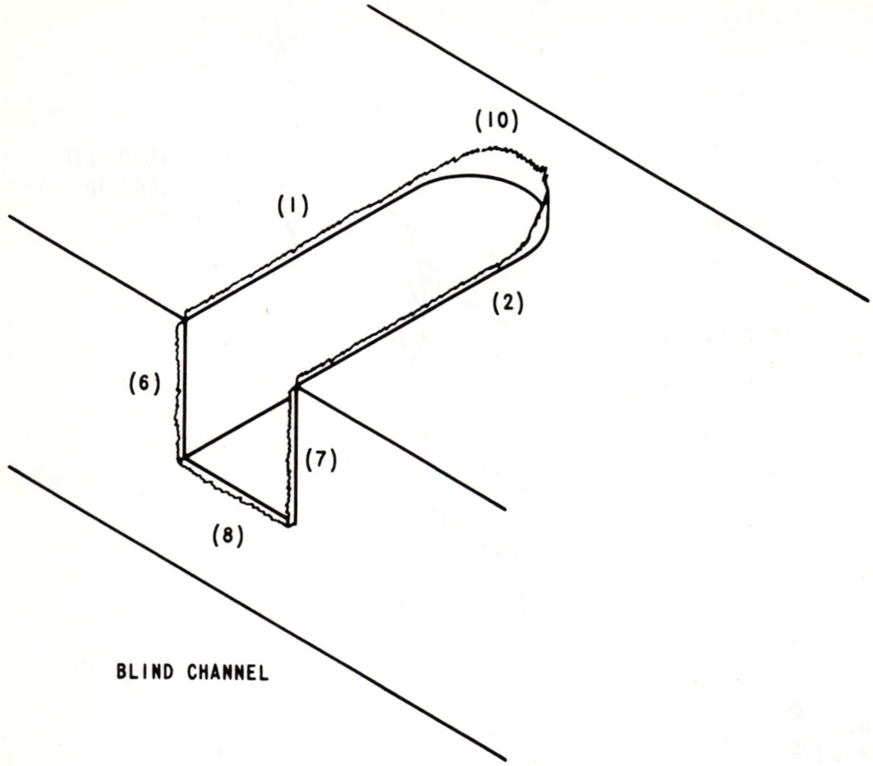

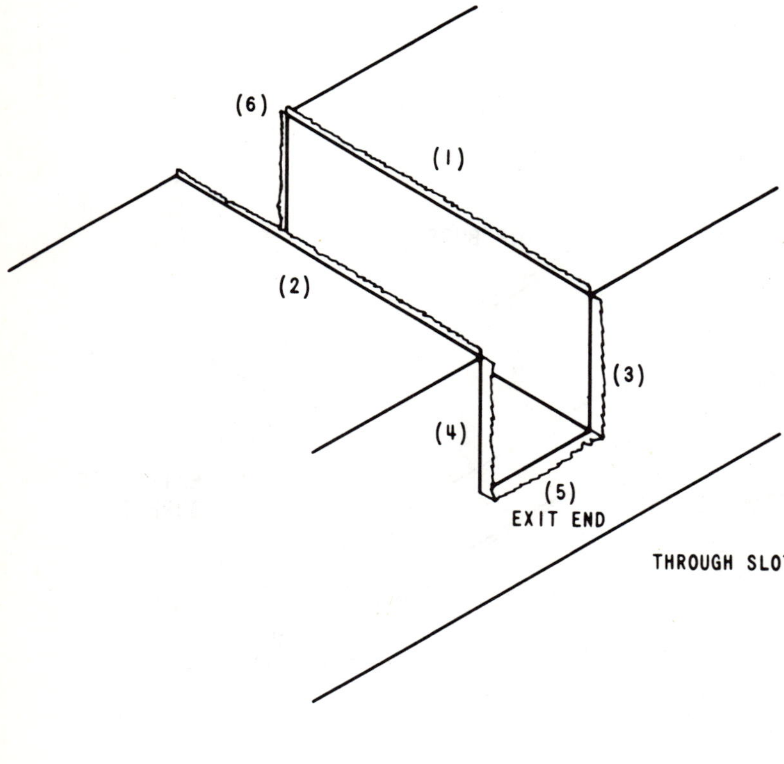

FIGURE 4-23: IDENTIFICATION OF BURR LOCATIONS IN END MILLING

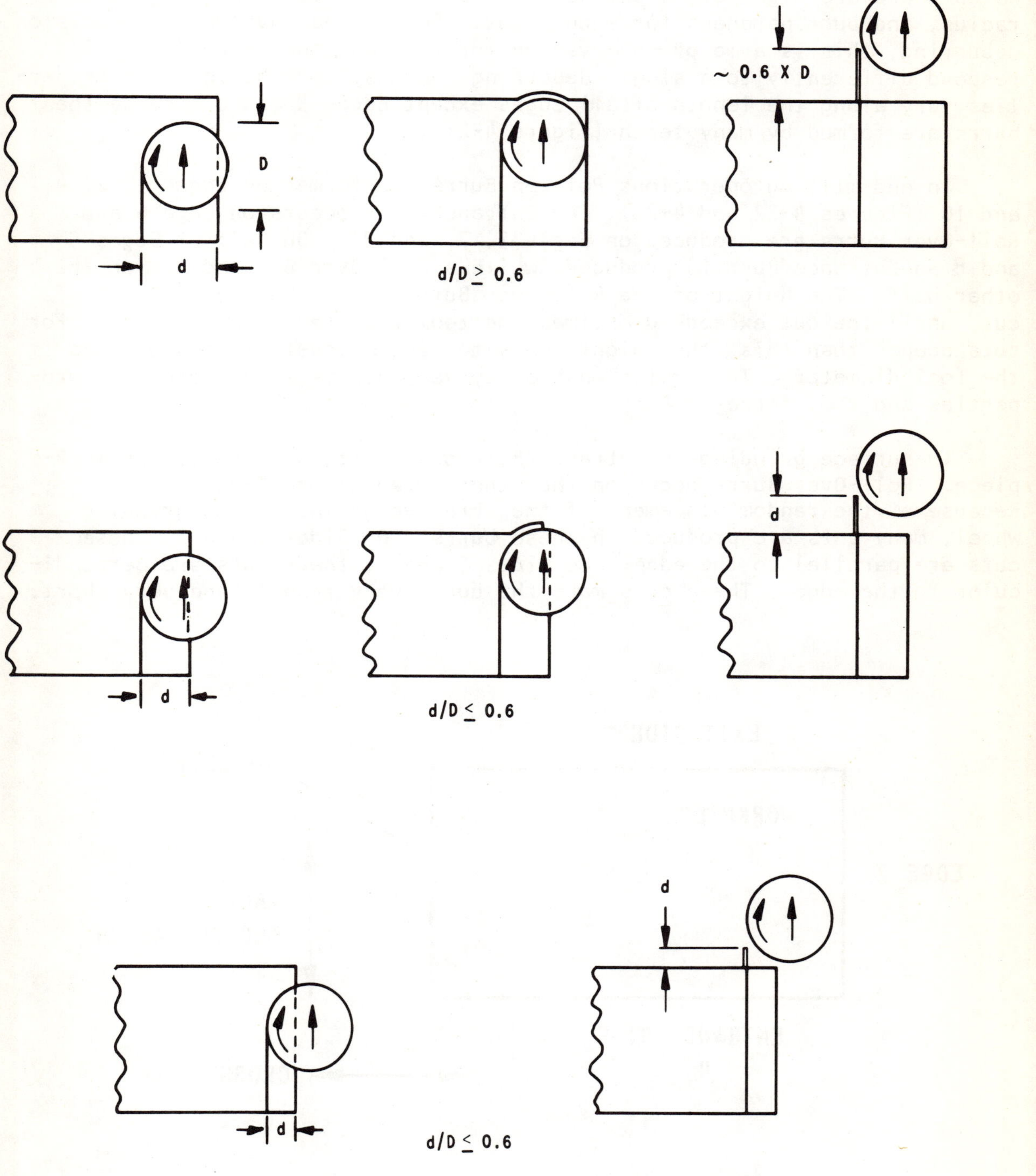

FIGURE 4-24: EFFECT OF RADIAL DEPTH OF CUT ON ROLL-OVER BURR

exits from the part.* The height of the burr is equal to the depth of the cut (Figure 4-20) or less. An Entrance Burr occurs at Edge 8. A hybrid Poisson/Tear Burr occurs at Edges 6 and 7. A Tear Burr occurs at Edges 1, 2, 3, and 4. If the cutter has a helix angle, the properties of the burrs at each edge will be different from those on the other edges. In other words there are eight different sets of burr thickness, burr height, burr radius, and burr hardness for such a cut. For anyone involved in precision deburring, this is a major observation for it means that each edge will respond differently to a single deburring process. Furthermore the properties vary along the length of all edges except Edges 5 and 8 because the burrs are formed by many teeth (Figure 4-21).

In end milling operations Poisson Burrs are formed on Edges 1, 2, 4, and 10 (Figures 4-22 and 4-23). An Entrance Burr occurs on Edge 6 and Roll-Over Burrs are produced on Edges 3, 7, and 9.* On half of Edges 5 and 8 an Entrance Burr is produced, while a Roll-Over Burr occurs on the other half. The height of the Roll-Over Burr equals the radial depth of cut, until the cut exceeds 0.6 times the tool diameter (Figure 4-24). For cuts deeper than this, the height tends to remain constant at 0.6 times the tool diameter. This relationship may vary however with workpiece properties and tool forces.

In surface grinding an Entrance Burr occurs at one edge of the workpiece. Roll-Over Burrs occur on the other three sides (Figure 4-25). Because of the random placement of the abrasive grains in the grinding wheel, many cuts are produced in these burrs. On Sides 1 and 2, these cuts are parallel to the edge. On Sides 3 and 4, these cuts are perpendicular to the edge. These cuts make the burrs very fragile and very short.

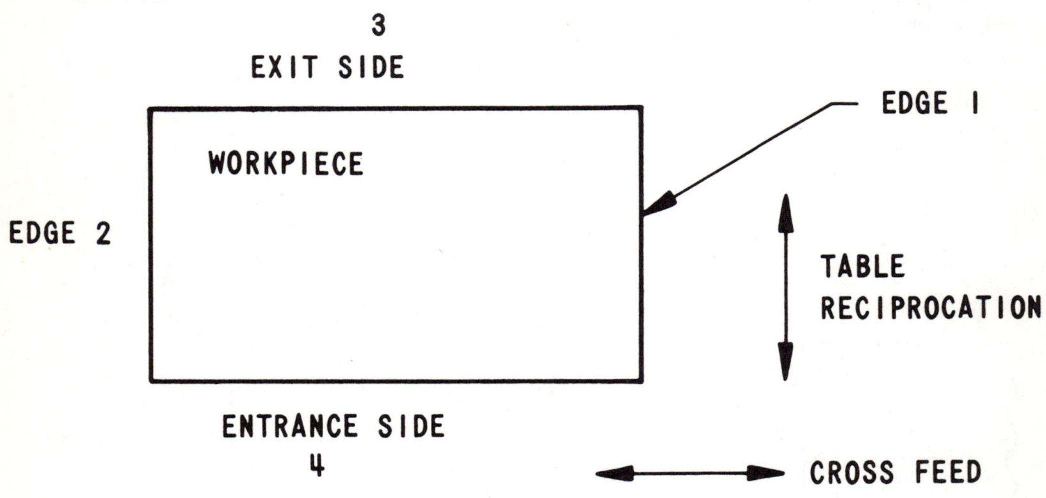

FIGURE 4-25: BURR LOCATION IN GRINDING

*For conventional cuts where the cutter enters the side and Edges 6, 7, and 8 are formed.

Predicting Burr Size

Poisson Burr

The thickness and height of burrs on machined parts can be predicted by the formulas expressed in this chapter if the cutting forces, material properties, and machining conditions are known. Equations 1, 2, and 3 however are difficult to use as a production aid because they require a knowledge of the pressure existing at the cutting edge radius. This pressure cannot be conveniently or accurately measured because of the small surface it acts upon. It is possible to obtain a rough estimate of this pressure by noting that

$$P = \frac{T}{bc} \tag{13}$$

Where T is the cutting force acting at the cutting edge (Figure 4-26) and b and c are defined in Figures 4-27 and 4-28. The contact length c can be expressed by

$$c = \frac{\pi r}{114.59} (57.3 + \gamma + \alpha_o) \tag{14}$$

where the true rake angle γ and the relief angle α_o are expressed in degrees.

Although the force T represents only part of the cutting force, it can be estimated empirically by varying the cutting edge radius and measuring the increase in cutting forces F_x, F_y, and F_z (Figure 4-29). By this means it is possible to derive the force T. This has been done by two researchers for specific workpiece materials.[10,11]

Three factors are significant in this analysis and require further explanation. First, Equations 1, 2, and 3 were developed for a perfectly plastic material. Few materials approach this ideal case. When large values of strain are involved even the normal "ideal" materials vary significantly from this assumption. In addition the theory assumes a state of stress at the cutting edge similar to a pressurized cylinder. The actual state of stress at the point is undoubtedly more complex. As a result of these assumptions it is possible that P/σ_p can be somewhat larger than the ideal value of 1.157. Similarly the assumption that all the plowing force T occurs at the cutting edge radius is an approximation which is close but not entirely accurate.

Secondly, flank wear can greatly affect the size of burrs produced. Thus although the cutting edge radius may be small, a separate mechanism on the flank can produce a large burr.

The third aspect of interest is the size of the cutting edge radius. The 0.005 inch cutting edge radius used in the sample calculation may seem grossly out of place to many individuals. A review of cutting tool materials, however, will soon banish this concern. While high speed steel (HSS) tools typically have near sharp edges, the nature of carbides and ceramics prevent them from approaching a similar sharpness. The sintered nature of carbide limits the sharpness obtainable on them; ceramic tools cannot have sharp edges without inducing catastrophic tensile stresses at the edge.

As discussed by Albrecht[12] the typical radius on a high speed steel tool is 0.0002 to 0.0006 inch. Siekmann[13] notes that radii of 0.004-0.005 inch on carbide tools produce longer tool lives in AISI 1045 steel than sharper edged tools. Mayer and Stauffer[10] recommend a radius of 0.002 inch or less for carbide tools. Agnew[14] presents several recommendations including radii of up to 0.007 inch for carbide tools. For ceramic tools in cast irons Siekmann recommends a 0.005 inch or larger cutting edge radius, and a 0.002 to 0.003 inch radius for most steels.

Most researchers have found that a built-up-edge (BUE) occurs on many if not most cutting conditions. This BUE effectively increases the cutting edge radius. As noted by Albrecht[12] even a small BUE can greatly increase the effective radius.

Entrance Burr

Strain hardening data is available for most materials to provide a qualitative comparison of Entrance Burr size. Quantitative comparisons cannot be made at the present time except on an empirical basis.

Roll-Over Burr

Roll-Over Burr properties can be estimated easily from the equations presented. Actual burr thickness however will be approximately one third of that predicted because of redundant work occurring in bending.

Tear Burr

While some mathematical theory exists to describe Tear Burr formation[2] it is not sufficient to predict actual burr size.

Burr Measurement

In the past, measuring burrs meant measuring burr height (length). As a result of recent studies, however, it is obvious that burr thickness is the single most important size parameter for burr removal. In most situations the combination of thickness and height, rather than just one, is important. Unfortunately measuring these properties is not as straightforward as one might expect.

The Foundations of Burr Measurement

There are four basic foundations to build on when measuring burrs:

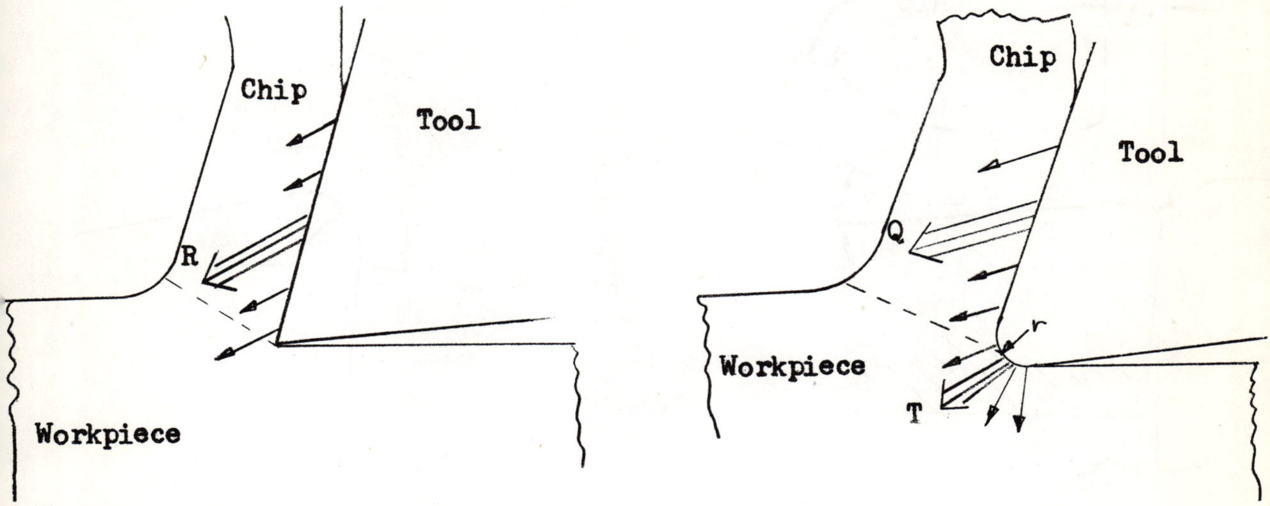

FIGURE 4-26: FORCE DISTRIBUTION ON SHARP AND ROUNDED TOOL EDGE

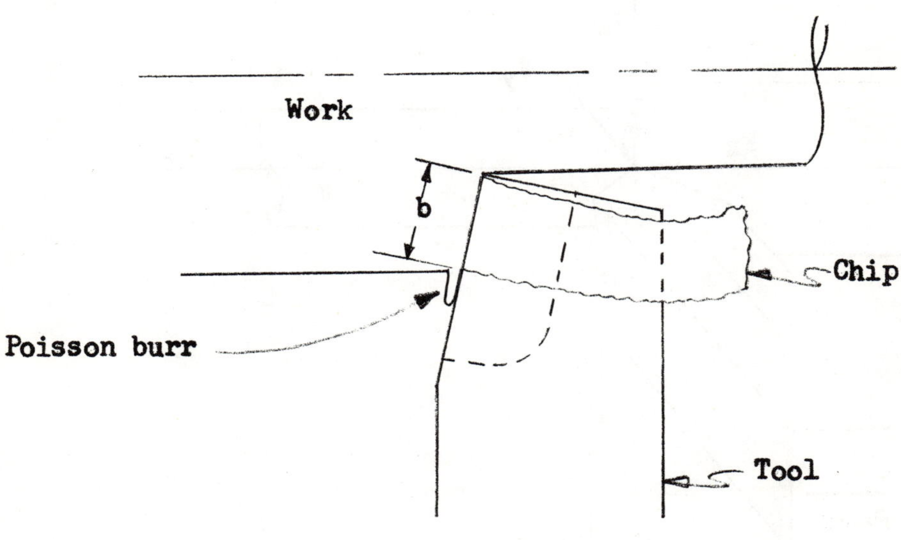

FIGURE 4-27: LOCATION OF POISSON BURR

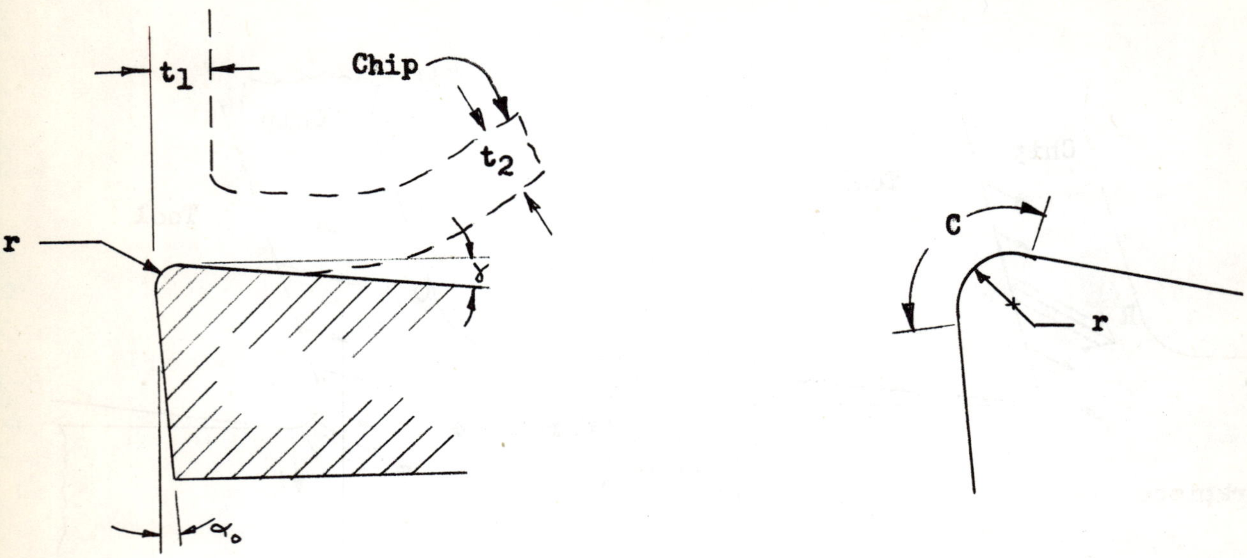

FIGURE 4-28: CONTACT LENGTH AT CUTTING EDGE

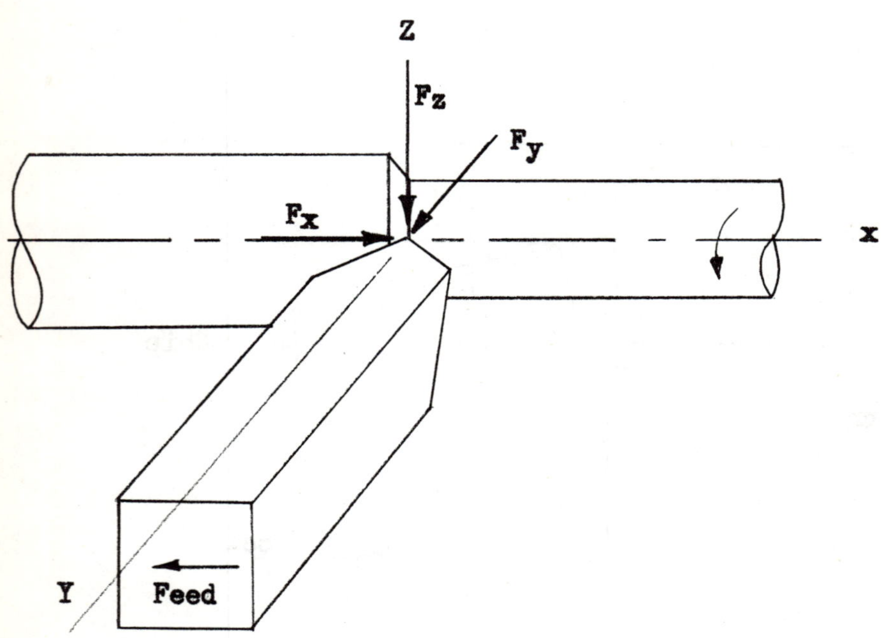

FIGURE 4-29: FORCE COMPONENTS

- Burr properties depend on which of four metal cutting mechanisms produced the burr.

- Burr height is not the sole property: thickness, shape, and toughness are equally as significant.

- The ability to obtain repeatable burr measurements depends on the definition of each property.

- Repeatable results can only be assured by measuring several samples.

Burr Properties

What makes a burr difficult to remove? The answer varies with each situation but the following list is typical.

- Burr Thickness
- Burr Location
- Burr Toughness
- Burr Height (Length)
- Workpiece Tolerances
- Burr Radius

For most applications the above list is in approximately the correct order of significance. It is burr thickness at the root of the burr, not burr height, which makes a burr most difficult to remove. This is easy to see by looking at two extremes. Grind a piece of 303Se stainless steel, then take a dull end mill and cut a channel in the same workpiece. Compare the burrs. At 0.001 inch per pass, a 0.010 inch downfeed will produce a 0.007 inch high by 0.0002 inch thick grinding burr on one edge. One of the milling burrs will be 0.003 inch thick and 0.005 inch high. Experience tells us that the grinding burr can be readily removed with a finger nail. The milling burr requires considerable effort to remove, yet it is shorter. A similar comparison can be drawn between hardness and height.

For a specific machining operation, is burr height directly relatable to burr thickness? From a theoretical standpoint the answer is yes, but from a practical standpoint, the relationship is not readily evident. Nowhere in the literature has an empirical relationship yet been described. A few authors maintain that burr heights will be roughly 20 times burr thickness. While for one material, one machining operation, and one point in tool life this may be true, it is by no means universally true. Although studies are currently underway to determine such relationships between burr height and thickness, it will be several years before sufficient data is available to establish a universal relationship. As a result it behooves burr researchers to measure all the burr properties which are meaningful.

Burrs assume many shapes. As shown in Table 4-1 they can be triangular, rectangular, or spherical in cross section, and they can be uniform or ragged (irregular), or feather-like in height. Each of these shapes indicate a different metal cutting phenomenon and as such the shape should be documented in any study.

TABLE 4-1: TYPICAL BURR GEOMETRIES

Operation	Typical Burr Shape
Blanking	Triangular cross section
Fine Edge Blanking	Thin rectangular cross section
Milling	Rectangular or triangular cross section, and feathery, uniform or ragged height
Drilling	Triangular or rectangular cross section, uniform or ragged height
Turning	Rectangular cross section
EDM Burrs	Spherical
Laser Burr	Globular

Figure 4-30 illustrates the cross section of a typical burr. As seen there, a radius exists on the back side of the burr. This radius determines how burr thickness varies with distance from the surface and as such is a significant burr property although it has only been discussed by one researcher.

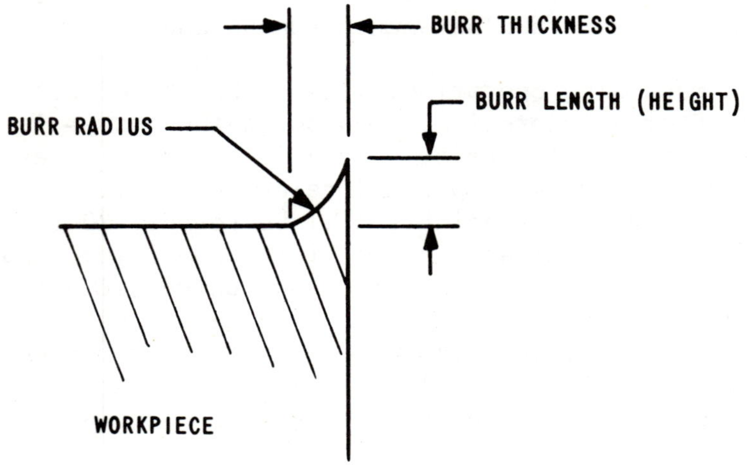

FIGURE 4-30: CROSS SECTIONAL VIEW OF TYPICAL BURR

Burr toughness is not measured directly. Burr hardness, however, is readily measureable and at least qualitatively can be related to burr toughness. For a given material, higher hardness indicates lower toughness.

Burr hardness by itself reveals little about deformation. It is important to also know the hardness of the workpiece. The burr hardness can then be related back to workpiece hardness and the strain produced by burr formation. Since workpiece hardness can vary noticeably on a single workpiece it is important to measure hardness in the general vicinity of the burr.

Measuring burr hardness requires preparation of metallurgical mounts. Although this is time consuming, these mounts also allow the investigator to measure four other properties:

. Burr thickness at base
. Burr height
. Burr shape
. Grain structure (macro patterns and metallurgical phase)

The flow lines and the crystalline phase in the burr are further measures of metal cutting mechanisms. A knowledge of burr crystal structure is also useful in chemical and electrochemical deburring. In some metals, work hardening changes the crystal structure to a phase which is more readily attacked by chemicals than the normal state. This allows the work hardened burr to be dissolved faster than the parent metal.

Defining the Proper Statistic

In measuring burr height, both Wang[16] and Kienzle[17] emphasize that measurements of maximum burr height should not be used. In measurements of blanking burrs, an average burr height should be used. Wang notes that the repeatability of the average burr height is 50 times greater than the maximum burr height. It is not surprising, therefore, that investigators who attempt to study tool geometry influences by monitoring maximum burr height find little correlation. This same effect can be observed in measurements of drilling and milling burrs. The maximum burr height should not be used as a burr measurement. Both mean and median measurements will provide more repeatable and explainable results.

The measurement of burrs requires two principal considerations:

. A definition of a burr (what is a burr and what is not); and
. An understanding of what the information will be used for.

A firm understanding of what constitutes a burr is essential if any practical use is to be obtained from the data. Both the experimenter and the individual making the measurements must understand which projections are burrs. Is a burr something which easily breaks off? These definitions must be understood and expressible by and to inspectors in the plant and individuals in other plants and institutions. Without clear definitions the measurement and classification of burrs will lead to some erroneous conclusions.

The techniques used to study and measure burrs may vary with the intended use of the data. For example, an in-depth study of burr formation may necessitate recording the complete profile of burrs. In this case the height of a burr would be continuously measured around the diameter of a hole. The thickness of a burr might be similarly recorded as a machine traverses the circumference of the hole. This information could then be compared with analytical predictions of burr properties.

For a general understanding of how burr properties vary with machining parameters, the average burr height, thickness, and hardness may be more functional than a continuous recording. With this information, one can obtain a working appreciation for the impact of machining conditions on burr properties, and the same information can be used to document the capabilities of each deburring process.

If the purpose of the study is to determine how to prevent the formation of high fragments then the maximum burr size would have to be recorded. This type of information would be useful to those who have to prevent burr roll-over, which is common in vibratory deburring.

In the first situation, it would not be reasonable to use a mechanical dial indicator to measure burr height fluctuations. The fluctuations are too sharp and the indicator ball will not ride the crest of the burr. In the second case, one may wish to remove all the loose burr fragments before making measurements. For this type of study, one has to decide whether easily removable fragments truly represent the phenomenon of interest and whether they in any way influence the deburring process. In the third situation, a dial indicator could bend the high burr fragments enough to provide meaningless data.

In any approach, it is essential in reporting the data to describe the method used to measure the burrs so that subsequent investigators and inspectors have a common base for comparing results.

Burr Variability

Burr properties vary considerably from part to part and even within the same part. As a result, it is essential that several measurements be made at each condition tested. Wang's study of blanking burrs[16] indicated that 24 measurements on each slug are required to provide a positive "average" measurement. While 24 measurements per part may seem excessive it does reemphasize the one dominant property of burrs--they vary considerably in size. As an example, an abusive facing milling operation in 303Se stainless steel produced an average burr height of 0.0095 inch (Table 4-2). Measuring 100 edges, however, would reveal heights varying from 0.0 to 0.020 inch ($\bar{x} \pm 2\sigma$).

Blanking is considered to be one of the more repeatable manufacturing processes. If multiple measurements are required to define a usable blanking burr property, it is obvious that multiple measurements are man-

TABLE 4-2: COMPARATIVE BURR SIZES GENERATED BY MILLING OPERATIONS

Material	Condition	Burr Thickness, in.	Burr Height, in.	Hardness (Knoop) Parent Material	Hardness (Knoop) Burr
Aluminum	Abusive Face Milling	$\bar{\chi}$ 0.0078	0.0138	115	129
		σ 0.0011	0.0045	4	9
	Gentle Profile Milling	$\bar{\chi}$ 0.0023	0.0015		
		σ 0.0003			
Low Carbon Steel	Abusive Face Milling	$\bar{\chi}$ 0.0046	0.0255	230	275
		σ 0.0010	0.0191	7	17
	Gentle Profile Milling	$\bar{\chi}$ 0.0014	0.0024		
		σ 0.0002	0.0009		
Stainless Steel 303Se	Abusive Face Milling	$\bar{\chi}$ 0.0049	0.0095	305	428
		σ 0.0020	0.0053	0	56
	Gentle Profile Milling	$\bar{\chi}$ 0.0020	0.0013		
Beryllium Copper	Abusive Face Milling	$\bar{\chi}$ 0.0045	0.0189	249	268
		σ 0.0008	0.0017	7	22
	Gentle Profile Milling	$\bar{\chi}$ 0.0007	0.0016		
		σ 0.0005	0.0011		

Abusive Face Milling--Dull cutter and .025 inch depth of cut.

Gentle Profile Milling--Burr produced by the bottom of a 1/4 inch diameter end mill 2730 RPM, 9-3/16 IPM.

Knoop Hardness Number with 100 Gram Load (249 = Rc 24; 115 = RB 73)

$\bar{\chi}$ = average measurement

σ = standard deviation of measurements

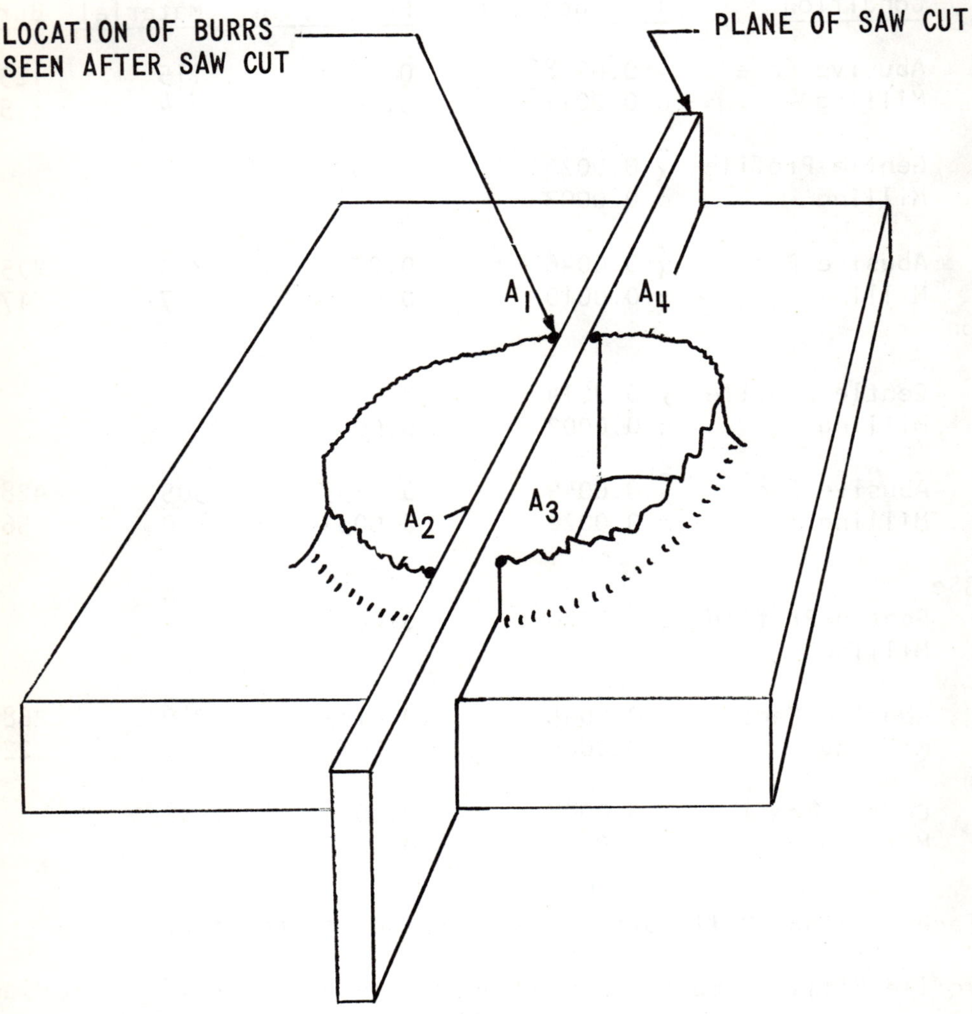

FIGURE 4-31: LOCATION OF BURRS SEEN AFTER SAW CUT

datory when measuring drilling or milling burrs.

In studies of burrs produced by small drills essentially two averages are taken. After cross sectioning each hole, two views of the same burr are revealed (A1 and A2 in Figure 4-31). The thickness of these two burrs are pooled to give an average burr thickness. This is repeated on three to ten specimens and the average of the averages is used. Although not immediately evident, the average within a specimen provides a statistically more reliable quantity than using the individual measurements. This same technique can be used to advantage on other burr properties.

Most tools including dies tend to produce a highly erratic burr during the first few cuts. For this reason, any test involving burr measurements should utilize tools which have been "worn" in.

Blanking, Piercing, and Stamping Burrs

Conventional blanking typically produces a triangular shaped burr. While the triangular shape is also produced by other operations, the mechanism forming the burr is essentially unique to blanking. Kienzle[17] and Seidenberg[18] illustrate the shape of these burrs. While no data is available on blanking burr thicknesses, it is reasonable to assume for this type of operation that thickness is proportional to height. Thus burr height can be used as an effective measure of burr thickness for burrs produced in blanking.

Blanking burr height can be measured by mechanical indicators[16,18,19] or by optical microscopes.[18] While burr thickness can be measured by slow traversing profilometers[16], metallurgical mounts showing burr cross section provide a more positive method. The profilometer, however, allows one to make several thickness measurements on a specimen while only two measurements can typically be obtained from metallurgical mounts. As pointed out earlier, however, one advantage of the metallurgical mount is that it can also be used to obtain burr hardness.

Burr size can also be measured by casting a silicone rubber impression of the burr. These RTV molds can then be sliced in thin slices to provide a cross sectional view of the burr. Repeating this after deburring provides a permanent record of before and after conditions.[20] While this process is time consuming, it does provide an accurate picture of edge conditions. Light section microscopes and scanning electron microscopes (SEM) are also useful tools for measuring burr size.

In most production applications burr hardness should be measured such that an average hardness is obtained. This hardness should be measured at the juncture of the burr and the parent material (Figure 4-32). The necessity for using an average value can be appreciated by viewing Figure 4-33. As in the bending of beams this strain is linearly proportional to the distance from the neutral axis. The strain in the roll-over burr shown there varies considerably across the thickness. Since hardness and strain are directly related, hardness will also vary across the burr cross section. While the roll-over burr is only one of four types of burrs, a similar

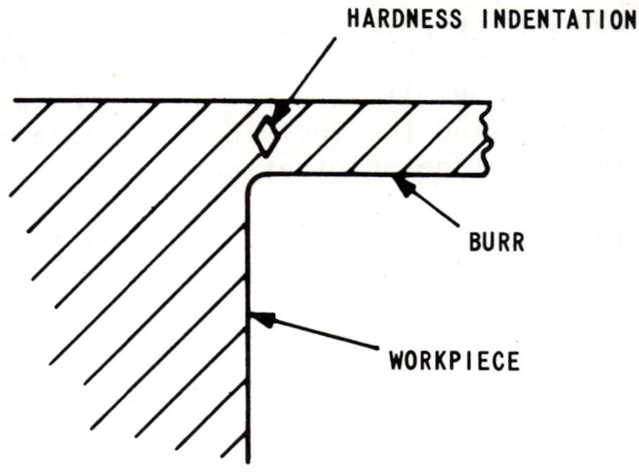

FIGURE 4-32: LOCATION AND ORIENTATION OF HARDNESS MEASUREMENT

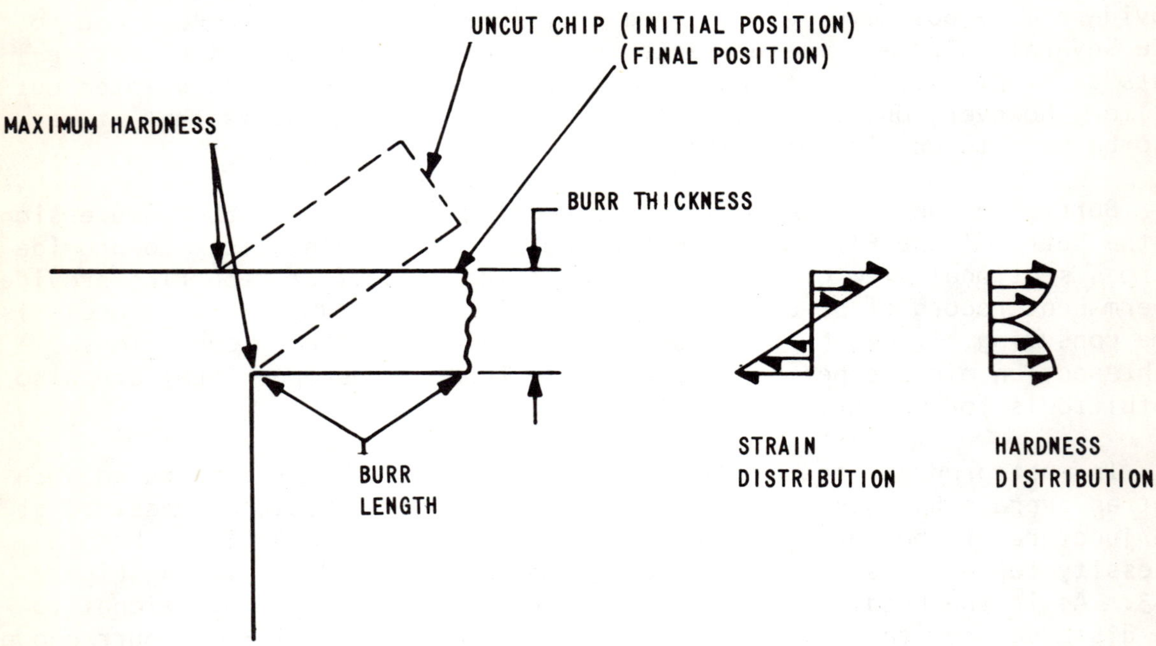

FIGURE 4-33: STRAIN AND HARDNESS DISTRIBUTION IN ROLL-OVER BURR

hardness variation will be observed in the other types. Burrs in all materials tend to be harder than the parent material. This is a result of work hardening. By the nature of blanking and machining operations, both the workpiece and the scrap are strained in the vicinity of the burr and this straining hardens the sheared area. The majority of workpiece materials tend to be work hardened to some extent.

Kienzle[17] recommends measuring burrs on the slug rather than the perforated part for the sake of simplicity and uniformity. Theoretically the burr on the perforated part should be similar to the one on the slug.

Drilling Burrs

The burrs formed at the entrance of a hole tend to be similar in size and shape to blanking burrs. They are basically triangular, reasonably uniform in size, and thus measurable by the same techniques used for blanking burrs.

Drill exit burrs, however, fall into four basic categories:

- Small uniform burrs (Figure 4-34)
- Exceedingly large but uniform burrs (Figures 4-35 and 4-36)
- Large ragged burrs (Figure 4-37 and 4-38)
- Burrs connected to a cap of material (Figure 4-39)

Most small burrs can be measured by the techniques already discussed. Occasionally when a drill is very dull it will extrude a large burr. This burr, which looks like the lip produced on sheet metal flanged holes, can easily be measured in a metallurgical mount with little need to take multiple length measurements. As this type of burr is an indication of poor machining technique it should seldom be seen in production.

The majority of drill exit burrs are very uneven, ragged burrs. Obtaining repeatable measurements is difficult on these burrs. Because of the flexibility and unevenness of these burrs, mechanical height indicators are not usually feasible. Because the height of these burrs may vary by a quarter of an inch, it is important to use a functional definition of burr height. Some of the possible definitions include:

- The minimum burr height
- The median burr height
- The maximum burr height
- The average (mean) burr height

As indicated before the maximum burr height should not be used. It is not a repeatable statistic. The minimum burr height is probably the best attribute to measure if one is limited to a single height measurement per hole. The median burr height is the most repeatable attribute. This can be measured quickly by optical means if the inspector is allowed to exercise his judgment in determining the median. An average burr height can be used but the occasional large "spikes" which inevitably occur will greatly influence the results. For tests which warrant high accuracy and

FIGURE 4-34: SHORT CURLED DRILL EXIT BURR IN 17-4 PH STAINLESS STEEL

FIGURE 4-35: LONG, SEMI-UNIFORM DRILL EXIT BURR IN 6061-T6 ALUMINUM (CROSS SECTIONAL VIEW)

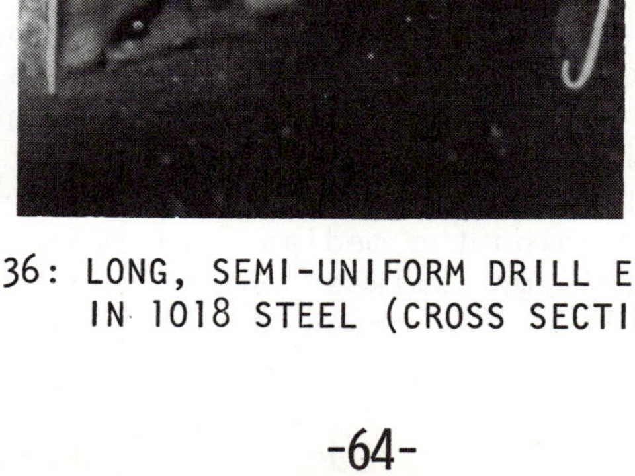

FIGURE 4-36: LONG, SEMI-UNIFORM DRILL EXIT BURR IN 1018 STEEL (CROSS SECTIONAL VIEW)

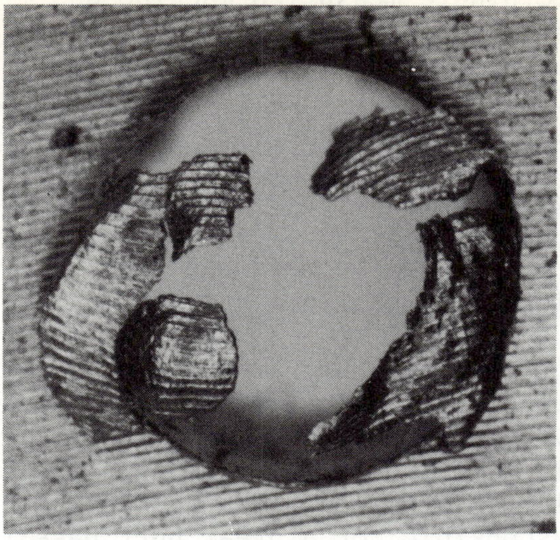

FIGURE 4-37: RAGGED DRILL EXIT BURR IN 1018 STEEL

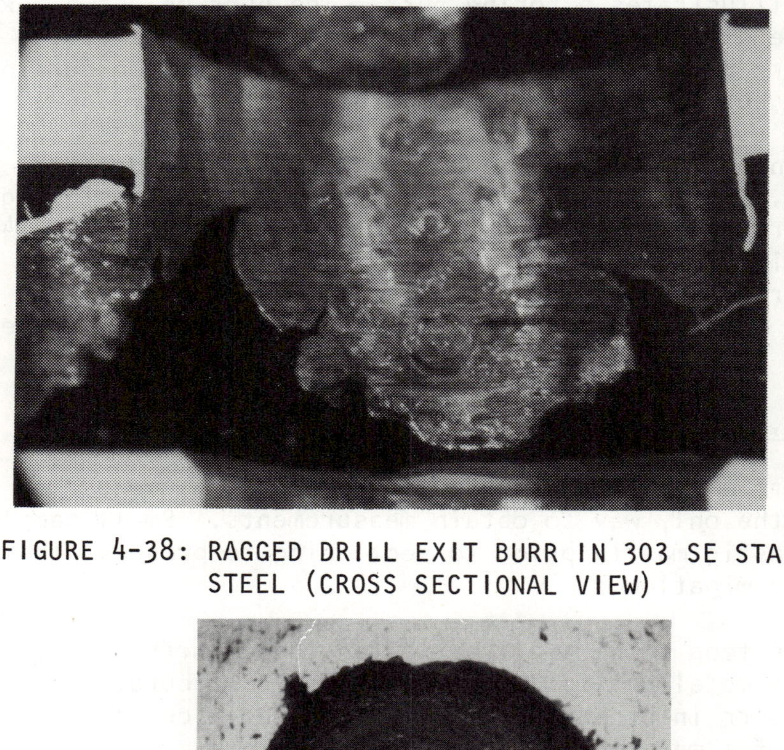

FIGURE 4-38: RAGGED DRILL EXIT BURR IN 303 SE STAINLESS STEEL (CROSS SECTIONAL VIEW)

FIGURE 4-39: CAP OF METAL PRODUCED IN 1018 STEEL

speed, a combination coordinate measuring machine and minicomputer can be used to determine average or median values.

After height measurements have been made, metallurgical mounts can be used to obtain the remaining burr properties.

Some drills in some materials leave a "cap" of material attached to one side of the hole. While this requires some additional effort to remove, the cap itself should not be measured. A study of the cap, however, may provide some additional insight into burr formation.

The presence of a continuous burr around the exit side of the hole as in figures 4-35, 4-36, and 4-40 indicates high feed forces and dulled cutting edges. In many deburring methods, it is more difficult to remove burrs with continuous longitudical profile because these cylindrical burrs being produced with dull tools have a high inherent stability. Moreover, these dull tools are responsible for a great burr base thickness.

Figure 4-41 illustrates a boring burr. The burr height h_o exactly corresponds to the difference between the drilling and the boring radius, that means the burr height is 0.25 mm (0.010 inch). At the burr peak one can recognize the burr of the drilling operation.

Countersinking of workpieces made of tough material does not always involve deburring. Strictly speaking, deburring by countersinking replaces a big burr by two small burrs (Figure 4-42). Figures 4-41 and 4-42 were taken with the help of a scanning electron microscope. This microscope is extraordinarily appropriate to study burr formation as the depth of focus in comparison to a light microscope is 500 times deeper.

Grinding, Turning, and Broaching Burrs

Grinding burrs are easily broken off and, as such, measuring them optically may be the only way to obtain measurements. Small samples can be potted in a plastic matrix prior to sectioning to preserve the burr for a metallurgical examination.

Turning burrs tend to be small but often these burrs are curled over. As a result only a metallurgical mount will provide accurate data on these burrs. The exit burr in broaching also is frequently curled and thus requires the use of a metallurgical mount.

EDM Burrs

These small, spherical burrs can be measured either with indicators or metallurgical mounts. From a production standpoint, several measurements made along an edge with an indicator should be adequate since the burrs are spherical, and thus width can easily be deduced from height measurements.

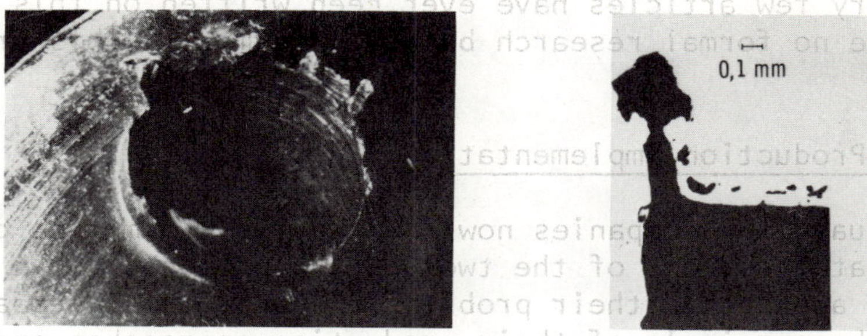

FIGURE 4-40 DRILL EXIT BURR CONTINUOUS AROUND HOLE CIRCUMFERENCE

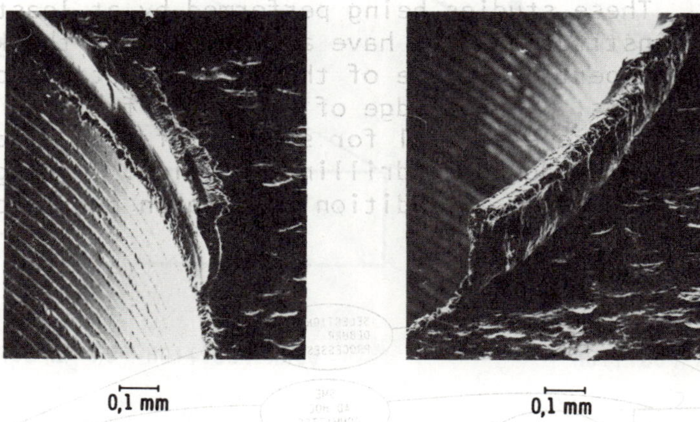

FIGURE 4-41: BURR PRODUCED BY BORING AFTER DRILLING

FIGURE 4-42: BURR PRODUCED BY COUNTERSINKING

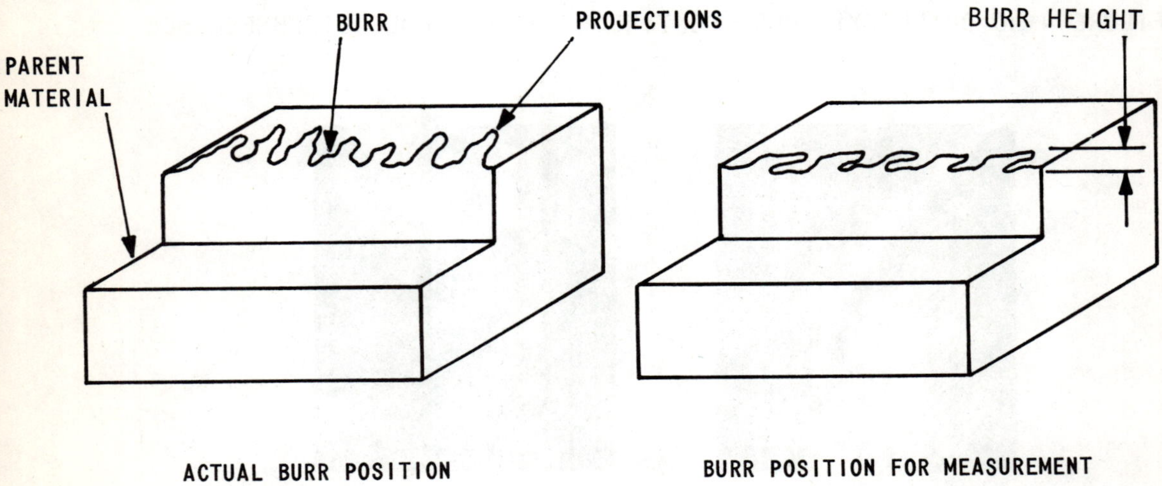

FIGURE 4-43: BURR HEIGHT FOR END MILLING BURRS

Comparison of Results from Two Measurement Approaches

To illustrate the impact the measurement technique can have on burr measurements, a simple test involving drill burrs was performed. A mechanical dial indicator was used to measure the height of both exit and entrance burrs at four places on each side of the 0.125 inch (3.175 mm) diameter holes. Each reading was verified by passing the indicator over the same spot three times. In the second method, a 0.010 inch (254 µm) wide saw was passed through the center of all specimens measured by the first method. Both sides of the cut specimens were mounted in metallurgical mounts and polished. Burr heights were then measured optically to the nearest 0.0001 inch (2.5 µm). Since a single cross section of a hole presents two views of each burr for each half of the hole, this technique provided four views of each burr (Figure 4-31).

In both methods six holes were inspected for each of the 12 combinations of drill geometry and workpiece material studied. Workpieces materials included 1018 steel, 303Se and 17-4 stainless steel, and 6061-T6 aluminum. The drills studied included four-facet, eight-facet, and radial lip points. (Eight-facet drills were prepared by dubbing the corners of four-facet drills.) The heights of burrs on both the entrance and exit sides of the drilled hole were measured.

A statistical analysis of variance indicated that statistically the two methods produced different results on the entrance burrs of five of the combinations studied (Table 4-3). In all cases in 6061-T6 aluminum, the metallurgical mount resulted in the longest burr readings. Apparently the dial indicator bent the soft, thin tips of the burrs enough to lower the measurement obtained. The same phenomenon could be responsible for the difference seen with the eight-facet drill in 1018 steel. Although statistically significant, there was little actual difference in the results of the eight-facet drill in 17-4 PH stainless steel. It is of interest that on 303Se workpieces both techniques produced equivalent results. The fact that these burrs were shorter (and thicker)[21,22] than those in other materials is also significant.

TABLE 4-3: EFFECTS OF MEASUREMENT APPROACH ON ENTRANCE BURR HEIGHT

Drill	Workpiece Material			
Geometry	1018	303Se	17-4 PH	6061
Four-Facet	0	0	0	M
Eight-Facet	M	0	I	M
Radial Lip	0	0	0	M

0 = results independent of measurement technique
I = a statistically significant difference and an indicator reading larger than that from mounts
M = statistically significant differences between methods and that measurements from mounts tend to be larger

TABLE 4-4: EFFECTS OF MEASUREMENT APPROACH ON EXIT BURR HEIGHT

Drill Geometry	Workpieces Material 1018	303Se	17-4 PH	6061
Four-Facet	0	I	0	0
Eight-Facet	0	I	0	0
Radial Lip	0	I	0	0

The dial indicator resulted in higher exit burr measurements in 303Se than those made optically from mounts (Table 4-4). The explanation of this phenomenon requires a closer look at the actual geometry of drill burrs. As previously indicated, drill exit burrs can be:

- Very short and uniform (sharp cutting tool--Figure 4-34)
- Very long and uniform (drill is extruding metal--Figures 4-35 and 4-36)
- Extremely ragged (typical in most metals--Figures 4-37 and 4-38)
- Short with a cap of metal (Figure 4-39)

In the first two cases either method of measurement should produce roughly similar results. However, when a mechanical dial indicator is used to measure ragged burrs, the ball at the end of the indicator is trapped by the sides of some segments and cannot fall to the lowest point. This wedging of the ball prevents it from showing the actual burr height at that one point. The saw used for metallurgical mounts sees no such restriction. If the saw cuts near one of the low spots, one obtains the actual burr height at that point. Since the 303Se had these large ragged burrs, the dial indicator could not reach the low points, therefore, the average results were higher than those from the metallurgical mount which could include the low values.

Two additional observations are worth noting. In many holes there is at least one high burr segment (Figure 4-37). This segment can be slightly larger than the radius of the hole. If this maximum value of burr height is used it can provide some misleading information about the typical burr size. For example, the eight-facet drill produced an average exit burr height of 0.0020 inch (50.8 μm) in 6061-T6 aluminum when measured by the dial indicator.[20] One segment of the burr on some of these holes was 0.06 inch (1.5 mm) high, or 30 times higher.

The second observation is that because of the ragged nature of most burrs, the standard deviation of burr height measurements is proportional to the mean value of the measurements. Thus in general as drill exit burrs get higher, they also become less consistent. This can greatly affect the repeatability of deburring efforts.

The fact that standard deviation is a function of the mean value indicates that burr height does not follow a true normal distribution. When burrs are small or uniform in height, the deviation from normality is not sufficient to require special data normalizing procedures. For the purposes of the tests described in this paper the use of normalizing procedures was not deemed necessary.

This test involved small drills and conditions which produced generally small burrs. By varying the machining parameters or allowing the tools to dull, one could obtain results different from those shown in Tables 4-3 and 4-4.

The results of this brief test do indicate that the technique used will affect the results. The fact that the differences are material-dependent is also significant. One cannot make all-inclusive statements about the effects of the technique, but for the conditions and materials studied the indicator readings produced entrance burr readings at least no larger than those obtained from mounts. When statistical differences existed for the exit burrs the indicator results were equivalent to or larger than those taken from mounts.

Effects of Experimental Design

The statistical design and analysis of tests on burr properties is sensitive to the data obtained. Wide variations in burr properties will obscure some noteworthy trends. In another study of drilling burrs it was conclusively shown that replication of measurements or samples results in a more accurate picture of process effects.[20]

Measuring burrs is time consuming. When the object of a study is to determine how process variables affect burr properties it is essential to utilize an efficient test program. Factorial design approaches such as described by Wang[16,23], Davies[24], and Hicks[25] will greatly reduce the number of specimen which must be analyzed. In addition this experimental approach will unveil combinations of variables which greatly affect the results but which are not evident in traditional tests.

Measured Burr Properties

As indicated by the equations presented, any changes which increase the cutting force will create a larger burr. Thus one can logically expect such parameters as feed rate, tool size, tool geometry, tool sharpness, and coolant to affect the size of burrs measured. Tests have verified this, although, again, it is important to note that not all burrs on a cut will respond the same way to a change in cutting variables. Thus changing the helix angle on a drill will reduce exit burr height but increase exit burr thickness. Increasing the feed rate in an end milling operation may increase the height of burr at Edge 1 but not affect the one at Edge 6.

Tables 4-2 and 4-5 through 4-8 present some typical burr properties produced by sharp cutters at conditions typically used to produce precision miniature parts. These are typical properties calculated over a moderate range of process variables. For measurements at specific speeds, feeds, geometries, etc., the reader is urged to review the many reports available (2, 8, 16, 17, 20, 22, 26-56).

Under ideal conditions most burrs are on the order of 0.004 inch (101.6 μm) or shorter (Table 4-5). They tend to be 0.002 inch (50.8 μm) thick except for grinding burrs which are thinner.

Burrs produced in hole making operations tend to be higher than milling burrs except for fine edge blanking (Table 4-6). Burrs produced in turning operations are strongly influenced by side cutting edge angle (SCEA) and rake angle (Table 4-7). This is logical since these parameters influence the cutting forces. As seen in Table 4-8, reaming produces burrs which are typically twice as repeatable as drilling burrs. The dramatic influence of tool sharpness is illustrated in Table 4-2. Not only do abusive cutting conditions produce burrs three to ten times larger than do gentle conditions but the repeatability of these burrs also worsens by a similar amount (Table 4-2).

Geometry and tool sharpness play a major role in determining burr size as seen in Figures 4-44 through 4-46. The formation force F_g is that component of the resultant cutting force F_Z, which is vertical to the exit edge. The burr formation force F_g however depends on the angle between edge and cutting force direction F_S.

The angle between edge and direction of cutting force not only influences the burr size but also the shape of the longitudinal profile. As shown in Figure 4-45, an angle between edge and cutting force direction within a range of 0° and 135° causes the following burr forms:

0° - 45° continuous burr with constant height
45° - 75° burr with steplike longitudinal profile
75° - 105° burr with saw-toothlike shape
105° - 135° burr peaks in cutting direction

Increasing tool wear rate involves increasing burr size as F_g is increasing.

Every chip section marked with roman numerals in Figure 4-46 is cut off by a cutting edge. As with a dull tool edge, the burr formation force F_g is higher than with a sharp edge. Also the chip section is larger which is no longer cut off but only bent over.

Figure 4-47 illustrates the milling burrs occuring at the acute edge angles of a helical gear. Acute edge angles cause very thick burrs as indicated in Chapter 2. The base thickness and burr radius is large in these conditions which make deburring difficult. The deburring costs for this workpiece shown in Figure 4-48 exceed the machining costs because of the acute angle effect.

Figures 4-49 and 4-50 represent a broaching and a grinding burr as taken by a scanning electron microscope.

Process and Property	Burr Location 1	2	3	4	5	6	7	8	9
Side Milling*									
Burr Height (inch)	0.0028	0.0020	0.0140	0.0073	0.0600	0.0001	0.0004	0.0002	--
Burr Thickness (inch)	0.0014	0.0016	0.0020	0.0016	0.0010	0.0014	0.0015	0.0011	--
Burr Hardness**	354	348	350	343	350	357	355	342	--
End Milling***									
Burr Height (inch)	0.0010	0.0020	0.0850	0.0001	0.0024	0.0002	0.0600	0.0007	0.0029
Burr Thickness (inch)	0.0007	--	0.0008	--	0.0009	0.0003	--	0.0003	--
Grinding†									
Burr Height (inch)	0.0053	0.0039	0.0032	0.0057	--	--	--	--	--
Burr Thickness (inch)	0.0007	0.0008	0.0006	0.0005	--	--	--	--	--

*4 inch diameter, 1/2 inch wide, HSS, 18 teeth, zero to 10 degrees helix angle at 98 rpm, and feed rates of 0.00057 to 0.002 iprt
**Knoop hardness number with 100 gram load. Parent material hardness was Knoop 325.
***0.125 to 0.375 diameter, 4 flute end mills at 1240 rpm and feeds of 0.0005 to 0.0033 iprt
†PA46HBV40 wheel at down feeds of 0.0005 to 0.0015 inch per pass

(0.0001 inch equals 2.54 μm)

TABLE 4-5: TYPICAL PROPERTIES OF BURRS PRODUCED IN 303SE STAINLESS STEEL (R_c32)

TABLE 4-6: TYPICAL BURR PROPERTIES PRODUCED BY COMMON PRODUCTION PROCESSES IN 303SE, 302, OR 304L STAINLESS STEEL

Process and Tool Size	Entrance Height (inch)	Burr Thickness (inch)	Exit Height (inch)	Burr Thickness (inch)
Drilling				
0.025 diameter*			0.0041	0.0021
0.082 diameter**			0.0060	0.0012
0.125 diameter	0.0033	0.0022	0.0063	0.0020
Reaming				
0.140 diameter***	0.0021	0.0005	0.0039	0.0008
Ball Broaching				
0.062 diameter†	0.0044	0.0015	0.0080	0.0010
Fine Edge Blanking			0.0020	0.0009

*118 degree point, HSS drill at 0.0005 ipr and 2000 rpm
**135 degree point, solid carbide drill, feeds and speeds not recorded
***HSS reamers, 6 flutes, radial tooth loads of 0.001 to 0.0045 inch, and feed rates of 0.001 to 0.003 iprt
†For interference fits of up to 0.0007 inch

(0.0001 inch equals 2.54 μm).

TABLE 4-7: BURR PROPERTIES PRODUCED IN TURNING 303SE STAINLESS STEEL*

SCEA (degrees)	Rake Angle (degrees)	Burr Height (inch)	Burr Thickness (inch)
17.5	15	0.0006	0.0020
	- 4.3	0.0010	0.0030
0	15	0.0040	0.0030
	- 4.3	0.0120	0.0060
-17.5	15	0.0800	0.0030
	- 4.3	0.1000	0.0080

*The side cutting edge angles were obtained by rotating a Kennemetal DVVNN-123 from its normal orientation (17.5 degrees SCEA) to a position in which the insert nose entered the cut before the side cutting edge (-17.5 SCEA). This latter condition simulates an exaggerated face and turn tool. Kennemetal VNMP-332 and VNMS-331 solid carbide inserts were used at feed rates of 0.0021 to 0.0065 ipr and a surface velocity of 225 feet per minute.

(0.0010 inch equals 25.4 μm).

		Burr Properties							
		Entrance Burr				Exit Burr			
Workpiece Material	Process	Thickness (inch)	(μm)	Height (inch)	(μm)	Thickness (inch)	(μm)	Height (inch)	(μm)
303Se SST	Drill	0.00124	31.5	0.00169	42.9	0.00038	9.6	0.00400	101.6
	Ream	0.00049	12.5	0.00094	23.9	0.00127	32.2	0.00017	43.2
1018 Steel	Drill	0.00239	60.7	0.00170	43.2	0.00100	25.4	0.00230	58.4
	Ream	0.00026	6.6	0.00031	7.9	0.00013	33.0	0.00109	27.7
6061-T6 Aluminum	Drill	0.00087	22.1	0.00092	23.4	0.00200	50.8	0.00232	58.9
	Ream	0.00112	28.4	0.00293	74.4	0.00051	13.0	0.00557	141.5

*Values shown are standard deviation

TABLE 4-8: REPEATABILITY OF DRILLING AND REAMING BURRS*

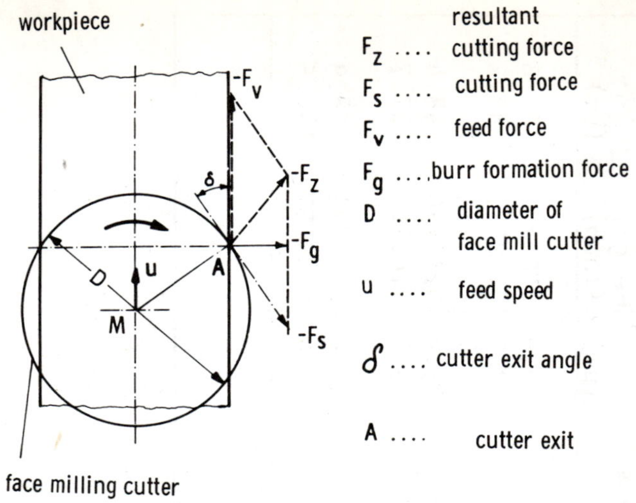

FIGURE 4-44: FORCES IN FACE MILLING OPERATIONS

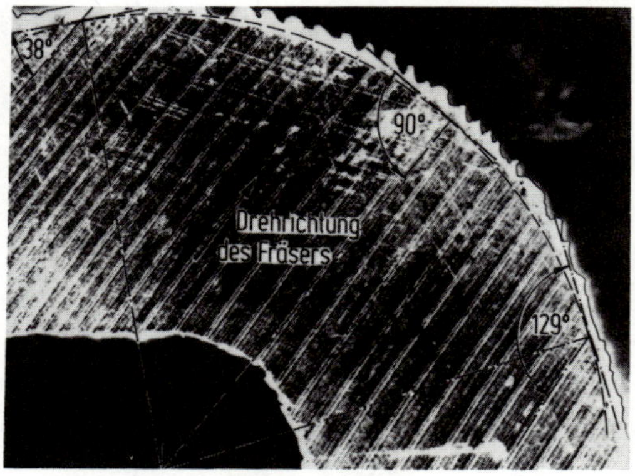

0°.....45 continuos longitudinal profile
45°.....75 steplike longitudinal profile
75°....105 saw - toothlike long profile
105°....135 burr peaks in cutting direction

FIGURE 4-45: PLAN VIEW OF BURRS AS A FUNCTION OF TOOTH EXIT ANGLE IN FACE MILLING

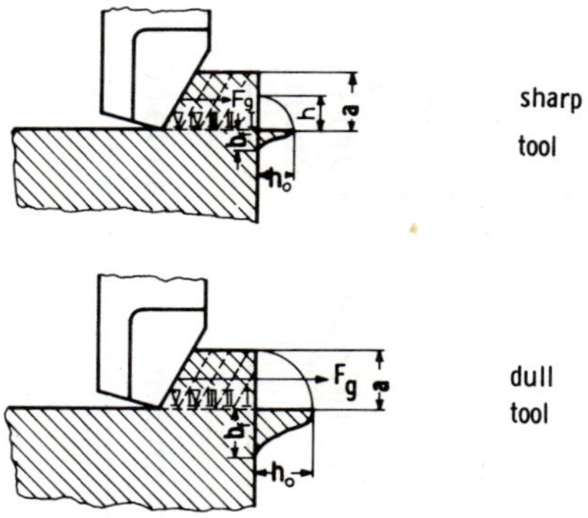

F_g ...... burr formation force
I ...V.. chip section
a .... depth of cut
h_o ... burr height
b_f ..burr base thickness

FIGURE 4-46: EFFECT OF DULL TOOL IN FACE MILLING OR TURNING

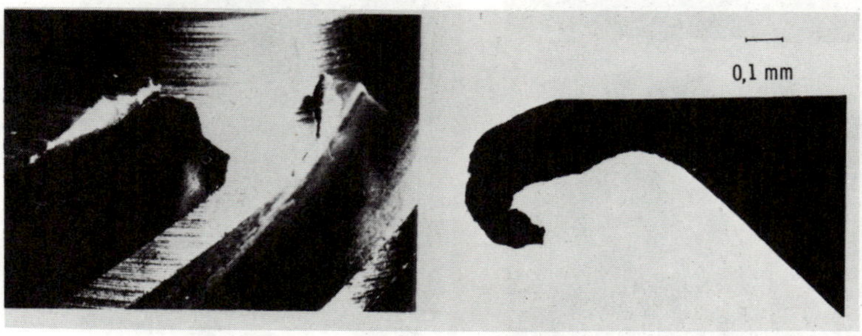

FIGURE 4-47: MILLING BURRS AT AN ACUTE EDGE IN A HELICAL GEAR

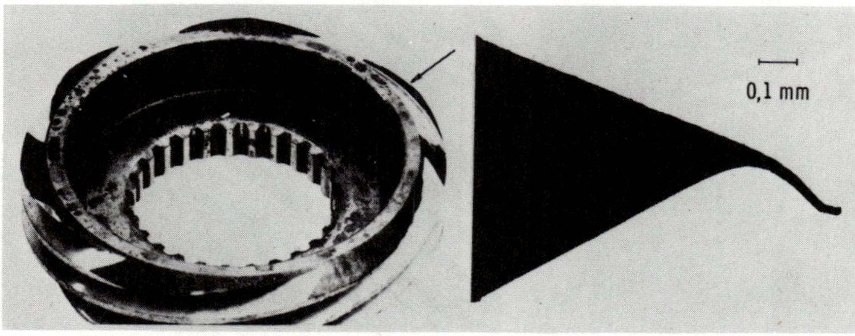

FIGURE 4-48: MILLING BURR PRODUCED ON AN ACUTE EDGE

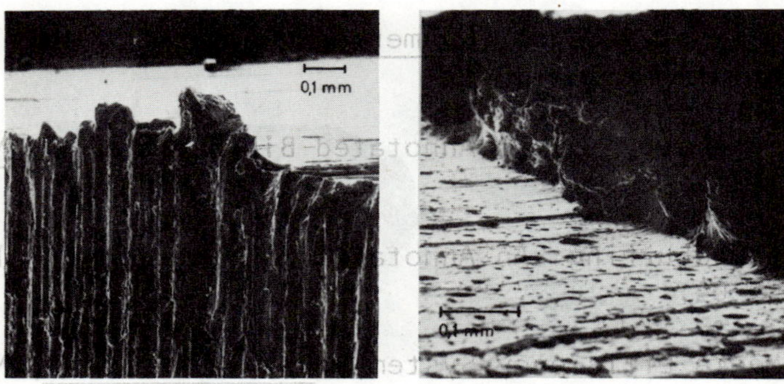

FIGURE 4-49: BROACHING BURR

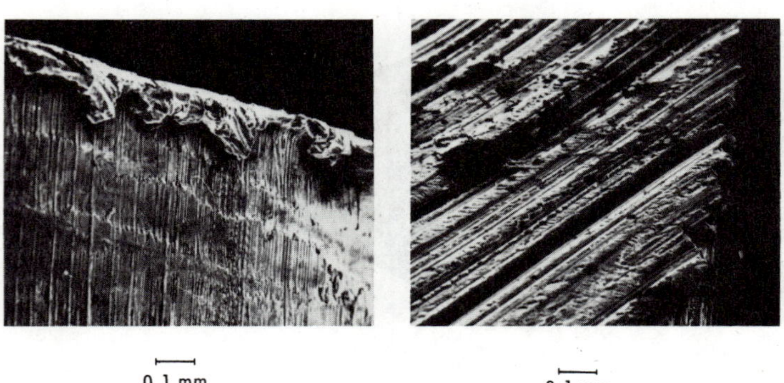

FIGURE 4-50: GRINDING BURR

Burr Size in Stamping Operations

The majority of work performed on documenting burr properties has been for the blanking and piercing operations. Wukusik and Zeno's 1958 study[23] appears to be one of the earliest. Since that time, studies or reports containing quantitative data have appeared almost yearly for these types of operations. The majority of these reports illustrate either the effect punch-to-die clearance has on burr height or the increase in burr height which occurs as the die wears.

Dr. Seidenberg's 1967 papers[18,55] on the measurement of blanking burrs represents one of the world's first notable attempts to obtain convenient yet meaningful burr measurements. In 1969, Wang, Taraman, and Wu reported a study[16] of stamping burrs in cold rolled steel. In their study, die radius affected burr height more than clearance, punch speed, die taper, die throat length, punch slide taper, or punch bottom taper. Their study is also one of the first to emphasize the ragged nature of burr height. As they indicate, the use of average burr height values produces much more repeatable data than maximum burr height.

Hugo's 1971 article[29] represents one of the most production oriented articles on burrs produced in stamping operations. Using average burr height measurements, he concludes that the burr height at the beginning of a run is greatly influenced by the internal construction of the die button. Burr height is minimized when a straight land is used. It is not affected by the clearance. When a tapered relief is used instead of a straight land, the burr height at the beginning of a run is inversely proportional to the clearance. The burr height increases in direct proportion to the number of punch strokes made. For a die button with tapered relief, burr height decreases as clearance increases. For the lamination steel studied, the optimum clearance for minimum burr height was 20 percent of material thickness. Hugo also notes that the burr on the bottom of the workpiece hole tends to be considerably shorter than the one on the slug. In addition to the quantitative empirical data, Hugo presents conceptual models of the mechanisms of burr formation in stamping operations.

Buhler and Pollmar[19] present a qualitative discussion of burr formation. Their paper describes the formation of the burr shape and the influence of the workpiece material. They too include a number of curves showing burr height as a function of number of parts stamped.

A brochure published by Dayton Progress Corporation[27] provides two curves of burr height as a function of clearance. It is obvious in these curves that burr height is a function of clearance and that there is one value of clearance which produces the shortest burr. For the results shown, this point occurs at ten percent of material thickness. The second curve indicates that for a 0.003 inch maximum allowable burr height in cold rolled steel, ten percent clearance (in contrast to a five percent clearance) increased die life from 71,000 to 221,000 hits.

While the recent work in the area of stamping burrs has provided a notable improvement in the understanding of die geometry effects on burr height, it is obvious that some of the results do not agree with those found in other tests. In addition three aspects of our knowledge of stamping burrs are missing. Firstly, no quantitative model of burr formation in stamping operations has yet been offered. That is to say that no mathematical equations have been developed from an analytical base which describes burr formation. Until this is done researchers will have to provide empirical results for each material of interest. Secondly, no mathematical models, analytical or empirical, have been published that will predict burr size as a function of tool geometry, material, and number of press strokes. Thirdly, no work has been reported on the burrs produced by fine edge blanking. Two notable aspects of the work which has been performed on stamping burrs are that the stamping industry has established an accepted definition of what a burr is and that it has developed convenient methods and techniques to measure burr size. Both of these are lacking in the metal cutting field.

Burr Size Produced by Other Machining Processes

With the exception of drilling, little work has been performed on the machining processes. Some work has been reported by the author on side milling, end milling, turning, and grinding of 303Se stainless steel.[2] As reported there, helix angle, radial depth of cut, feed rate, and certain combinations of these greatly affected the height of the ten burrs produced by end milling. As previously indicated, the side cutting edge angle (SCEA) and depth of cut affected the height of turning burrs. The SCEA and feed rate also affected the thickness of turning burrs. Increasing the down feed per pass in surface grinding increased the height of two of the four burrs produced in this process.

In reaming initially burr free holes, the author has observed that the amount of stock removed by the reamer and the reamer geometry are the variables which have the greatest impact on burr properties.[3] Feed rate had little effect on burr height or thickness. Austenitic stainless steel had thicker burrs than brass or aluminum.

Some empirical and analytical results of burrs produced by ballizing have also been reported by the author.[3]

Seidel[56] recently reported an approach which can be used to determine the tendency for any brass alloy to form machining burrs. This system requires only a knowledge of the mechanical properties of the alloy of interest.

McBride's[8] thesis describes the effects of Electrical Discharge Machining (EDM) parameters on the rescast burr produced by the EDM process.

Recent Research on Drilling Burrs

In drilling, burr size is a function of many variables including the type of workpiece alloy, the heat treatment, size and shape of the drill, the cutting conditions, tool wear, built up edge and machine tool and drill vibrations. To identify how burrs from drilling could be minimized several recent studies have been performed.[57] These studies utilized aluminum workpieces. Because of the bulging which occurs in aluminum, burr thickness was used as the factor most characteristic of burr size.

For these studies, burr thickness was measured by placing the drilled workpiece on a fine adjustment stage with the heavy exit burr facing up. The inside diameter is measured with the dial indicator and the diameter at which the burr ceases to be visable is measured with a 50X microscope on the same staging device (Figure 4-51). When the microscope is focused on the flat workpiece surface then moved over to the burr, the start of the burr is that point at which the surface becomes out of focus. The burr thickness is half the difference of the inner and outer diameter. To improve accuracy the burr thickness was defined as the average thickness observed with two readings taken perpendicular to each other. While burr heights were taken using the microscope they were not reported in this study.

The drill used in this study was a 10 mm (0.4 inch) diameter drill with a 27 degree helix angle and a web eccentricity of 0.04 mm (0.0016 inch). The drill material was a Japan Industrial Standard (JIS) SKH9 alloy (similar to M2 tool steel). The workpiece materials included 1050-0, 2024-T3, and 6061-T3 aluminum.

Effects of Cutting Conditions

Although it has been said that burrs become smaller at high cutting speeds[58], burr thickness was essentially not affected by cutting velocity in this experiment. Speeds were varied up to 50 m/min (167 ft/min).

Increasing feed rates does increase burr thickness as shown in Figure 4-52. Applying coolant, however, can significantly decrease burr thickness (Table 4-9).

TABLE 4-9: EFFECT OF CUTTING FLUIDS ON THICKNESS OF BURR

Cutting fluid	Carbon-tetra-chloride	Inactive compound oil with extreme pressure agents	Water soluble oil (emulsion type mixed with 30 parts of water)	Dry
Thickness of burr Δr	0.18 mm 0.0071 in	0.20 mm 0.0079 in	0.38 mm 0.0150 in	0.45 mm 0.0177 in

Point angle 118°, Chisel edge angle 130°,
Lip clearance angle 10°, cutting speed 18 m/min (60.1 ft/min),
Feed 0.13 mm/rev (.0051 inch/rev), cutting fluid was supplied by conventional method.

As seen there, cutting oils compounded with extreme pressure agents are almost as effective as carbon tetrachloride. The success of carbon tetrachloride is due to its ability to prevent rewelding of chips.[59] In deep hole drilling, it is essential to use oil hole drills in order to assure that cutting fluid reaches the drill lips.

The recommended lip clearance angle of 12 degrees for aluminum workpieces[60] appears to be optimum also from the standpoint of burr thickness. Lower clearances increase thickness while larger clearances provide no additional benefit for burr size (Figure 4-53).

Helix angle plays a major role in determining burr size (Figure 4-54). With a helix angle of 40 degrees, exit burrs were less than 0.2 mm (0.008 inch) thick. With a standard 27 degree helix, burr thickness was twice this thick. The reason for this apparently is that high helix angles reduce the friction of the rake surfaces of the drill. This in turn reduces the width of the plastic deformation zone. Okoshi[58] found similar results in his study of rake angles used in aluminum and by Okushima in his study of brass.[61] It should be noted, however, that high helix angle drills are not suitable for drilling thin sheets as they cause the drill to pull into the workpiece too fast. In deep hole drilling, high helix angles can cause chip packing in the flutes.

Point angles and corner angles also affect burr size (Figure 4-55). In order to obtain data on angles near 180 degrees, a fishtail point was used. The angle included between the lips at the outer corner of the fishtail point drill, projected upon a plane parallel to the drill axis and parallel to the two cutting lips is equivalent to the point angle of the chisel point drill. This is called the "equivalent point angle." Accordingly, the equivalent point angle can become larger than 180 degrees. Table 4-10 describes the drills used in this study.

TABLE 4-10: DRILL SHAPES FOR EXPERIMENT OF THE INFLUENCE OF POINT ANGLE

Drill point geometry	Point angle (deg.)	Remark
Chisel point drill	70, 80, 90, 100, 118, 130, 140, 150	(1) Web eccentricity 0.03-0.04 mm (0.0012-0.0016 inch)
Fishtail point drill	160, 180, 200, 235	(2) Height difference between chisel point and outer corner of lip for fishtail point drill 1.4-1.5 mm (0.56-0.60 inch)

Helix angle 27°, Lip clearance angle 10°.

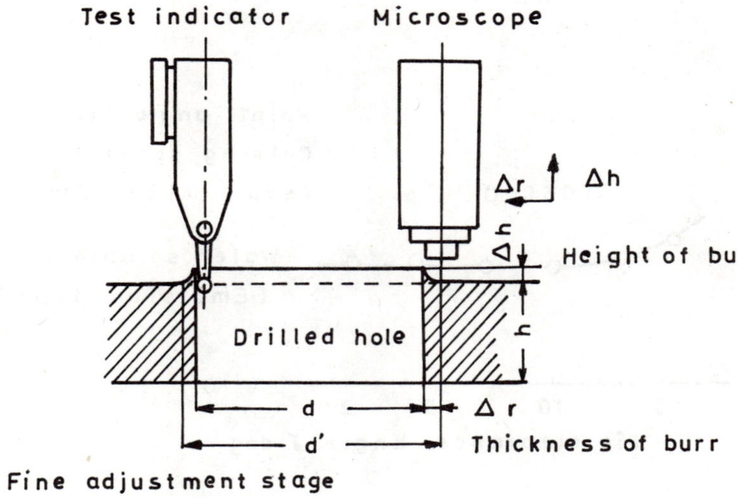

FIGURE 4-51: MEASURING METHOD FOR BURR SIZE

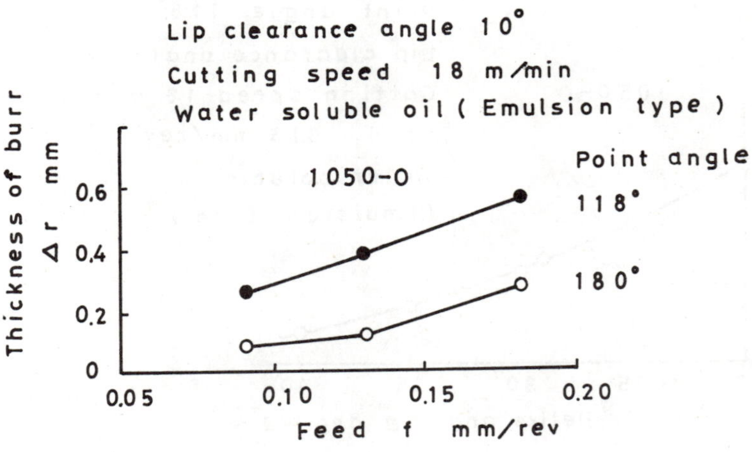

FIGURE 4-52: EFFECT OF FEED ON THICKNESS OF BURR

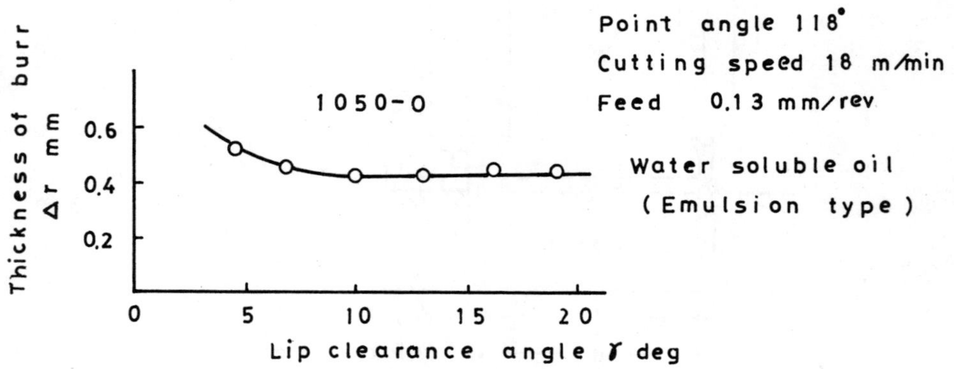

FIGURE 4-53: EFFECT OF LIP CLEARANCE ANGLE ON THICKNESS OF BURR

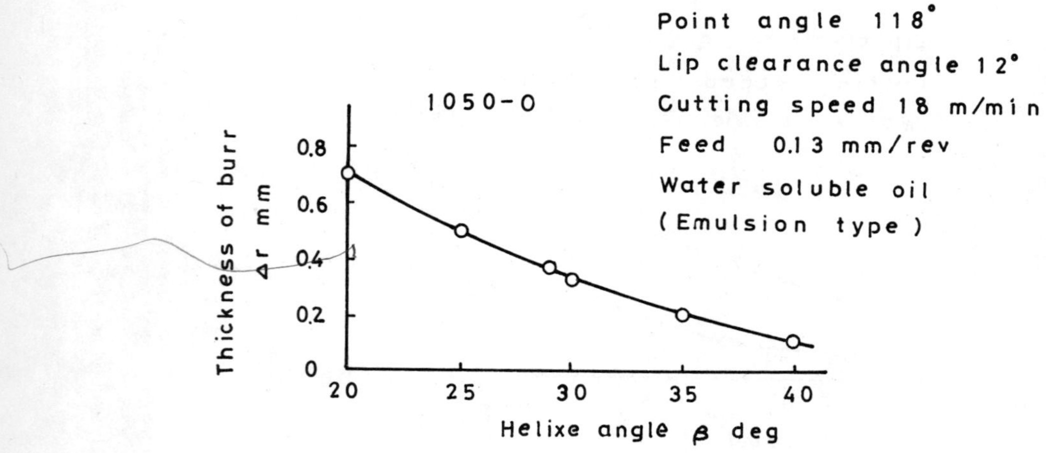

FIGURE 4-54: EFFECT OF HELIX ANGLE ON THICKNESS OF BURR

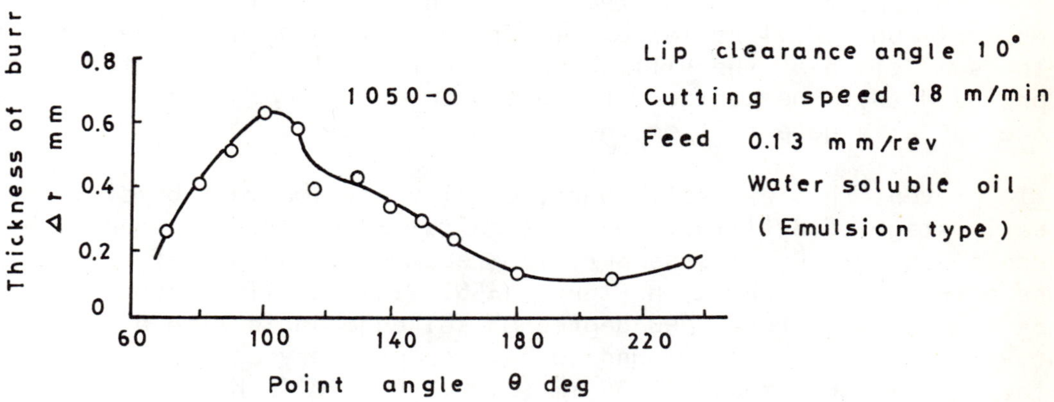

FIGURE 4-55: EFFECT OF POINT ANGLE ON THICKNESS OF BURR

As seen in Figure 4-55 increasing the point angle to 100 degrees increases burr thickness. Above that angle thickness decreases to a minimum at about 180 degrees. The recommended point angle in aluminum is 118 to 140 degrees[57] which results in comparatively large burrs.

Minimizing Burr Size in Drilling

When the above results are considered, there are two approaches to diminish the thickness of burr with respect to the point angle. The first is to decrease the equivalent point angle at the outer corner, and the other is to make it a large angle such as 180 degrees. In the case of the small equivalent point angle, the zone of plastic flow diminishes at the outlet of hole due to the small chip thickness. On the other hand, in the case of the equivalent point angle of 180 degrees, the radial component of cutting resistance becomes small. Therefore, either method reduces burrs. The latter is employed widely in drilling thin sheet metals as with a fish-tail point drill.

Drills with chamfered or dubbed corners such as used for cast iron are examples of the first approach as is the radial lip drill geometry. The radial lip drill can reduce the chip thickness at the drill corners to a very small value which produces the same effect as slowing the feed rate.[62]

It takes a long time for this type of drill to constrain the movement of drill by the margin, contrary to the case of the fishtail point drill. Therefore, in drilling thin sheet metals, this type of drill is accompanied by vibration. However, this drill is effective for the drilling of deep holes.

Another significant aspect of drilling burrs concerns the burr produced on the entrance side of the hole. In soft metals, if the drill vibrates, this burr will be large. If drill walking occurs, the burr shape has the same pitch as the rifling marks in the hole.[63] Therefore, it can be concluded that the contact pressure between the margin and the hole surface has a large effect on the entrance burr.

To prevent this excessive burr, drill vibration must be prevented in the early stage of drilling. Using a spiral point drill is one technique to accomplish this.[64] Hasegawa[57] also recommends the use of a stepped ground point such as shown in Figure 4-56. This drill is ground in two setups in order to prove a 90 degree chisel edge angle with a sharp point.[65] Moreover, it is ground to have an ordinary clearance angle leaving the chisel land of 0.03-0.05 mm (0.0012-0.0020 inch) width in order to remove the chip easily. This type of drill point is able to prevent the walking.[66]

Figure 4-57 shows the effect of the radius on the stepped ground drills on hole accuracy and burr size. These drills had a radial lip configuration at the outer corners. As seen in Figure 4-57 increasing the radial lip radius decreases hole oversize when a soluble cutting oil is used but has little effect when a kerosene and lard mixture is used. Burr height, however, is greatly reduced when a 2 mm (0.080 inch) radius is ground on the drill. The burr in the hard alloy 2024-T3 is much smaller than that in the softer 6061-T3 alloy and is not shown for that reason.

The stepped ground drill point gives a much smaller hole oversize condition than conventional 114 degree chisel point drills. Burr sizes are not significantly different however.

The stepped ground point geometry shown in Figure 4-56 was applied to an oil hole drill. At 18 m/min (60 ft/min) and a feed rate of 0.13 mm/rev (0.0052 inch/rev) a burr height of only 0.07 mm (0.0028 inch) was produced in 6061-T3 using soluble oil. For these conditions hole oversize was 0.04 mm (0.0016 inch) and the surface finish was 6.5 μm (260 microinch).

While several groups have performed research on burrs produced by drilling, only four have published their results. Six studies, all apparently sponsored by the same group, have been reported on the burrs resulting from drilling aluminum.[46-51] Fleming reported a study of drilling and reaming 300M steel and titanium alloys.[52] The author of this paper recently reported two studies[22,45] of drilling aluminum, low carbon steel, 303Se and 17-4 PH stainless steel. Phillips has described ongoing tests in drilling laminated aerospace materials.[53,54] Each of these studies is summarized in the following paragraphs.

Zaima's study of drilling aluminum[46] indicated that the thickness of

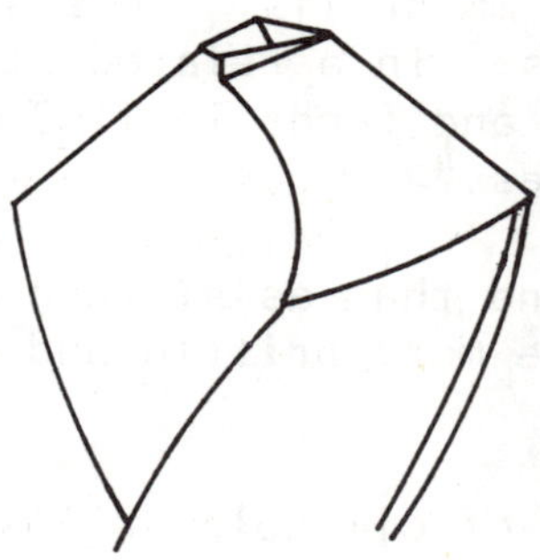

FIGURE 4-56: STEPPED GROUND POINT DRILL

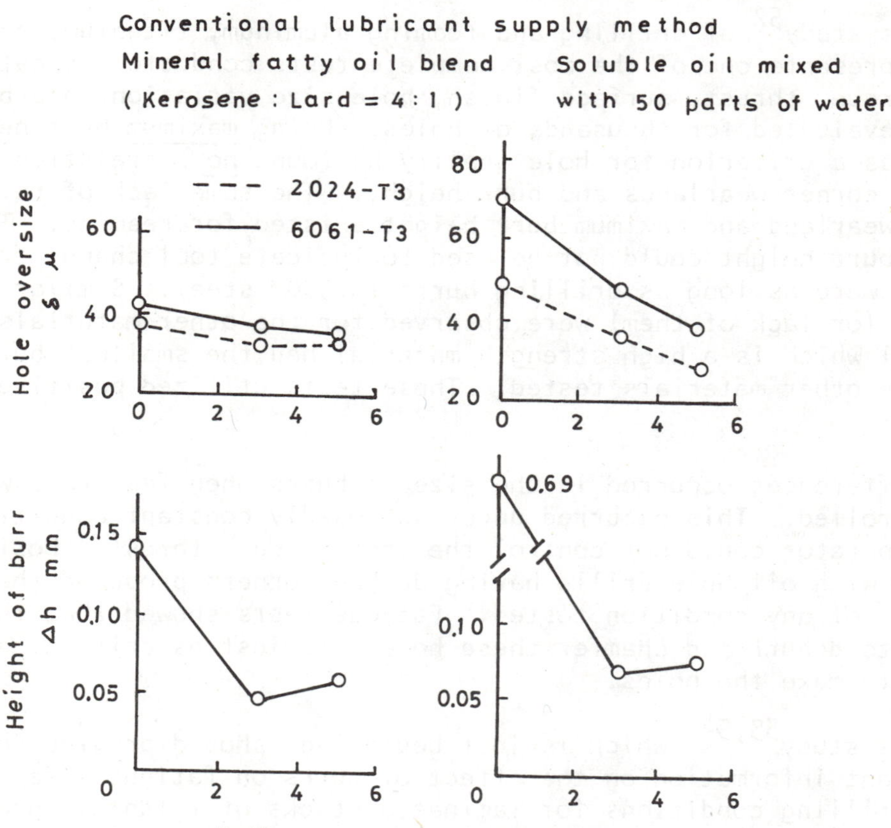

FIGURE 4-57: EFFECTS OF RADIUS OF LIP ON HEIGHT OF BURR AND HOLE OVERSIZE

these burrs is a function of feed rate and lip clearance. Cutting speed does not affect thickness.

Saito and his associates also found burr height to increase with feed rate.[48] In this study it was noted that drilling can produce two burr shapes in aluminum. Although they present a number of response surface curves, it is obvious that no simple relationship exists between burr height, feed rate, and drill diameter. Burr height was roughly proportional to drill diameter up to 6 mm drills. This study involved aluminum plates of 0.5 to 1.0 mm thickness. In a related study, Saito and his associates found that low speeds and feeds in 1100 aluminum produce a more acceptable burr than higher values.[49]

Sato and his associates found that as silicon content is increased in aluminum, drill exit burrs become more brittle and are thus more easily removed.[50]

Yuki and Hasegawa's study[51] of oil hole drills indicated that the radial lip design with a 3 mm radius at the drill corners produced a smaller burr in aluminum than conventional drills. Lubricants with 20 percent lard content produced much shorter burrs than those with lower lard content. Forced lubrication provided the smallest burr. Both 2024-T3 and 6061-T6 aluminum were studied in this investigation.

Fleming's study[52] of drilling and reaming aluminum, titanium, and 300M steel represents one of the most complete tests conducted to date. Tool wear, torque, thrust, surface finish, hole size variation, and burr heights were evaluated for thousands of holes. Using maximum burr height measurements as a criterion for hole quality he found no correlation between drill corner wearlands and burr height. The same lack of relationship between wearland and maximum burr height existed for reaming. Therefore maximum burr height could not be used to indicate tool change frequency. Reaming burrs were as long as drilling burrs in 300M steel. Similar relationships (or lack of them) were observed for the other materials tested. The 300M steel which is a high strength material had the smallest burr heights of any other materials tested. These tests utilized positive power feed.

Large differences occurred in the size of burrs when feed rate was manually controlled. This occurred under supposedly constant conditions because the operator could not control the breakthrough thrust. Positive feed drilling with oil hole drills having dubbed corners produced the smallest burrs of any condition tested. Fatigue tests showed that the methods used to deburr and chamfer these holes are just as critical as the methods used to make the holes.

Phillip's study,[53,54] which is just beginning, should provide some very significant information on the effect of burrs on fatigue life and on the optimum drilling conditions for laminated stacks of titanium, precipitation hardening stainless steel, aluminum, and other materials. This study will include an analysis of the effectiveness of adaptive and quasi-adaptive control in minimizing burr size.

In another study of drilling, the author of this book noted that the use of backup material is the most effective method for minimizing burr size.[22] Solid plates of backup were less effective than plates which had a hole in them the same diameter as the drill to be used. Reducing feed rates reduced exit burr height. The thickness of exit burrs was proportional to the average burr height in stainless steel, but no such relationship was observed for other materials. In this study burr height, thickness, repeatability, and shape were analyzed for four materials and both sides of the holes. Both drilling and reaming were studied.

To the author's knowledge, only two inventions are being marketed for the purpose of minimizing machining burrs. The radial lip drill is one of these. With proper design, it has been shown to effectively produce smaller drilling burrs than other drills. The second product is a spring loaded drill guiding mechanism[67] for use on portable drilling tools. The drill guide prevents drill walking and workpiece "cork-screwing" up the drill flutes. The spring loaded portion minimizes the thrust at drill breakthrough. By minimizing the thrust, the burr height is apparently minimized.

Summary

Burr formation can be explained from easily understandable models. The effects of tool wear, feed rates, and material properties can be seen in the equations defining many of these models. The fact that many different burrs can occur on a single cut emphasizes the complexity of the situation. Burr sizes can vary widely on a single part even on a single edge.

The measurement of burrs must be approached as scientifically and carefully as hole size or cutting forces. Failure to exercise such effort can easily mask significant relationships. Careful consideration of property definitions and measurement techniques will provide a realistic base for using burr measurements.

References and Sources of Additional Information

[1]Hugh O'Neill, The Hardness of Metals and Its Measurement, Sherwood Press, Cleveland, Ohio, 1934.

[2]LaRoux K. Gillespie, The Formation and Properties of Machining Burrs, Master of Science Thesis, Utah State University, Logan, Utah, 1973.

[3]LaRoux K. Gillespie and P. T. Blotter, "The Formation and Properties of Machining Burrs," Journal of Engineering for Industry (ASME Transactions Series B), February, 1976, pp. 66-74.

[4]D. Vilenski and M. C. Shaw, "The Importance of Workpieces Softening on Machinability," Annals of the C.I.R.P., Vol. 18, 1970, pp. 623-631.

[5]Joseph Datsko, Material Properties and Manufacturing Processes, John Wiley and Sons, Inc., New York, 1966.

[6]M. E. Merchant, "Metal Cutting Research," Machining Theory and Practice, American Society of Metals, Metals Park, Ohio, 1950, pp. 5-44.

[7]Serope Kalpakjian, Mechanical Processing of Materials, D. Van Nostrand Company, Princeton, New Jersey, 1967.

[8]James Nielson McBride, The Magnitude of Burrs Caused by Electrical-Discharge Machining, Master of Science Thesis, Utah State University, Logan, Utah, 1969.

[9]LaRoux K. Gillespie, "The Measurement of Burrs," SME Paper MR 74-993, 1974.

[10]J. E. Mayer, Jr. and D. J. Stauffer, "Effects of Tool Edge Hone and Chamfer on Wear Life," SME Paper MR73-907, 1973.

[11]Keiji Okushima and Yoshiaki Kakino, "Study on the Generating Process of Machined Surface," Bulletin of the Japan Society of Mechanical Engineers, 12 (49), 1969, pp. 141-148.

[12]P. Albrecht, "New Developments in the Theory of the Metal-Cutting Process: Part 1. The ploughing process in metal cutting," ASME Transactions (Journal of Engineering for Industry), November, 1960, pp. 348-358.

[13]H. J. Siekmann, "Why Hone the Cutting Edge," Automatic Machining, January, 1959.

[14]John Agnew, "The Importance and Methods of Carbide Edge Preparation," SME Paper MR73-905, 1973.

[15]F. Schafer, "Deburring in Production Engineering," Werkstattstechnik, vol. 63, 1973, pp. 692-696 (in German).

[16]K. K. Wang, Khalil Thalil Taraman, and S. M. Wu, "An Analysis of Punching Variables by Two-Level Fractional Factorial Design," ASME Paper 69-WA/Prod-27, 1969.

[17]Otto Kienzle and Werner Kienzle, "Tool Wear in the Cutting of Thin-Gauge Steel Sheets," (Translation of Stahl und Eisen, Vol 78, No. 12, June 12, 1958, pp. 820-828). ASTME Research Report No. 22, May 1,1959.

[18]H. Seidenberg, "Schittgrad und Messverfahren," Werkstatt und Betrieb, Vol. 100, 1971, No. 6, pp. 471-475.

[19]Hans Buhler and Fedor Pollmar, "Die Gratbildung Beim Schneiden Von Feinblech," Bander Belche Rohre (Dusseldorf) Vol. 12, No. 3, 1971, pp. 105-111.

[20]F. Schafer, "Deburring Processes in Perspective," SME Paper MR75-482.

[21]LaRoux K. Gillespie, "Effects of Measurement Technique and Experimental Design in the Analysis of Burrs," SME Paper MR75-985, 1975.

[22]LaRoux K. Gillespie, The Burrs Produced by Drilling, Bendix, Kansas City: BDX-613-1248 (Rev), December, 1974.

[23]K. K. Wang and M. F. Devries, "Investigation of Manufacturing Processes by Statistical Experimental Design Techniques," SME Paper MR69-263, 1969.

[24]O. L. Davies, Design and Analysis of Industrial Experiments, Hafner, New York, 1956.

[25]Charles R. Hicks, Fundamental Concepts in the Design of Experiments, Holt, Rinehart and Winston, 1964, p. 91.

[26]"Piercing of Low Carbon Steel," Metals Handbook, Vol. II, ASM, Metals Park, Ohio, 1965.

[27]"The Birth of a Hole," Dayton Progress Corporation Report 110, Dayton, Ohio, 1973.

[28]L. A. Holiga, "Punch-To-Die Clearance--Its Effect on Punched Hole Characteristics in Cold Rolled Steel," ASTME Paper 65-713, 1965.

[29]Harding R. Hugo, "How to Improve Metal Stamping Die Performance," Sheet Metal Industries, February, 1971, pp. 120-135.

[30]LaRoux K. Gillespie, The Effects of Reaming Variables on Burr Properties, Bendix Corporation (Kansas City Division) Report BDX-612-1083, March, 1974 (available from NTIS).

[31]LaRoux K. Gillespie, Properties of Burrs Produced by Ball Broaching, Bendix Corporation (Kansas City Division) Report BDX-613-1084, March, 1974 (available from NTIS).

[32]Leonard R. Allingham, "The Importance of Clearance in the Manufacture of Blanking Dies," ASTME Paper 622, 1964.

[33]C. S. Wukusik and R. S. Zeno, "Improving Punchability of Silicon Steel," Tool Engineering, Vol. 41, No. 12, December, 1958, pp. 63-70.

[34]Otto Kienzle and Kurt Buchmann, "Beeinflussung und Prufung des Verschleisses und der Gratbildung beim Schneiden von Stahlfeinblech," Archiv fur das Eisenhuttenwesen, Vol. 34, No. 6, June, 1963, pp. 443-445 (in German).

[35]H. Seidenberg, "Presseneinwirkungen auf Werkzeugverschleiss und Grathohe beim Schneiden von Feinblech im Gessch Lossenen Schnitt," Dr. -ing. Dissertation, TH Hannover, 1965 (in German).

[36] E. Sondershaus, "Die Beurteilung der Schneidbarkeit von Feinblech Durch ein Knabberverfahren," Diss. T. H. Darmstadt, 1966 (in German).

[37] H. Wentzel and H. Mehlhorn, "Erkenntnisse uber Gratbildung und Verschleiss von Schnittwerkzeugen," Fertig,-Techn. uber Betr., Vol. 14, 1964, pp. 236-240 (in German).

[38] H. Seidenberg, "Werkzeugverschleiss und Grathohe beim Schneiden von Feinblech," Stahl und Eisen, Vol. 86, 1966, pp. 1677-1688 (in German).

[39] E. Roeper, "Ursachen der Gratbildung und die Messung des Stanzgrates," Blech, Vol. 4, No. 4, 1957, p. 44 (in German).

[40] G. Oehler, "Zulassige Grathohe an Teilen aus Blech und Bandern," Blech, Vol. 15, 1968, pp. 462-463 (in German).

[41] Federico Strasser, "How Control of Burrs Aids Sheet Metal Stamping," The Iron Age, Vol. 185, No. 3, January 21, 1960, pp. 90-92.

[42] L. K. Gillespie, The Burrs Produced by Side Milling Cutters, (Topical Report) UNCLASSIFIED, Bendix Kansas City Division, BDX-613-1303, September, 1975.

[43] L. K. Gillespie, The Burrs Produced by Grinding, (Topical Report) UNCLASSIFIED, Bendix Kansas City Division, BDX-613-1372, February, 1976.

[44] L. K. Gillespie, The Burrs Produced by End Milling, (Topical Report) UNCLASSIFIED, Bendix Kansas City Division, BDX-613-1503, April, 1976.

[45] L. K. Gillespie, The Effects of Drilling Variables on Burr Properties, (Topical Report) UNCLASSIFIED, Bendix Kansas City Division, BDX-613-1502, March, 1976.

[46] Shigeo Zaima and Akiyasu Yuki, "On the Burr Formed in Drilling Aluminum Sheets," Kei Kinzoku (Light Metals), Vol. 17, No. 6, 1967, pp. 362-367 (in Japanese).

[47] Akiyasu Yuki, "Investigation of Burrs From Drilling," Res. Bull, Fac. Educat. Oita Univ. Nat. Sci., Vol. 3, No. 3, 1968, PP. 47-58.

[48] Yuzo Saito, Kinji Sato, Tsungo Saga, Mokoto Ogawa, and Shozo Mizusawa, "Drilling Machinability of Aluminum Sheet," J. Japan Inst. Light Metals (KEI Kinzoku), Vol. 20, No. 2, February, 1970, pp. 69-76 (in Japanese).

[49] Yuzo Saito, Kinji Sato, Tsuneo Saga, Makoto Ogawa, and Shozo Mizusawa, "Drilling Machinability of Aluminum Sheet," J. Japan Inst. Light Metals (KEI Kinzoku), Vol. 20, No. 8, August, 1970, pp. 373-380 (in Japanese).

[50]Kinji Sato, Tsuneo Saga, and Shozo Mizusawa, "Drilling Machinability of Aluminum-Silicon Binary Alloys," J. Japan Inst. Light Metals (KEI Kinzoku), Vol. 20, No. 1, January, 1970, pp. 26-35 (in Japanese).

[51]Akiyasu Yuki and Yoshio Hasegawa, "Effect of Lubricating Methods on Dimensional Accuracy and Surface Roughness of Drilled Holes in Aluminum Alloys," KEI Kinzoku, Vol. 23, No. 9, September, 1973, pp. 404-408 (in Japanese).

[52]C. M. Fleming, Precision Hole Generation Methods, McDonnell Aircraft Company Technical Report AFML-TR-73-135, Vol. I and II, March, 1973.

[53]Joseph L. Phillips, Multi-Layer Fastener Systems, Boeing Commercial Airplane Company Report IR-752-4 (I), July, 1974.

[54]Joseph L. Phillips, Multi-Layer Fastener Systems, Boeing Commercial Airplane Company Report IR-752-4 (II), October, 1974.

[55]Harald Seidenberg, "Neues Verfahren zur Messung der Grathohe an Schnitteilen," Werkstatt und Betrieb, Vol. 100, No. 10, 1967, pp. 801-805 (in German).

[56]Jurg R. Seidel, "Machinability Evaluation of Free Cutting Materials With Specific Reference to Copper Alloys," Proceedings of the International Conference on Manufacturing Technology, ASTME, Dearborn, Michigan, 1967, pp. 729-746.

[57]Yoshio Hasegawa, Shigeo Zaima, and Akiyasu Yuki, "Burr Drilling Aluminum and Prevention of it," SME Paper MR75-480, 1975.

[58]M. Okoshi, H. Watanabe, and Y. Watanabe, Journal of the Japan Society of Precision Engineering, Vol. 11, No. 9, 1945.

[59]E. Usui, A. Gujral, and M. C. Shaw, International Journal of Machine Tool Design and Research, Vol. 1, 1967, p. 187.

[60]F. W. Wilson (Ed.), Tool Engineer's Handbook, McGraw-Hill, 2nd Ed., 1969.

[61]K. Okushima and K. Hitomi, Journal of the Japan Society of Precision Engineering, Vol. 24, No. 8, 1958.

[62]R. M. Dyke, Cutting Tool Engineering, May, 1968.

[63]S. Zaima, A. Yuki, and M. Iio, Journal of the Light Metals, Vol. 18, No. 3, 1968.

[64]E. Ernst and W. A. Haggerty, Transactions ASME, 80, 1059, 1958.

[65]A. Yuki and Y. Hasegawa, Journal of the Light Metals, Vol. 19, No. 10, 1969.

[66] A. Yuki and Y. Hasegawa, The Journal of the Light Metals, Vol. 23, No. 9, 1973.

[67] "New Drill Guide Improves Efficiency in Drilling With Portable Tools," Cutting Tool Engineering, January/February, 1974, pp. 18-19.

CHAPTER 5

BURR MINIMIZATION

Since the prevention of burrs is not an economically realistic approach for most companies, it is essential to consider how burr size can be minimized. In utilizing this approach to minimizing deburring costs it is convenient to consider burr minimization from two viewpoints:

1. Physically limiting burr size and toughness
2. Making the burr easy to remove

Controlling Burr Size

It is almost trivial to point out that small burrs are easier to remove than large burrs, and yet few individuals have made any attempt to analytically determine what maximum burr size should be allowed for their machining and deburring conditions. For small parts having tolerances of ± 0.002 inch (±50.8 μm) burrs 0.003 inch (76.2 μm) thick by a similar height can be readily removed by loose abrasive processes. For very miniature parts having tolerances of ± 0.0002 inch (±5.08 μm) burr size must be maintained at 0.001 inch (25.4 μm) thick or thinner. For different processes and geometries different limits exist.

Burr minimization requires a quantitative knowledge of how burrs vary with machining variables. Minimum cost deburring also requires a knowledge of how deburring costs vary with burr properties. While Chapter 4 summarizes the existing knowledge of how machining variables affect burr size, no data has yet been published on deburring costs as a function of burr size. Chapter 8 however does indiciate, for a particular situation, how burr size affects deburring time.

Several general rules about burr minimization are known:

- Sharp tools minimize all burr properties
- Workhardening materials produce thicker burrs than non-workhardening materials
- Burr hardness is higher than the parent material hardness for workhardening material
- Supporting the machined edges minimizes burr size
- Ductile materials form larger burrs than brittle materials
- High feed rates typically (but not always) produce thick burrs

Table 4-2 illustrates the differences in burr properties produced by sharp and dull cutters. If a choice is available in materials, choose a non-workhardening material to minimize burr thickness and the hardness differential between the burr and workpiece. While the last rule in the above list is universally known the actual economics of this approach are often ignored. Several companies have found that "burr" type cutters such as Metal Removal Company's "Master Cut" greatly reduce the burr thickness and length produced in milling operations.

Burr Placement

As indicated previously, burr minimization in its broadest interpretation includes more than just minimizing burr size. It involves more than a knowledge of how all variables affect a given operation. It includes other approaches to making the burr easy to handle.

Making the burr easy to remove involves two aspects:

- . Locating the burr in the best position
- . Controlling burr size by appropriate selection of machining sequences

Figure 5-1 provides an excellent example of burr placement. The cut-off burr in the right hand view is masked by the large diameter of the part. In the left hand view the burr is exposed to the action of vibratory deburring and can be readily removed provided tolerances are adequate. Figure 5-2 is another example of burr placement. Production routing sheets should indicate where the cutter exit burr should be. In this case a burr at the step in diameter would double the time required to deburr this part.

Correct burr placement can pay big dividends on milled parts because of the large number of edges produced. On intersecting features placing the burr on the most accessible edge can reduce burr costs by 50 per cent. The correct placement of burrs can change a normally hand deburr situation to a much more economical and repeatable vibratory deburr operation. In some operations burr placement can be controlled by a simple N/C tape change. In some cases climb milling rather than conventional milling will provide better location. Since tooling may be involved it is important to consider burr placement in the preliminary processing steps.

The decision of which machining sequence to use can affect deburring time. As an example the part in Figure 5-3 would typically be turned then the three flats would be milled. This creates two undesirable facets though. First milling typically makes a heavy burr, and secondly deburring would require a separate operation. The company which produced this part chose to mill the flats before turning the stem diameter. This produced a smaller burr and allowed the operator to brush the burrs off with a wire brush as a part of the lathe operation.

When the manufacturing engineer considers the machining sequence he wants to follow he should consider the "burr" as significant a factor as the desired surface finish and tolerance. The following checklist has been prepared to help in processing and in troubleshooting burrs:

ASK YOURSELF

1. Does the burr have to be removed?
2. Can the part be redesigned
   a. for less machining
   b. for less deburring
3. Will the burr be cut off in a later machining operation? If the machining sequence were changed would the burr be cut off?
4. Is the burr accessible?
   a. Should I change the sequence of operation?
   b. Should I change the direction of the cut?
5. Can I choose a cutter that gives a smaller burr?
6. Do I know the feed rate which gives the smallest burr?
7. Can I use a subsequent heat treat to make the burr brittle?
8. Would a change in coolant or method of application make the burr brittle?
9. Does the burr have to be removed <u>now</u>?

VIBRATORY DEBURRING OPERATIONS CAN BE EXPEDITED IF THE CUTOFF BURR IS LEFT ON THE ROUNDED END OF THIS SCREW MACHINE PART, RATHER THAN ON THE FLAT END.

FIGURE 5-1: PLACE THE BURR FOR EASIEST REMOVAL

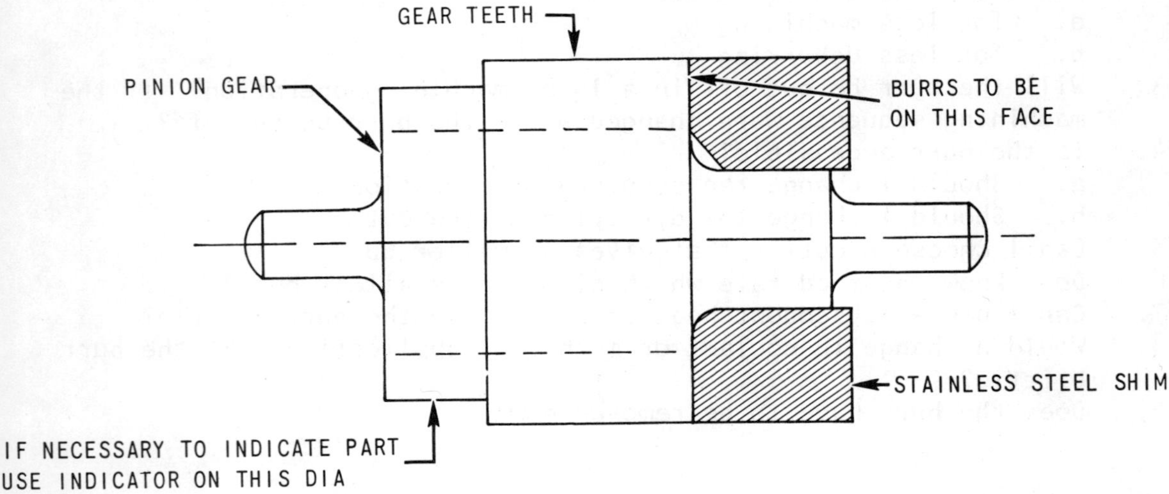

FIGURE 5-2: IDENTIFY DESIRED BURR LOCATION AND UTILIZE BACKUP MATERIALS TO MINIMIZE BURRS

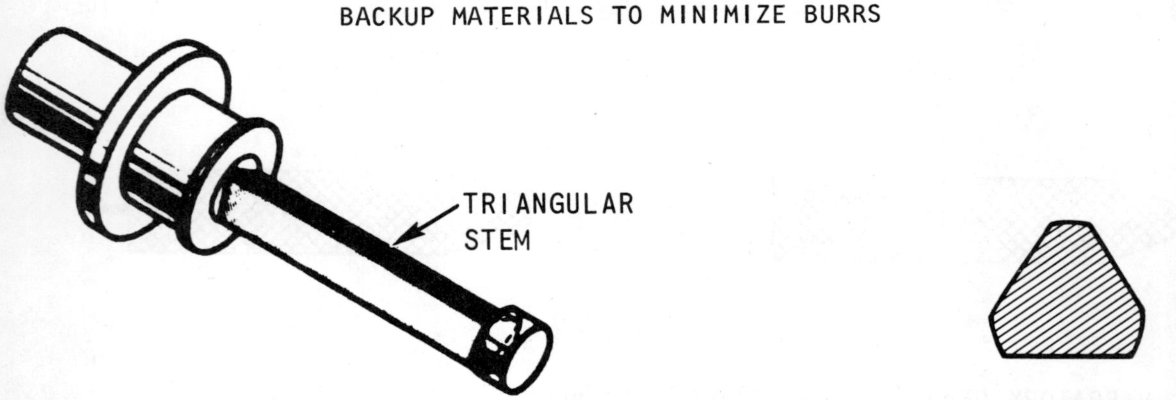

TURN STEM AFTER MILLING FLATS
WIRE BRUSH ON LATHE TO REMOVE BURRS

ADVANTAGE

THIS SEQUENCE OF OPERATIONS MINIMIZES THE BURR
AND
DEBURRING IS PERFORMED AS PART OF TURNING OPERATION
(IT IS RAPID AND DOESN'T REQUIRE EXTRA SETUP OR HANDLING)

FIGURE 5-3: EFFECT OF PROCESS ON DEBURRING

Summary

Burr minimization is always possible. For effective utilization of this approach, however, it is essential to have a knowledge of how burr size affects deburring costs. Changing feeds and speeds will reduce burr size, as will selecting appropriate cutters, using backup material, and choosing workpiece materials which have low ductility and strain hardening exponents. Burr removal can be expedited by placing burrs in a position they can easily be removed from. Utilizing different machining sequences can also minimize the amount of deburring required.

References

[1] Federico Strasser, "How Control of Burrs Aids Sheet Metal Stamping," Iron Age, Vol. 185, No. 3, January 21, 1960, pp. 90-92.

[2] "Prevention of Burrs," Tooling (England), September, 1975, pp. 23-26, 29.

[3] F. Schafer, "Entgratgerechtes Konstruieren und Fertigen," Proceedings of Fachtagung Entgraten '75, IPA, Stuttgart, West Germany, November, 1975 (in German).

CHAPTER 6

PLANT WIDE APPROACHES TO MINIMIZING DEBURRING COSTS

While most of the techniques described in this book can be implimented by individual engineers and supervisors, effective cost reduction can best be effected by a coordinated plant wide program. The reason such an approach is needed is due to the apparent complexity of the problem and the need to involve individuals from several departments.

As an example consider the following aspects of minimizing deburring costs:

1. Design to minimize deburring costs
2. Practice burr prevention
3. Minimize burr size
4. Improve deburring processes

The design aspect requires input from product designers and draftsmen and, if standards are involved, of inspection and manufacturing. Burr prevention requires nontraditional machining techniques which requires input from manufacturing engineers. Minimizing burr size requires input and cooperation from machinists, methods engineers, and manufacturing engineers. Improving deburring processes requires an analysis of existing capabilities and the cooperation of deburring workers. In addition to these already named, machining department foremen, quality control engineers, safety engineers, and production control individuals may also be involved.

Background

Deburring is one of the few areas of manufacturing which has been typically approached by trial and error rather than treated in a rational cause and effect manner. It is a twentieth century process relegated to historical precedent rather than scientific evaluation.

In looking at other manufacturing processes it can be seen that much of the success in optimizing productivity and profits is the result of two principal factors:

1. Having a data base which defines pertinent characteristics of the process.
2. Having a rationale for comparing alternatives.

In turning operations, for example, one can depend upon handbooks or manufacturers' literature to indicate tool life; recommend appropriate speeds, feeds, and geometries; and to provide relationships between tool life, depth of cut, material hardness, and other parameters. The literature also abounds with analytical techniques to use to optimize production or to minimize costs. These same two key factors have been available for all manufacturing processes but one--deburring.

Consider for a moment the three primary elements in deburring:

- The burr
- The deburring process
- The resulting edge condition

When did you last see a table, a graph, or a formula which quantitatively related these three elements? The fact is, such information has only recently been developed for burrs. As a result many, if not most, companies are still relying on precedent, intuition, and trial and error. Comprehensive guidelines are not generally available to assure that deburring is optimized. If deburring is not optimized then probably the manufacturing process as a whole is similarly not optimized.

Deburring--An Integral Element

Deburring is the one manufacturing process common to every metal cutting and blanking facility in the world. It is a $2-6 billion annual expense to U.S. industry. It represents from 1 to 5 percent of most companies' direct cost of manufacturing. Just the labor required to operate the 33,000 barrel tumbling and vibratory deburring machines in the U.S. requires at the minimum an estimated $30,000,000 a year.

On close tolerance parts, allowance must be made for the stock loss which accompanies deburring. In fine edge blanking where tolerances of 0.0005 or less are common, it is essential for the die to be designed to leave extra stock which will be removed by the deburring process. Since vibratory or tumbling processes do not remove stock uniformly around complex contours, careful consideration of the deburring effects is essential.

Each deburring process is limited in the size of burr which can be removed without affecting part dimensions significantly. The feeds, speeds, and tool geometry used to machine the parts all affect the size of burr produced. Thus trade-offs are required between machining parameters and the deburring process. Ignoring the machining effects can only result in increased deburring costs and can easily result in higher total costs of fabrication.

In many situations it is essential to have the best process on the first run. It is not economical to cut-and-try various deburring processes on components worth $200 apiece. Similarly one cannot afford to damage low volume, long lead time parts by experimenting with deburring parameters. Some contracts specify that no change will be made to a manufacturing process once satisfactory parts have been produced. In this situation, even though a company develops a less expensive deburring technique, the company is committed to using equipment and labor which could better be applied elsewhere.

In each of the situations mentioned, deburring constitutes an integral element in the manufacturing process. While in most cases it is not a critical element, it does cost money. With a better understanding of burrs and deburring process capabilities, it is possible to minimize some of the problems and costs associated with deburring.

Several factors can be identified which have contributed to the cut-and-try approach to deburring rather than a rational approach. These are:

. Failure to distinguish between deburring and finishing
. Influence exercised by large volume, single product companies
. Lack of knowledge about burr properties
. Inability to define actual geometry requirements
. Proliferation of deburring processes
. Diversity of deburring requirements

Deburring versus Finishing

Deburring is only one of several finishing processes (Table 6-1). Because most components need more than one finishing process there is a logical desire to utilize processes which will accomplish all finishing in one unit. In those instances where a high polish is required, the polishing is the most important process. Thus while deburring may have been completed after a one hour run, two additional hours are required to obtain the required lustre. In these types of operations, which are typical of many parts, there has been no need to take a close look at deburring. As a result while the knowledge of polishing increases, nothing is added to the knowledge base of deburring.

TABLE 6-1: FINISHING PROCESSES

Polishing	Deburring
Radiusing	Mechanical Hardening
Plating	Descaling

High Production versus Short Runs

Most manufacturers of deburring equipment will develop a process tailored to their equipment and the user's product if a large volume of parts are to be run. In the automotive industry for example several companies manufacture equipment specifically for brush deburring gears. While this relationship between user and supplier eliminates the direct cost of developing a technique in-house, it also results in a situation in which no data base is produced.

Although the equipment manufacturers are in an excellent position to establish such information, they have not done so. The reasons for this are obvious--first they want to maintain their competitive edge, and second they are results-oriented rather than developers of comprehensive information.

All of this, of course, works to the disadvantage of companies producing only short runs and which have neither the time nor money to develop or have others develop processes for them. They need dependable, usable information immediately.

Lack of Knowledge About Burrs

One of the major reasons deburring is still treated as an art rather than a science is that few people know anything about burr properties. The whole subject of burrs and deburring appears to most individuals to be a terribly complex subject (Figure 6-1). With a little study, however, the picture becomes clearer (Table 6-2). There are five factors which influence deburring. The material properties and the machining parameters determine the burr properties. The burr properties and the part definition in turn dictate the allowable deburring processes.

With the exception of burrs produced by blanking and piercing, until recently no one had taken a close look at burr properties. The few authors who discussed burr size used only burr height as the feature which identified the burr properties. More recent studies indicate that burr thickness, hardness, and shape at the workpiece surface are more significant properties. In general for most situations, burr height does not have any apparent relationship to burr thickness and only a secondary relationship to difficulty of removal.

It has also only been recently that any attempt has been made to analyze how burrs form or to predict their properties. The studies that have been performed indicate that burrs are produced by four basic mechanisms. Complex plasticity equations are required to provide the quantitative description of burr formation. As a result there is no single nor simple formula which describes burr formation.

Another serious problem is the lack of an industry wide accepted definition of a burr. To some, a burr is any material extending past a ninety degree corner. To some "extruded" metal at an edge is a burr, to others it is not. To some a "little burr" is not a burr, to others any burr which does not interfere with the next assembly is not a burr. With no universally accepted definition of a burr, it is not surprising that deburring still does not have a data base.

Proliferation of Deburring Processes

Today industry has 20 principal deburring processes to choose from (Figure 6-2). Each of these processes works best on certain materials, on a limited size of burrs, on specific geometries, and within specific workpiece tolerances. As a result, out of a large jig saw puzzle of possibilities, often only one process appears feasible (Figure 6-3). With additional knowledge the feasibility of other processes would also be evident. Seven of these processes have been commercially introduced within the past 12 years. The relatively recent introduction of these processes is one of the reasons little information is available on these processes.

Material Properties	Machining Process	Burr Properties	Deburring Processes	Part Definition
Hardness	Type	Hardness	Type	Tolerances
Toughness	Tool Design	Toughness	Cycle Time	Edge Requirement
Thermal Conductivity	Machining Rates	Width	Burr Size	Size
Specific Heat	Coolants	Length	Cost	Geometry
Density	Machining Sequence	Abrasion Resistance	Repeatability	Function
Chemical Composition	Tool Wear	Thermal Conductivity	Lead Time	
Abrasion Resistance	Cost	Chemical Composition	Quantities	
Metallurgical Phase		Specific Heat	Burr Location	
Yield Strength		Mass		
Ultimate Strength		Repeatability		
Melting Point				

TABLE 6-2: FACTORS INFLUENCING DEBURRING

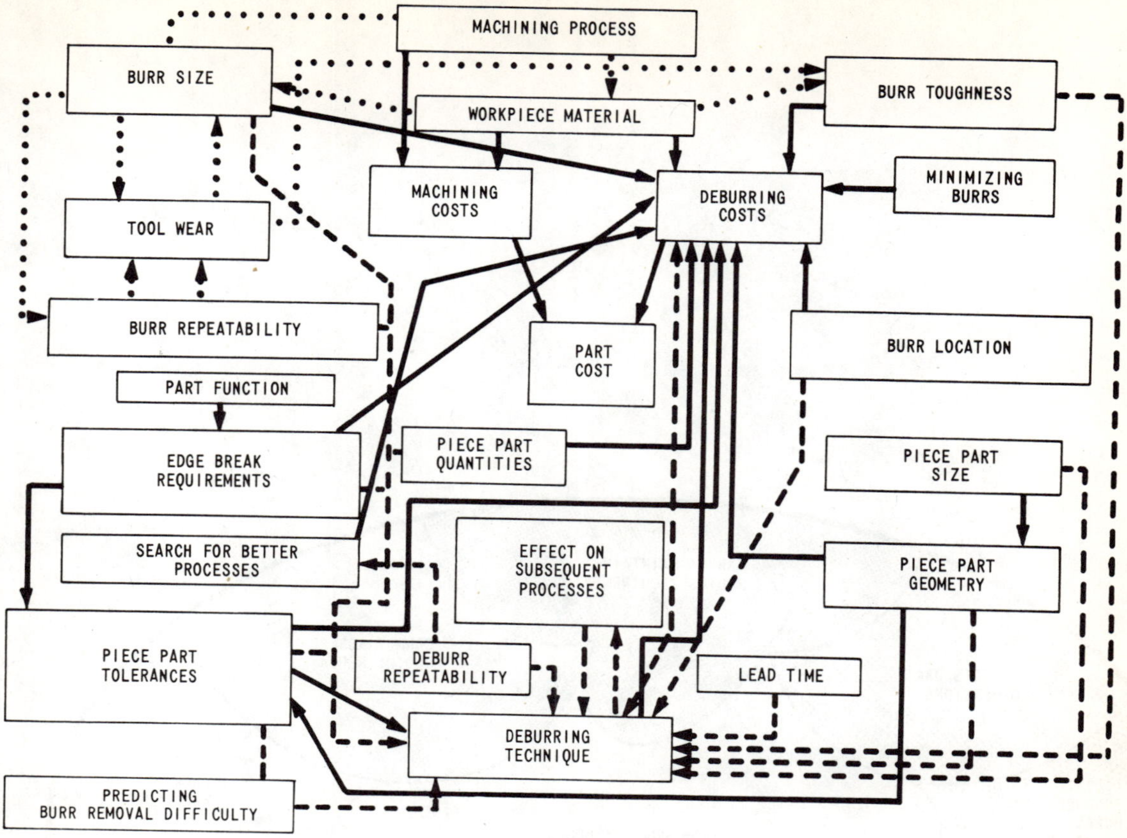

FIGURE 6-1: DEBURRING DECISION MODEL

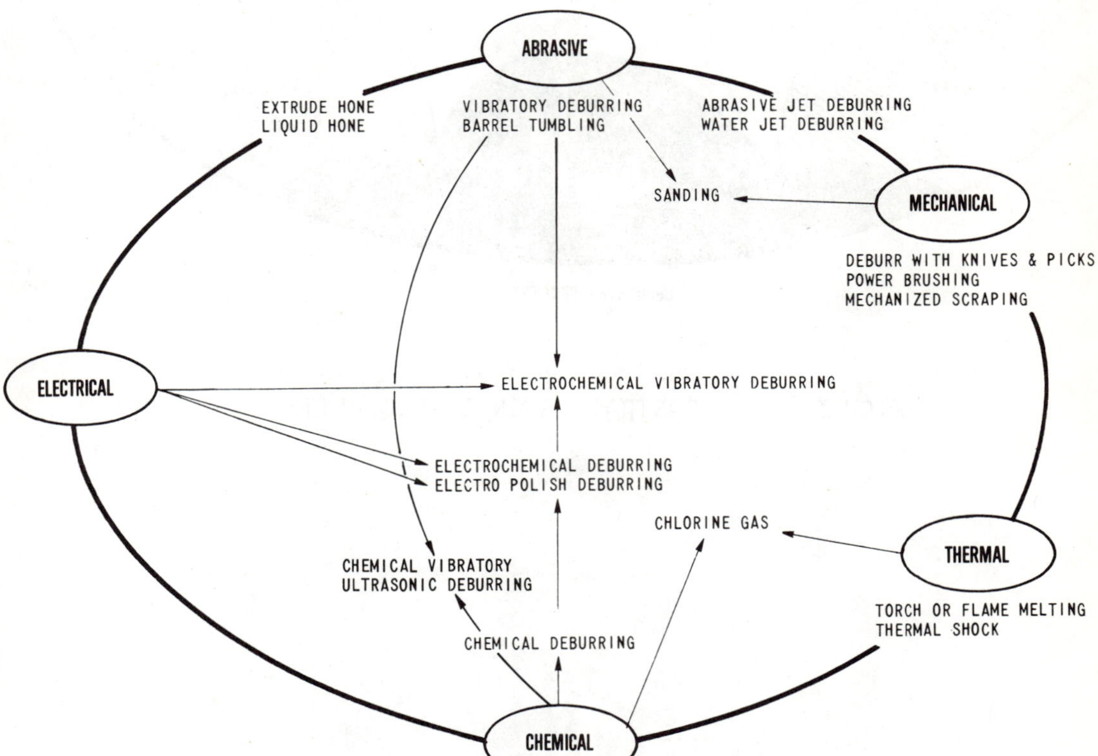

FIGURE 6-2: PRINCIPAL DEBURRING PROCESSES

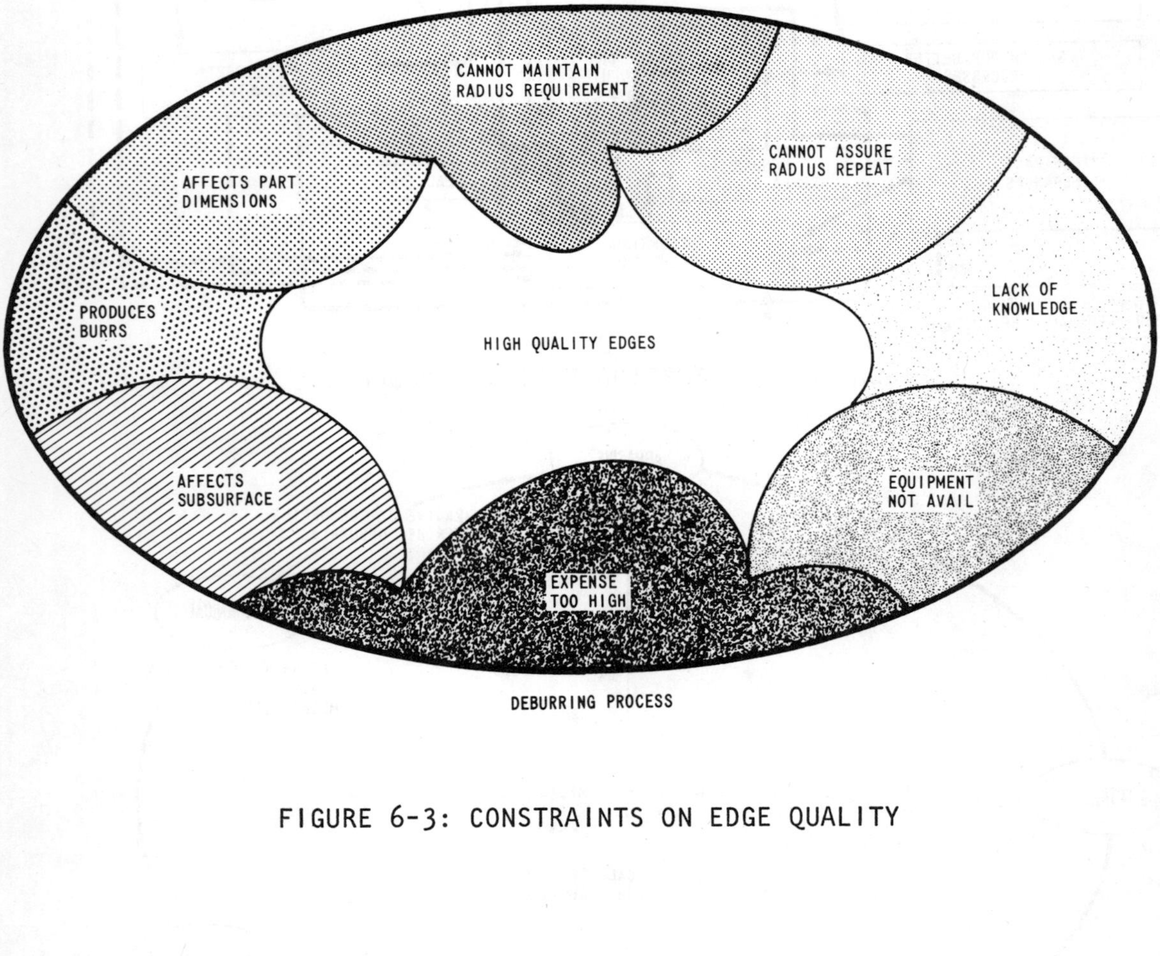

FIGURE 6-3: CONSTRAINTS ON EDGE QUALITY

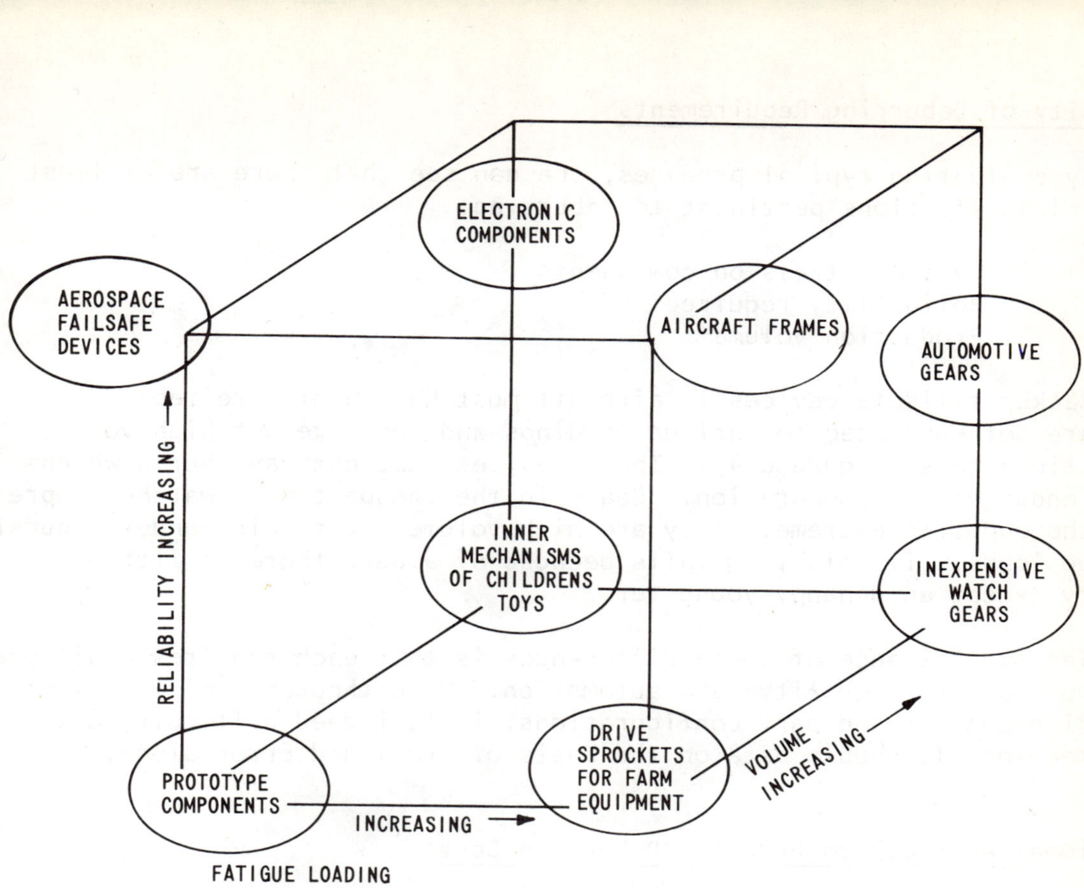

FIGURE 6-4: VARIABLES WHICH INFLUENCE DEBURRING REQUIREMENTS

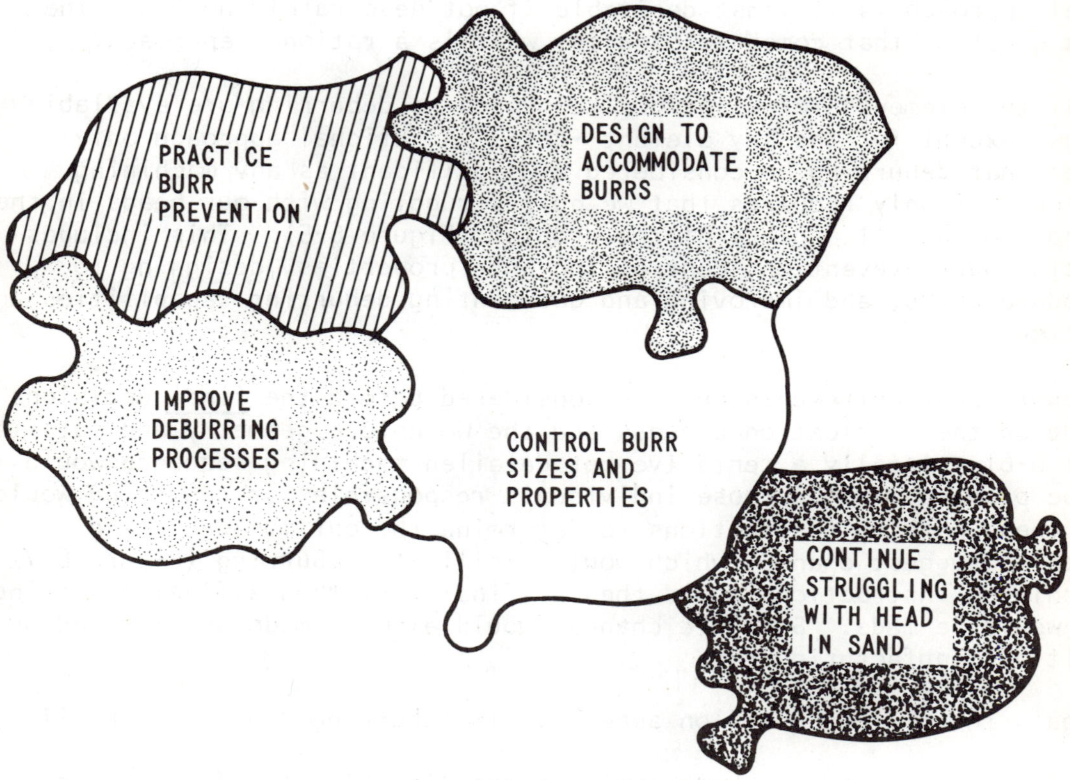

FIGURE 6-5: APPROACHES TO BURRS

Diversity of Deburring Requirements

By evaluating typical produces, one can see that there are at least three classifications pertinent to deburring:

- . Type of stress on components
- . Reliability required
- . Production volume

Backup failsafe devices in aircraft must have a high reliability. They are not subjected to fatigue loadings and they are not high volume production items (Figure 6-4). These devices must not have burrs which would endanger their operation. Gears in the cheapest wristwatches represent the opposite extreme. They are high volume items which undergo quasi-fatigue loads. If this item fails because of a burr there is little penalty except an unhappy youngster.

The significance of these differences is that each requires a different level of deburring quality* and automation. With thousands of levels of deburring quality and part configurations, it is indeed difficult to provide meaningful, usable data on the basis of trail and error alone.

A Rational Approach to Reducing Deburring Costs

With 132,000 manufacturing plants in the United States spending at least two billion dollars annually to remove burrs, it is obvious that a rational approach is at least desirable if not desperately needed. The obvious question that comes to mind is, what is a rational approach?

All the elements for a rational approach to deburring are available and known except for the key element--data. A rational approach only requires that deburring be considered as significant as any machining operation. It only requires that we quit struggling with our heads in the sand and look at all aspects of the problem (Figure 6-5). This includes practicing burr prevention, controlling burr properties, designing to accommodate burrs, and improving and documenting deburring capabilities and limitations.

Deburring requirements must be considered before the final decisions are made on the fabrication process and the workpiece geometry and material (Figure 6-6). Ideally a tentative but detailed machining sequence would first be prepared. Then those individuals responsible for deburring would review these working instructions to determine the optimum deburring approach to suggest changes which would facilitate deburring (Figure 6-7). The machining instructions would then be finalized; then a final deburring review would be made. Suitable changes would also be made as dictated by the initial manufacturing run.

Again the obvious question appears, "Is deburring really worth all

*Deburring quality as used here includes the side effects of stock loss, impregnation, surface finish, and edge radius.

this special effort?" If the individuals responsible for specifying deburring requirements utilize an in-depth or systems approach, the answer is yes. Consider the check list shown in Table 6-3. These are the questions that must be answered and then implemented in both the initial and final reviews of work instructions. How many of these questions can the originators of the machining work instructions correctly and satisfactorily answer? What penalty in cost, performance, rejects, safety, and delays does one face if the wrong "guesses" are made?

When a single end mill produces eight different burrs on a single cut, each with a different height, thickness, hardness, and shape, is it reasonable to guess? Is it economical to process parts "the way we've always done it," or is cost reduction important?

TABLE 6-3: DEBURRING CHECK LIST

- What level of deburring quality is needed?
- Do all edges have to be burr free?
- How do we define acceptable edge conditions?
- Do fixtures have relief for burrs and sharp edges?
- Can we prevent some of the burrs?
- How can we economically minimize the burrs?
- Can we change machining sequences to make the burr easier to remove?
- Can we use the heat treat operation to make the burr brittle?
- Can we use different cutting tool geometry to make it easier to remove burrs?
- On what operations can the operator remove these burrs?
- Can we economically justify the optimum equipment?
- Can we design special hand deburring tools to speed up the operation?
- What change in product geometry or material will reduce deburring costs?
- How much stock will the deburring process remove from feature of the workpiece? How repeatable is this stock loss?
- What is the most economical process to remove these burrs?

Definitions and Standards

A company must adopt some universal definition(s) of a burr and how it should be measured. Without this common base the compilation of any data is essentially useless. The definition of a burr should be independent of the part's function, material, and method used to produce it. A definition such as "a burr is any material extending beyond the edges of a part which was not at that position originally and which is not an intentional feature of the part" would appear to be functional.* In all cases these burrs are material which has been pushed out of the way. (See Chapter 4.)

*The screw machine cut-off burr or any burr formed by allowing the workpiece portion of the stock to fall from the bar stock before the cut is completed is not a burr by this definition. Since it still may have to be removed one might want to consider definitions for two types of burrs--plastic flow burrs and cut-off burrs.

Providing a common definition of a burr also will necessitate developing within companies a definition of what edge conditions are allowable on specific parts. For example, on one part the designer might state:

> "Burrs are permissible on parts provided they are no longer than 0.003 inch, that they do not interfere with assembly, and that they will not break loose or become dislodged. A sufficient check for burr adherence is to rub a round nose steel rod 1.0 inch long by 0.020 inch thick across the edges in question. If this rod dislodges material the burr is not acceptable."

It is this definition of allowable edge condition which has most bothered inspection departments and hindered the scientific approach to deburring. The acceptance of common definitions will, in part, be dependent upon also having definitions of (1) how the burr should be measured and (2) what conditions are acceptable within the company.

Industry must also develop "standards" which can be used to compare data from different sources. These standards must be phrased such that differences between equipment, media, and operating conditions can be analyzed. Richerson[1] and Almco's[2] test for media efficiency represents the type of standards needed by industry as a whole.

There is certainly nothing restrictive about these standards. They merely provide a uniform technique for comparing results. In establishing standards it is also essential to establish an understanding of which variables are significant for each deburring process. For example, Table 6-4 lists most of the variables significant in vibratory deburring. Many individuals who have worked for years with this process have not yet realized the significance of some of these factors. It is for this reason that a broad cross-section of knowledgeable individuals is required to establish these standards, to identify significant variables, to propose test approaches, and insofar as possible to rank the significance of the variables.

Data Development

Both vendors and users of deburring equipment must begin supplying meaningful data, data which correlates the burr, the removal process, and the final result. To date suppliers of deburring equipment have been extremely reluctant to take the lead in providing facts. An often heard reaction is, "There are too many variables, it's impossible to make a science of deburring." F. W. Taylor faced the same reaction when he began his studies of metal cutting in 1906. Today we have reams of data on metal cutting. Generating facts on deburring cannot be too difficult or manufacturers and suppliers of equipment would not still be in business. Their success is, in part, the result of building a certain data bank of experience.

Obtaining useful data and facts is a slow process. Like most ventures unqualified success comes after years of work, but useful results begin immediately. One of the first needed results is a general yet factual

TABLE 6-4: VARIABLES SIGNIFICANT IN VIBRATORY DEBURRING

Compound	Media	Machine	Workpiece	Loading
Composition	Size	Frequency	Material	Amount of Media
Grit Size	Shape	Amplitude	Burr Thickness	Amount of Abrasive
Grit Material	Material	Basic Design*	Burr Length	Amount of Water
Emulsifying Agent	Coarseness of Grit	Shape	Burr Toughness	Volume of Parts
	Composition		Geometry	Weight of Parts
	Weight		Burr Location	
			Size	
			Weight	

*Some machines have "dead" zones in which little deburring action occurs.

TABLE 6-5: TYPICAL BRUSH DEBURRING PERFORMANCE

Type of Wheel	Remove Burrs To	Provide Radius Of To	Typical Applications	Expected Wheel Life (Lineal Feet)	Typical Cost of Wheel & Media (per Lineal Foot)	Effect on Dimensions

definition of the capabilities of each process--a definition which describes the burr as well as the results. The next need is for some form of visual guide which relates deburring parameters to the results, a guide which indicates the next logical step to take. A flow chart such as Figure 6-8 is one approach. The next phase might be the preparation of a series of tables which provide more detailed information (Table 6-5). To provide a scientific basis for selecting deburring parameters, scientific testing will be required. Fortunately such statistical experimental design techniques as factorial designs, response surface methodology, and evolutionary operations (EVOP) are available to provide inexpensive yet comprehensive data.

Eventually a sufficient data base will exist that formulas can be developed which will predict the optimum deburring process and parameters. Computer selection of deburring parameters has already been attempted but there is not a sufficient data base at this time to make it reliable.

A nucleus already exists for the development of data on deburring. Two extensive multilingual listings of literature on deburring are available.[3,4] A general summary of the capabilities of each of the 20 major deburring processes is also available. (Chapter 7) A start has been made on measuring burrs and determining how machining factors and workpiece geometry affect them. (Chapter 4) A basic quantitative theory of burr formation has been developed (Chapter 4) and a summary of common burr minimization techniques has been prepared (Chapter 5). Each of these studies represents only the initial thrust in a burgeoning new area, an area which must eventually also include plastics since there is no deburring data at all in that field.

Developing a rational approach to deburring requires a great deal of vision; vision to see the monetary return, to develop sound yet inexpensive test programs, and to see the relationships which have been obscure for decades. It requires a desire to reduce costs, to apply value engineering, systems engineering, and statistical approaches. A beginning has been made, needs have been identified, and goals have been outlined. Deburring can be approached as a rational process.

Reducing Deburring Costs Through Value Engineering

One of the basic approaches often employed to reduce costs on a plant wide basis is value engineering. Value engineering, conceived by Larry Miles of the General Electric Company, differs from cost reduction in that it assigns a predetermined value of cost to each of the functions of the product to insure that the value of each function is properly in proportion to the value of each other function. The predetermined values of cost may be established by past performance and judgment. The project engineer and designer must know the manufacturing costs as each new requirement, such as "remove all burrs" is added to the blueprint. When this information is assimilated, the project engineer and the designer will see design requirements which are generating excessive cost. Along with this, if the project engineer and the draftsman create the new project within the confines of an outline of predetermined cost by function, the total of

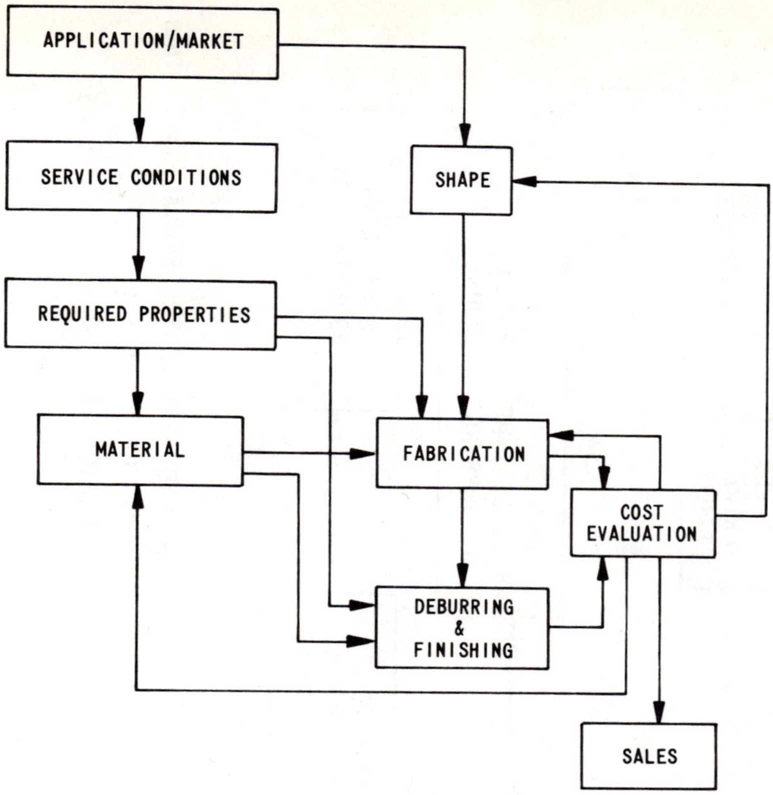

FIGURE 6-6: SEQUENCE OF EVALUATIONS FOR NEW PRODUCTS

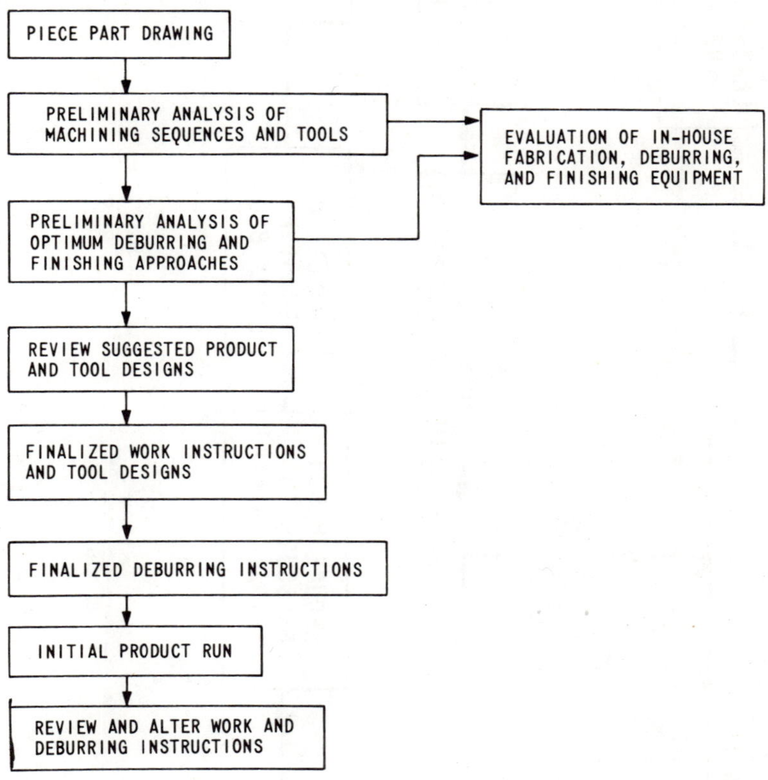

FIGURE 6-7: PROCESSING EVENT FLOW CHART

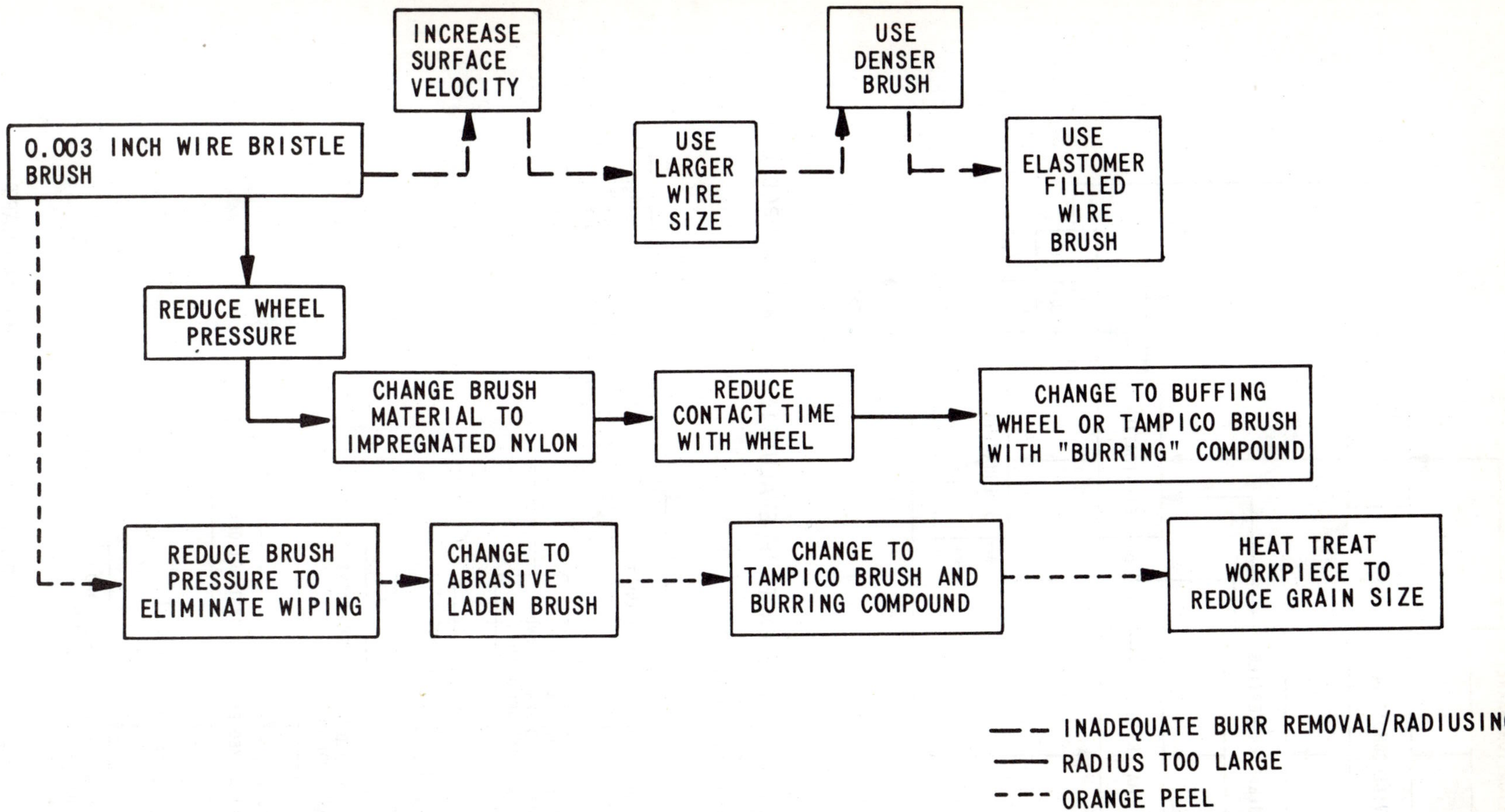

FIGURE 6-8: GUIDELINES FOR IMPROVING BRUSH DEBURRING OF PRECISION WORKPIECES

the cost of the functions will comprise a selling price which will be competitive.

What is Value?

A definition of value might be: Value is obtained when, based on the need, an action or thing provides a maximum function for a minimum of dollars expended.

Value is not a function of failure. In the classical example of the airplane wing bolt, the proposition is that the value of the bolt is equal to the value of the airplane and all of its crew because if the bolt fails the airplane will fall and all will be lost. Therefore, that we may conceivably put a million dollar bolt in the wing of an airplane is based on a false premise and the solution is incorrect.

The value of a report is the relationship of cost to create it, the materials to provide it, but mostly the use to which it is put. As may be seen, a very ornate and efficient computer report furnished to a supervisor who files it has no value. And also, this same computer report based on false information furnished to a supervisor who uses it likewise has no value.

So what is value? Value is obtained when, based on the need, an action of thing provides a maximum of function for a minimum of cost.

Making a Value Judgment

When making a value judgment we ask ourselves the following six questions:

1. What is it?
2. What does it do?
3. What does it cost?
4. What is it worth?
5. What else will perform the function?
6. What does that cost?

Each member of each value engineering team must ask himself: "Are the things that I am doing worth the cost to my group? Am I using a reliable and economical method for this function? Does anyone else perform this function that I might use his method?"

The Value Engineering Job Plan

The following is a brief description of the six phases of value engineering:

1. <u>The Information Phase</u>--This portion of the plan is the most important. It is in this phase that the individual determines the exact function or purpose. This phase continues through all of the

other phases. The six age-old questions: Who? What? When? Where? Why? and How? are asked and the proper answer is decided upon for each of these functions as analyzed. Some questions which might be asked at this time are: What is it? What does it do? What is the function that it performs? What does it cost? How is it made? What are the specifications, the tolerances, the limits, the requirements? Are temperature conditions a factor? Are there other usages for this function? What is the history of reliability of appearance, of delivery of this function? What is the rejection rate? What is the yield? How do we now obtain this function? How is it put together? And again, what does that cost? Poor decisions are made using incomplete information.

Several questions follow the Information Phase:

a. Does its use contribute value?
b. Is its cost proportionate to its usefulness?
c. Does it need all of its features?
d. Is there anything better for the intended use?
e. Can a lower cost method be used?
f. Can a standard part be used?
g. Is it made with proper tooling considering quantities?
h. Does the total cost of the item seem to be reasonable?
i. Will another supplier provide it for less?
j. Is anyone else obtaining it for less?

2. <u>The Speculative Phase</u>--The second phase of the value engineering program is called the speculative phase. During this phase, the information gained in the information phase is attached using the creative thinking technique to separate, to add, to combine, to change, and to alter. The questions in this phase are asked each with a challenge; to challenge the information that has been gained; to create a better, more simple answer to the function. Questions such as: Why is it this way? Why do we have this size? Why do we have this specification? Why do we use this material? The answers to these questions provide the speculative phase. Every possible suggestion is recorded and every possible suggestion is encouraged. There can be no criticism in this stage of the decision. Every idea is recorded from the most practical to the most impractical. Each member of the team is requested to make a suggestion which comes to his mind and to do exactly what the name of this phase says--speculate. Every suggestion will be recorded no matter how wild it seems. No one is permitted to disagree with an idea.

3. <u>The Analytical Phase</u>--During this, the third phase, the ideas advanced during the speculative phase are analyzed. The first question in this phase is: "How can this idea be made to work?" The analysis must be made on an accurate evaluation of the cost estimates of the idea; an analysis of the value of each idea; and this is the phase where we place a dollar sign on each of the ideas. Previous records and past history are used to put this dollar value, which is so important to each idea, on it. We must investigate reasons which appear obvious that an idea won't work and to overcome and eliminate

the objections which are not valid.

Every effort should be made to make the idea work. This phase is used to cull out ideas which will not be used.

4. The Planning Phase--The fourth phase of value engineering is the planning phase. During this phase, the aforementioned ideas with the greatest potential of return are selected. A look back into the information phase may be advisable at this time to rehash ideas or thoughts which may have come up since. But now we must plan a program backing up each of the ideas accepted with factual information and descriptions of the practicality of the suggestion. We must at this time reevaluate the function, consult with company specialists, and begin to plan our action.

5. Summary and Conclusion--The necessary implementation costs, scrap and rework and any other indirect costs which may be involved are analyzed and the idea is evaluated on the basis of total cost. This is the phase which implements the suggestion and future cost records of actual performances.

Another precept of the value engineering definition program is the functional approach. The function being performed is to be clearly stated in the information stage. When asking the questions, "What does it cost and what is it worth?" the answers are found by an evaluation of the functions. There are many ways to evaluate the function; for example, cost per pound, cost per piece, cost per order, or cost per year. There may be a cost increase to insure a reliable function; we find a correlation between cost and function. Since it is our intention to place a value on the function, we must first determine a definition for value.

Examples of Value Engineering

As examples of value engineering, consider the following approaches:

Operator Influences: Production operators can do a great deal to minimize burr removal costs. The most obvious item is to have them deburr parts while their machine is cycling on the next part. Even a partial deburr will save time in subsequent operations. Shops in which deburr operators and machine operators are in separate labor classifications should encourage machine operators to work with deburr operators in deciding when tools should be changed. Unfortunately in many shops the machine operator could care less about the burr he produces because he is not held accountable for the extra work required to remove gross burrs.

Some additional approaches to in-process burring include:

1. . Remove heavy burrs only (leave small burrs if they have to be removed mechanized processes will remove them cheaper)

. Provide operators with special design tools to remove heavy burrs. Special knives, reamers, countersinks, or cutters dramatically speed burr removal.

Machine off heavy burrs.
Thick burrs can often be cut or ground off faster than an operator can remove them with a knife.

2. Deburr only as required for later fixturing.
Save the majority of deburring for the final operation. This reduces the time operators spend in scrutinizing edges which have already been deburred. Some burr laden edges may be cut off in later operations; thus complete in-process deburring would have added some unnecessary expense. If the part is scrapped in a later machining operation deburring costs will not have been wasted.

3. Change burr location.
Put the burr where it is easy to remove, or where it will not have to be removed. (Figure 5-1)

4. Deburr on machine time.
Provide deburr aids (pictures, samples of adequately deburred parts, special deburring tools).
Add deburr steps in the machining cycle.
Often an extra machining pass can be used to partially deburr a part. It is cheaper to do this in the machining cycle than to have an operator pick up, fixture, deburr, and verify that edge. In addition it eliminates the flow time associated with an extra deburr operation, it is controllable, and it is independent of operator efficiency. Examples of this include brush deburring on screw machines and lathes, touching edges with sandpaper while a burr is still chucked on the lathe, and milling off a heavy roll-over burr.

There are four important rules to remember and twelve questions to answer on every part.

Rule #1 If you don't make a burr, you don't have to remove it.
Rule #2 Every conventional machining operation produces some burr.
Rule #3 If you don't have to remove a burr, don't remove it.
Rule #4 The machining conditions affect the deburring costs.

DEBURRING CHECK LIST

1. Does it have to be deburred?
2. Does it have to be deburred now?
3. What burr must be removed?
4. How much edge break is allowed (required)?
5. Does it have to be done by hand?
6. Will fixturing or hand tools help?
7. Can it be done on machine time?
8. Can two deburring techniques be combined to minimize total

deburring cost?

9. Will backup material make deburring more economical?
10. Can fixtures be designed with burr clearance?
11. How can the burr be minimized?
12. Can the part or process be redesigned to reduce deburring?

Job Shops for Deburring

Many companies frequently overlook the possibility of reducing costs by using job shop deburring facilities. While the immediate costs may seem high for such work, there are several advantages to this approach.

1. They often have equipment available which can do the work faster than companies which cannot afford such equipment.
2. They are accustomed to improvising inexpensive tooling to aid in deburring.
3. Being small, they typically have a faster turn around time than larger companies.
4. On very high quality parts, very skilled workers are required which are not available in many companies.
5. Since job shops see a wide variety of parts in a year's time, they learn to solve a greater variety of problems.

Some job shops specialize in such deburring process as TEM, abrasive flow, or electrochemical deburring. Many companies cannot justify the purchase of such equipment or the associated training required to efficiently use these processes.

Time spent in machining special tools to assist in burr removal is generally quite helpful. Most companies that do in-house deburring always have certain tools that will fit into areas where conventional deburring tools cannot reach. These come in the forms of a specially shaped knife or scraper, a rotary tool, or a special handle. Job shops continually make such tools in a relatively short period of time in order to save a considerable amount of bench time.

Some job shops have developed their own deburring machines and techniques for fixturing parts. These tools can make big savings in deburring times. While individuals in larger companies can do the same, there is less incentive than in many job shops.

The complete job shop is one that can perform deburring from gyro quality to commercial standard. This requires a double standard and close supervision to keep the separation enforced. The supervisory personnel must separate between precision and commercial acceptance and educate his people in these requirements. The job shop is probably the most complete deburring plant because it must be capable of performing such a wide variety of deburring. Their methods must produce a better part in less time in order to be competitive. To stay in this position, they must view all new products with special attention and constantly try to improve their capabilities. This continuing vigilance is often missing in companies having in-house capabilities.

Summary

Plant wide approaches to reducing deburring costs require time, training, and coordination between many departments. It requires an in-depth look at current approaches and costs. Such an approach can solve many of the roadblocks which face an individual trying to improve deburring on just the parts for which he is responsible. Many companies, however, have found that plant wide solutions offer major cost savings and quality increases.[5]

References

[1] David W. Richerson, "Testing and Selecting Wear Resistant Materials," Materials Engineering, January, 1972, pp. 46-49.

[2] "A Comparative Media Efficiency System," Precision Metal, May, 1974, pp. 32-34.

[3] L. K. Gillespie, "Deburring: An Annotated Bibliography," SME Bibliography Series, SME, 1974.

[4] L. K. Gillespie, "Deburring: An Annotated Bibliography, Vol. II," SME Paper MRR75-04, 1975.

[5] Bernard S. Miller, "Control Deburring Quality," Metalworking, November, 1965, pp. 40-48.

[6] Richard Golden, "Value Engineering Using a Functional Management Team Approach to Minimize Deburring Cost," SME Paper MR75-478, 1975.

[7] Robert R. Narins, "Deburring From a Job Shop Point of View," SME Paper MR75-479, 1975.

[8] Lawrence P. Miles, Techniques of Value Analysis and Engineering, New York, McGraw Hill Book Company, Inc., 1961.

[9] Alex F. Osborn, Applied Imagination, Charles Sribner's & Sons, New York, 1953.

[10] Robert E. Parsons, "Why Not Cut Costs by Value Engineering," Materials in Design Engineering, November, 1961.

CHAPTER 7

AN OVERVIEW OF THE DEBURRING PROCESSES

Providing comparisons between different types of processes is a difficult problem and yet it is essential for a full appreciation of process capabilities and limitations. It is particularly difficult for the deburring processes because so many factors have to be considered (Table 6-2).

The following chapter represents one approach to providing useful comparisons between different processes. The data presented was obtained from experiments, observation of production parts, and an analysis of over 1100 references. The data is expressly oriented to precision miniature parts in such materials as 303Se and 17-4 PH stainless steel, 6061-T6 aluminum, and beryllium copper. It is based on data obtained in 1973 on parts the size of wristwatch components or larger. Many of these parts had thin sections and tolerances of .001 inch or less.

A schematic of each process is presented along with a table, depicting its advantages as well as its limitations. Seven summary tables (Tables 7-16 through 7-22) list all deburring processes with an evaluation of each process based in the majority of cases on actual studies and experience. A more complete summary of most processes is given in subsequent chapters.

Twenty-one major deburring processes are discussed in this chapter. The majority of the processes are commercially available except for the use of chlorine gas which, due to the safety requirements and undesirable side effects, is still a laboratory process. There appear to be no plans in the foreseeable future to offer this process to industry. Also there are other processes such as electrochemical vibratory deburring which was developed in Japan which have not found much use in the United States. On the other hand, such processes as vibratory and barrel tumbling are widely used in this country because of their low cost and compatibility with mass produced parts.

The following is a list of the deburring processes and the corresponding figures and tables that illustrate and discuss each process.

Process	Figure	Table
Principal Deburring Processes	7-1	
Abrasive Jet Deburring	7-2	7-1
Chemical Deburring	7-3	7-2
Chlorine Gas Deburring	7-4	7-3
Electrochemical Deburring	7-5	7-4
Electropolish Deburring	7-6	7-5
Extrude Hone Deburring	7-7	7-6
Flame Deburring	7-8	

Process cont.	Figure	Table
Hand Deburring	7-9	7-7
Liquid Hone Deburring	7-10	7-8
Power Brushing Deburring		7-9
Power Sanding Deburring	7-11	7-10
Thermal Shock Deburring	7-12	7-11
Ultrasonic Deburring	7-13	7-12
Vibratory and Barrel Tumbling Deburring	7-14	7-13
Water Jet Deburring	7-15	7-14
Miscellaneous Deburring Processes		7-15
Comparison of Deburring Materials and Mechanisms		7-16
Comparison of Side Effects, Cycle Times and Required Secondary Operations		7-17
Comparison of Life of Materials and Safety Considerations		7-18
Comparison of Tooling, Equipment and Materials Costs		7-19
Comparison of Applicable Workpiece Materials and Part Geometry Limitations		7-20
Comparison of Burr Sizes Removed and Edge Radii Produced		7-21
Comparison of Side Effects and Repeatability		7-22

Using the Information

The figures and tables included in this chapter are designed to be brief and self-explanatory. Figure 7-1 illustrates the basic nature of the deburring processes. All these processes are either abrasive, utilize mechanical cutting or thermal, chemical or electrical techniques. Some such as electrochemical vibratory deburring utilize more than one removal phenomenon as indicated by the multiple arrows pointing to the process. Figures 7-2 through 7-15 illustrate the basic configuration of the individual processes, and Tables 7-1 through 7-15 provide details on capabilities and limitations. As an example, Figure 7-5 and Table 7-4 describe electrochemical deburring. As seen in Figure 7-5, this process requires a reservoir, pump, filter, and in general a multifacet machine. Electropolish deburring as seen in Figure 7-6, however, only requires a power supply, a solution, and a tank. Although they are very similar processes, they produce very different results as indicated in Tables 7-4 and 7-5.

Table 7-15 lists two processes for which little information was available in 1974.

Tables 7-16 through 7-22 provide some basic comparisons of capabilities and limitations. As seen there, abrasive jet deburring is applicable to both metals and plastics. It can remove burrs up to 0.005 inch (127 μm) thick and typically will produce a radius of 0.003 to 0.010 inch (76.2 to 254.0 μm) in stainless steel (SST) and in low carbon steel after removing the burr. It produces a matte finish on all surfaces around the edges to be deburred. Since the action is usually concentrated at edges rather than surfaces in deburring applications, it essentially removes no material from surrounding surfaces. The process is only effective if the stream of abrasives hits the burr. When edge radii produced by deburring must be controlled closely, 95 percent of the edges can be controlled within ±0.0015 inch (±38.1 μm) of a nominal radius on stainless steel parts. This value, however, assumes a reasonable size burr (less than 0.003 inch thick). When parts are subsequently welded or electroplated, one has to select abrasives which do not affect these processes.

As also indicated in these tables, abrasive jet deburring typically uses aluminum oxide grains as the blasting grain, although many other materials are available and are used in special industries. The basic deburring mechanism of this process is abrasion and the process does result in a matte finish on surfaces hit by the abrasive. Tooling costs are essentially nonexistent. In job shop situations, without automated equipment, it may take five minutes per part to remove all burrs. In many cases, considerably less time is required. Because loose abrasive particles adhere to the surface and in crevices, a subsequent cleaning process is often required. Abrasive jet deburring equipment typically can be purchased in 1974 for $40,000 or less. Small blasting systems can be obtained for $2500. Safety considerations include the need for controlling the dust which is produced in this process. With single nozzle heads used on a manual basis, up to four pounds (1.81 Kg) an hour of aluminum oxide would have to be replaced. On automated machines, this could increase by a factor of 20. The aluminum oxide abrasive can be purchased for $1.50 per pound ($3.31 per Kg) which amounts to $6.00 per hour at the four pound per hour rate. The process is typically used on steels and related hard materials.

Limitations of the Data

In utilizing this material, it is essential to consider several facts.

1. The information presented is for the most widely used form of the indicated process.

2. The information presented is based on the best data available in 1974.

3. The data given in Tables 7-16 through 7-22 was obtained in part from literature reviews, in part from estimates based on the nature of the processes, and in part from actual experiments. These tables are specifically applicable to parts smaller than hand size having tolerances of ±0.001 inch or less and thin cross-sections or diameters.

In using Tables 7-16 through 7-22, it is important to recognize that several variations extend the capabilities of the basic processes. If the reader is merely looking for a process which will work, it may suffice to review the material presented, select the most likely process or processes, and then contact equipment vendors or others familiar with these processes for final verification. If the reader is trying to obtain the optimum process, the material presented should be only a starting point. For this purpose, it is essential to explore in depth the capabilities and limitations of each process and each variation. Fortunately, bibliographies are available which summarize the available literature on deburring and reducing the costs of burrs.[1,2,3] Equipment vendors also are happy to discuss and verify process capabilities.

Summary

There is no rule of thumb which indicates the most applicable process to use. The material presented in this chapter and book, however, allows the manufacturing engineer to identify the most promising processes. Other chapters provide more in-depth information on a wider cross-section of parts. From this point it is advisable to submit parts to an individual evaluation by qualified vendors or in-house personnel. In some cases, it may be economical to utilize two or more processes to completely deburr the parts. This may involve hand deburring of large burrs first and then removal of all smaller burrs through vibratory or other means. The inability to predict the specific "optimum" deburring process is that the selected method depends not only on the material but also on burr properties such as hardness, thickness, length, etc. Unfortunately, burr properties at this time are generally unpredictable and unrepeatable and vary substantially from material to material and from one machining operation to another.

References

[1]L. K. Gillespie, Deburring: An Annotated Bibliography, SME Bibliography Series, 1974.

[2]L. K. Gillespie, Deburring: An Annotated Bibliography, Vol. II, SME Paper MRR75-04, 1975.

[3]L. K. Gillespie, Deburring: An Annotated Bibliography, Vol. III, SME Paper MRR76-07, 1976.

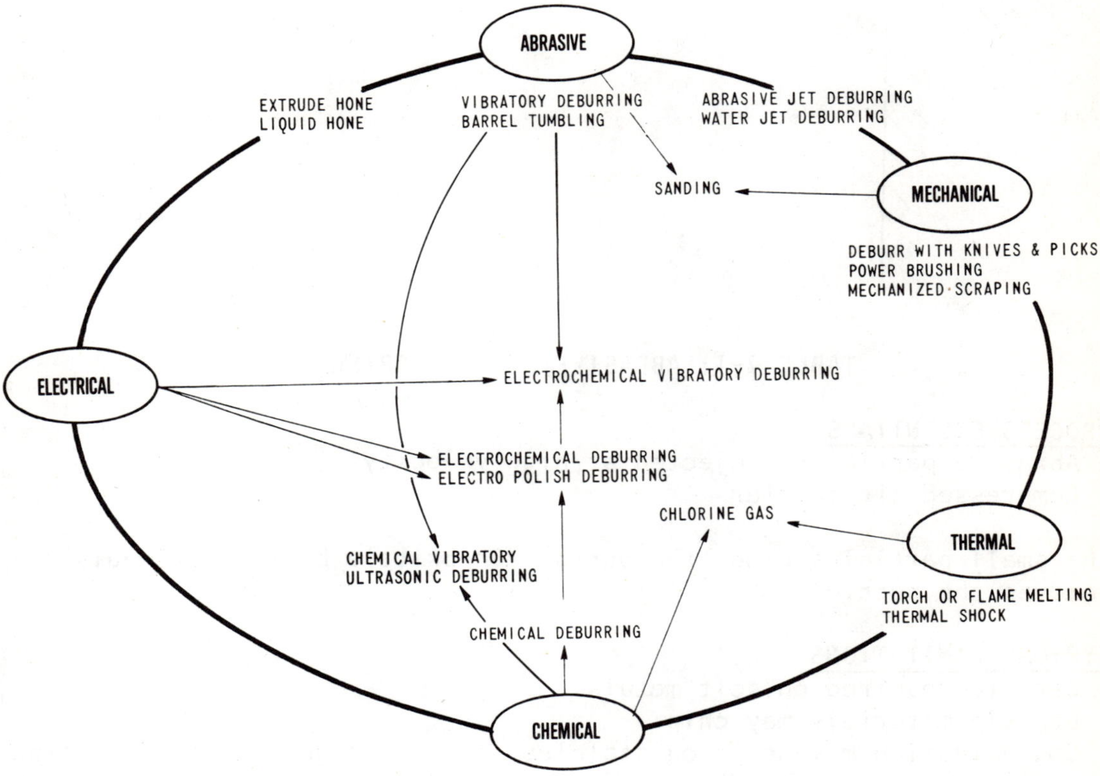

FIGURE 7-1: PRINCIPAL DEBURRING PROCESSES

FIGURE 7-2: ABRASIVE JET DEBURRING

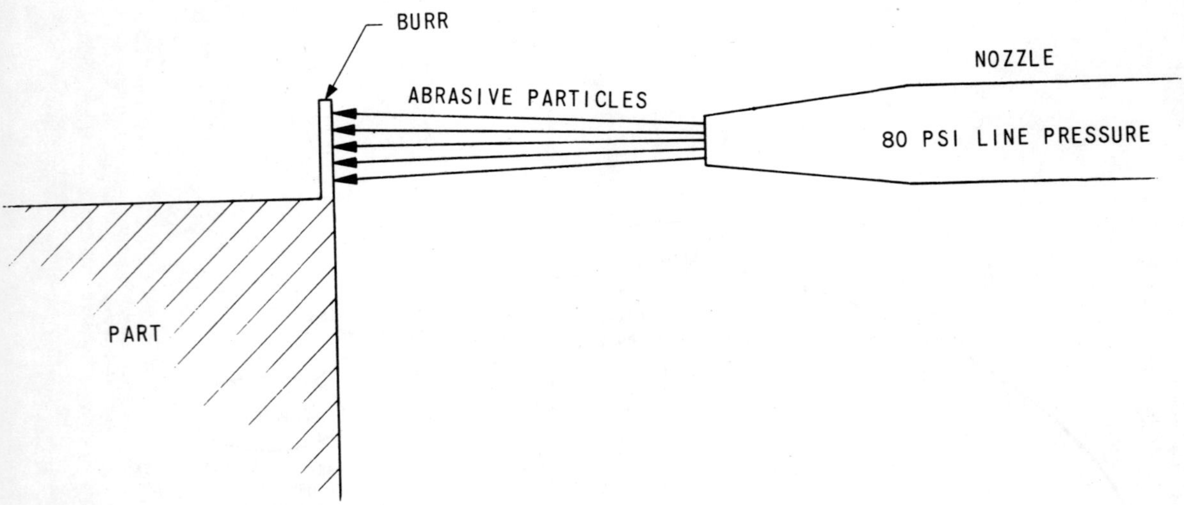

TABLE 7-1: ABRASIVE JET DEBURRING

PROCESS ESSENTIALS

. Abrasive particles projected at high velocity
. Compressed air carrier

The small particles erode the burrs. On brittle burrs the impulse removes the burr.

TYPICAL LIMITATIONS

. Care is required on soft metals
. Brittle materials may chip
. Contamination may occur on stainless steels (significant if plating or welding is required)
. Thin parts may distort
. Will degrade surface finish and precision dimensions
. Not suited to deep pockets

TYPICAL APPLICATIONS

. Miniature parts
. Hand size parts
. Plastics
. All metals
. Small burrs
. Brittle burrs
. Also used for deflashing, cleaning, texturizing, and improving fatigue strength

TYPICAL RESULTS

. 1-5 minutes per part

FIGURE 7-3: CHEMICAL DEBURRING

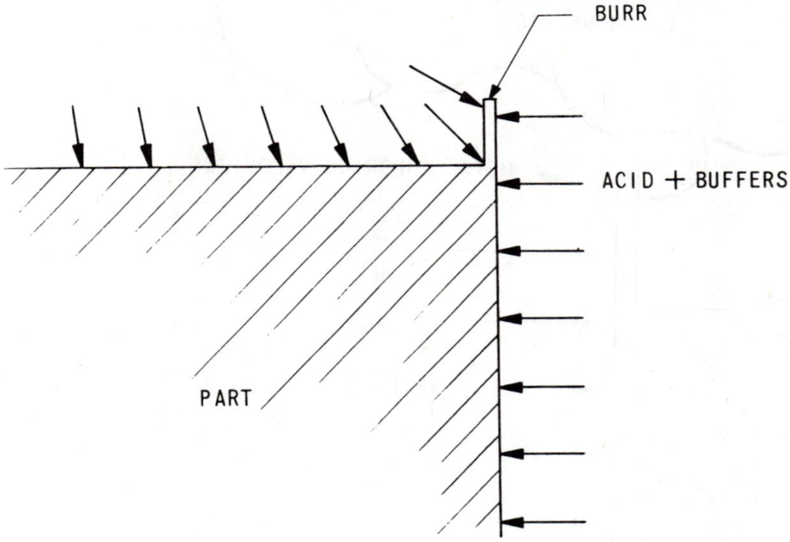

TABLE 7-2: CHEMICAL DEBURRING

PROCESS ESSENTIALS
. Chemical etchant
. Large tank

Chemical attack, resulting gassing, combine with tumbling to remove burr

TYPICAL LIMITATIONS
. Long burrs are softened chemically then removed by vibratory deburring
. Limited success on beryllium copper

TYPICAL APPLICATIONS
. Used on carbon steel
  300 series stainless
  Steel
  Aluminum
  Copper base alloys
. Small whisker burrs
. Screw machine parts

TYPICAL RESULTS
. Process is not widely used
. Finishes of 6 to 8 AA are quoted
. 5-30 minute batch cycle times are quoted
. 0.0005-0.003 edge radii
. Edge break is 2 to 3 times larger than amount of stock removed

FIGURE 7-4: CHLORINE GAS DEBURRING

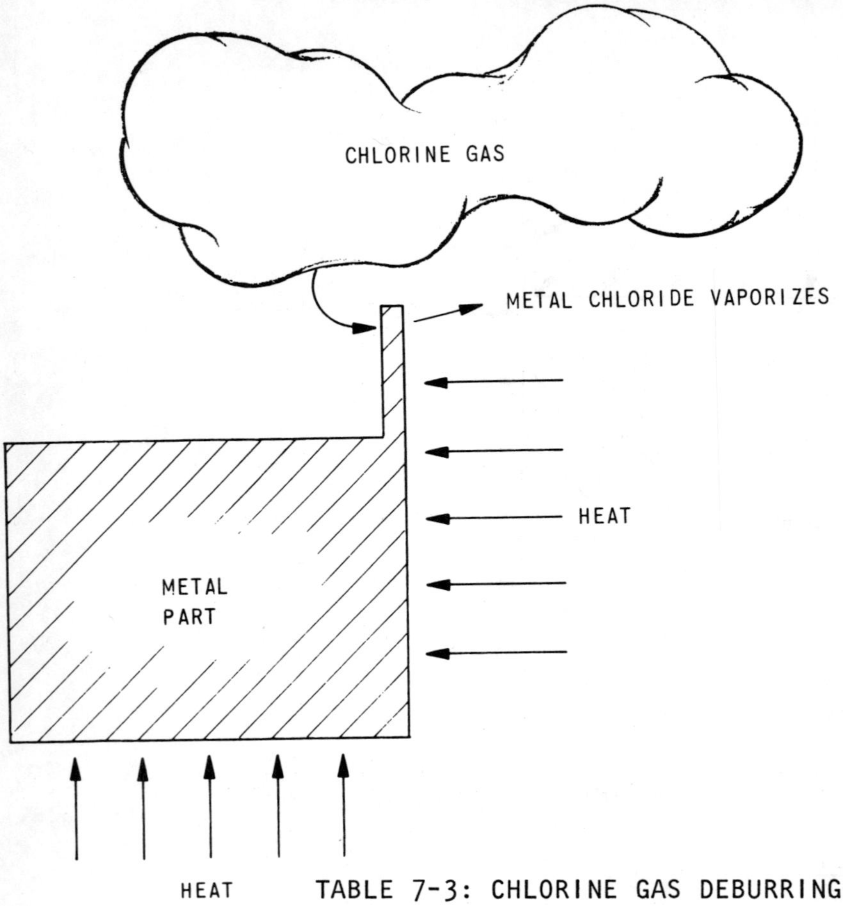

TABLE 7-3: CHLORINE GAS DEBURRING

PROCESS ESSENTIALS
. Heated part (or heated burr)
. Chlorine gas

The chlorine combines with the steel burr to form a chloride which immediately vaporizes

TYPICAL LIMITATIONS
. Irons or steels

TYPICAL APPLICATIONS
. Steel parts
. Also used for machining and surface finishing

TYPICAL RESULTS
. 0.004 size loss minimum
. 1/2 minute deburr time

Effects of chlorine gas temperature

Temp. F.	Steel Removed
1000	0.0004
2000	0.004
2500	0.01

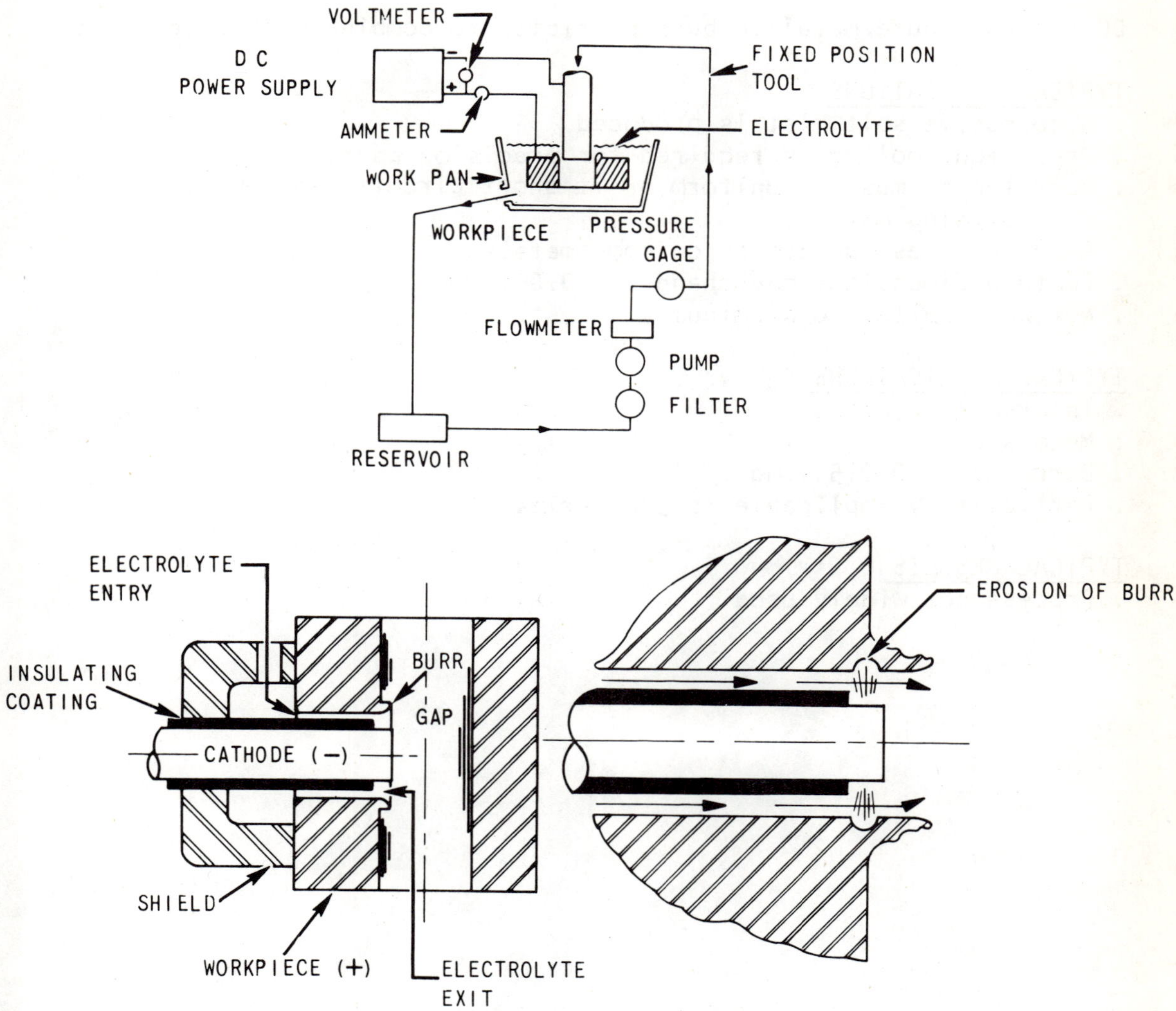

FIGURE 7-5: ELECTROCHEMICAL DEBURRING

TABLE 7-4: ELECTROCHEMICAL DEBURRING

PROCESS ESSENTIALS

. Salt electrolyte
. Electrical power supply
. Shaped electrode
. Fixturing

DC current cause metallic burr particles to combine with electrolyte

TYPICAL LIMITATIONS

. A corrosive salt mist is produced
. Precision tooling is required for precision parts
. Burr length must be uniform or it short circuits equipment
. Part pitting may occur
. A surface residue occurs on some metals
. Feature dimensions may change by 0.001
. Not well suited to aluminum

TYPICAL APPLICATIONS

. Intersecting holes
. Metals only
. Burrs up to 0.015 long
. Particularly applicable to 300 series SST

TYPICAL RESULTS

. Process not widely used

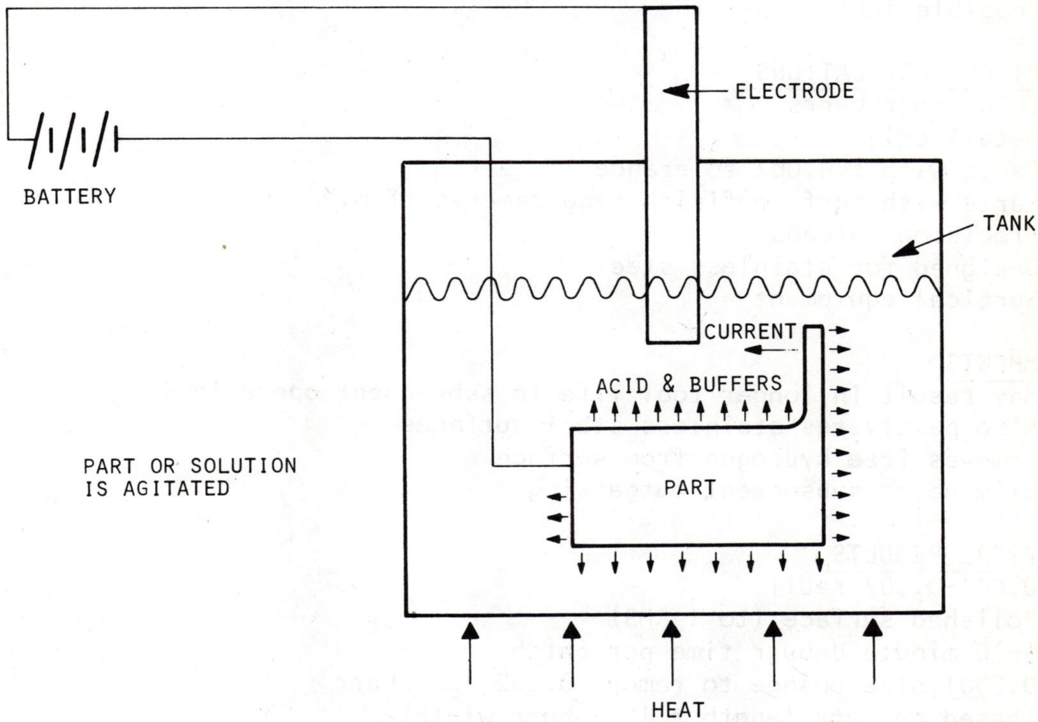

FIGURE 7-6: ELECTROPOLISH DEBURRING

TABLE 7-5: ELECTROPOLISH DEBURRING

PROCESS ESSENTIALS
. Acid electrolyte
. Heated solution
. Electrolytic tank
. Racking and agitation
. Requires electrical connection to each part

DC current causes burr to dissociate

TYPICAL LIMITATIONS
. Metals only
. Thin short burrs
. Possible surface pitting

TYPICAL APPLICATIONS
. Thin short burrs
. Metals only
. Parts with ± 0.002 tolerance
. Parts with surface finish requirements of 6.32 RMS
. Precision threads
. Designed for stainless steel
. Surgical equipment

COMMENTS
. May result in longer tool life in subsequent operations
. Also passivates stainless steel surfaces
. Removes free hydrogen from surface
. Eliminates subsequent outgassing

TYPICAL RESULTS
. 0.001-0.002 radii
. Polished surface (to 1 RMS)
. 5-10 minute deburr time per batch
. 0.0001 size change to remove 0.002 long burr
  (based on burr length = 10 x burr width)

FIGURE 7-7: ABRASIVE FLOW DEBURRING

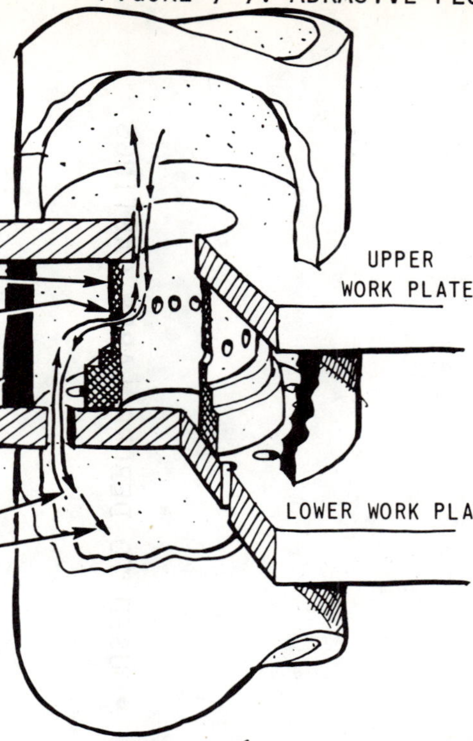

TABLE 7-6: ABRASIVE FLOW DEBURRING

PROCESS ESSENTIALS
. Putty-like abrasive is forced over burred edge
. Tooling (fixtures) direct abrasive flow
. Exceptional flexibility

The abrasive masticates the burr and radiuses the edges

TYPICAL LIMITATIONS
. Some size change occurs
. Large parts (1 ft^3) not generally applicable
. Parts generally require subsequent cleaning

TYPICAL APPLICATIONS
. Valve bodies
. Intersecting holes
. All metals
. Plastics
. Small holes
. Intricate, close tolerance parts
. Hand size parts
. Process is also used to polish mold dies

TYPICAL RESULTS
. Deburr time: 30-60 seconds
. Cycle time: 1 part per minute
. 0.001-0.005 radii on miniature parts
. 16AA microinch surface finish

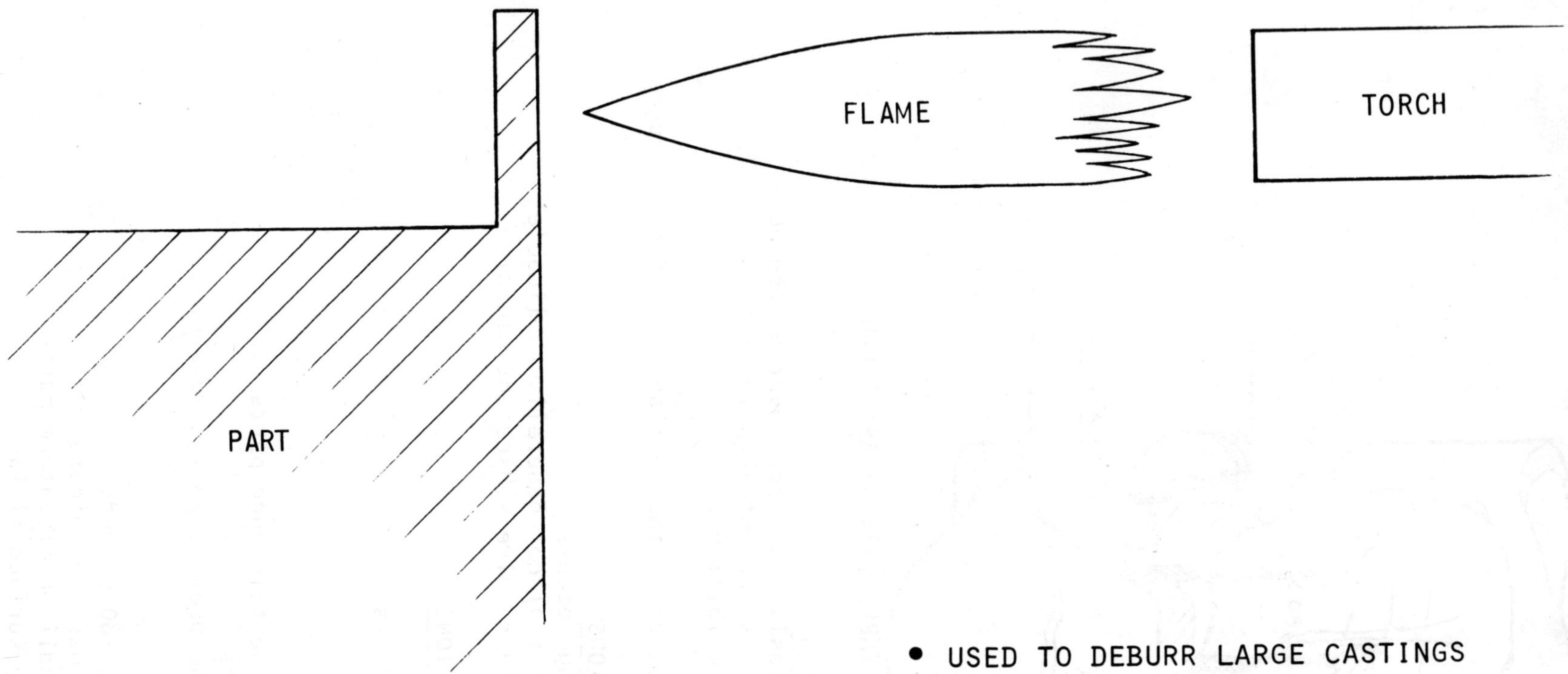

• USED TO DEBURR LARGE CASTINGS

FIGURE 7-8: FLAME DEBURRING

FIGURE 7-9: HAND DEBURRING

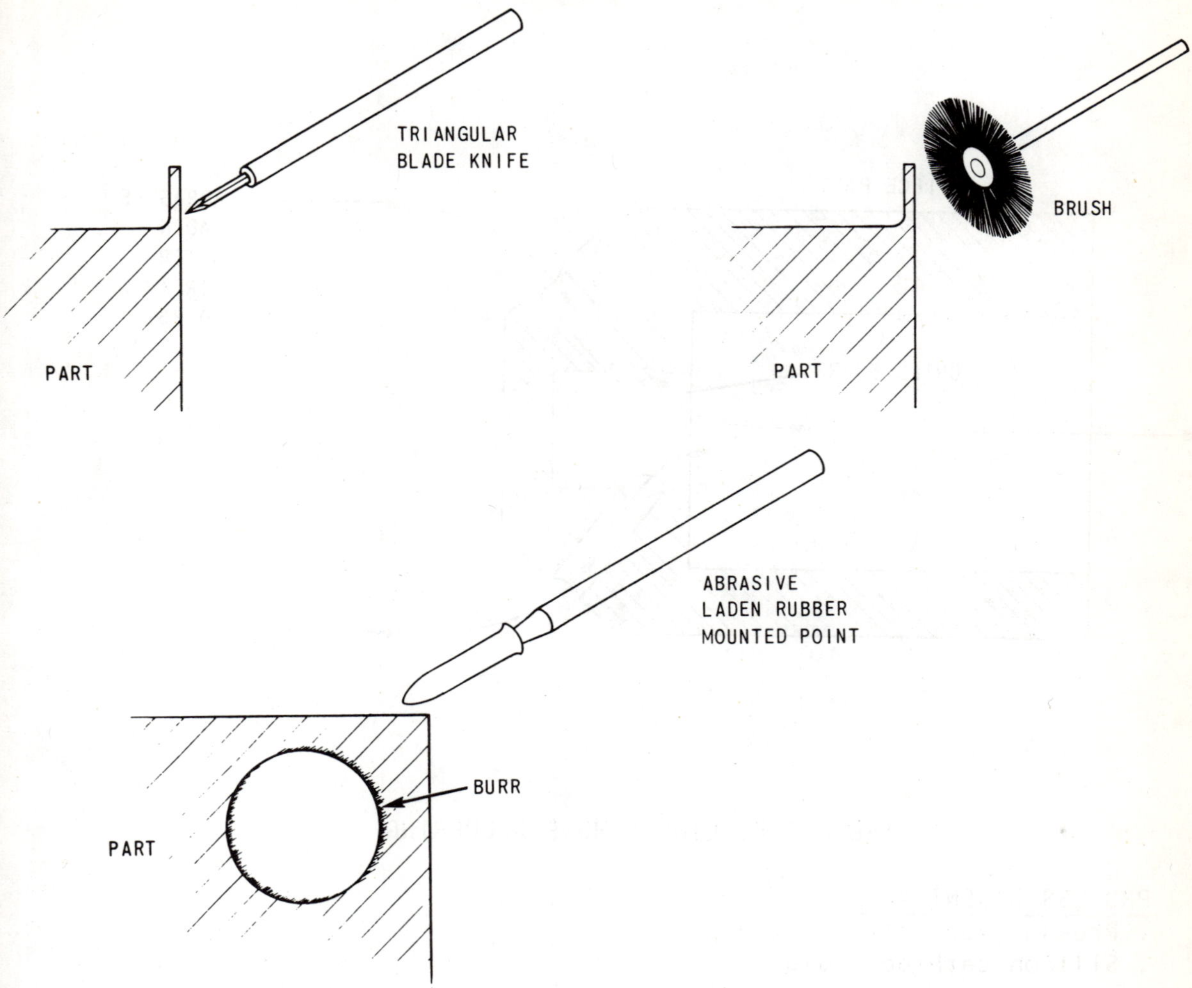

TABLE 7-7: HAND DEBURRING

PROCESS ESSENTIALS

. Knife
. Brush
. Mounted point

TYPICAL APPLICATIONS

. Hard to reach areas
. High quality parts
. Small quantities

TYPICAL RESULTS

. 0.002-0.015 radii

FIGURE 7-10: LIQUID HONE DEBURRING

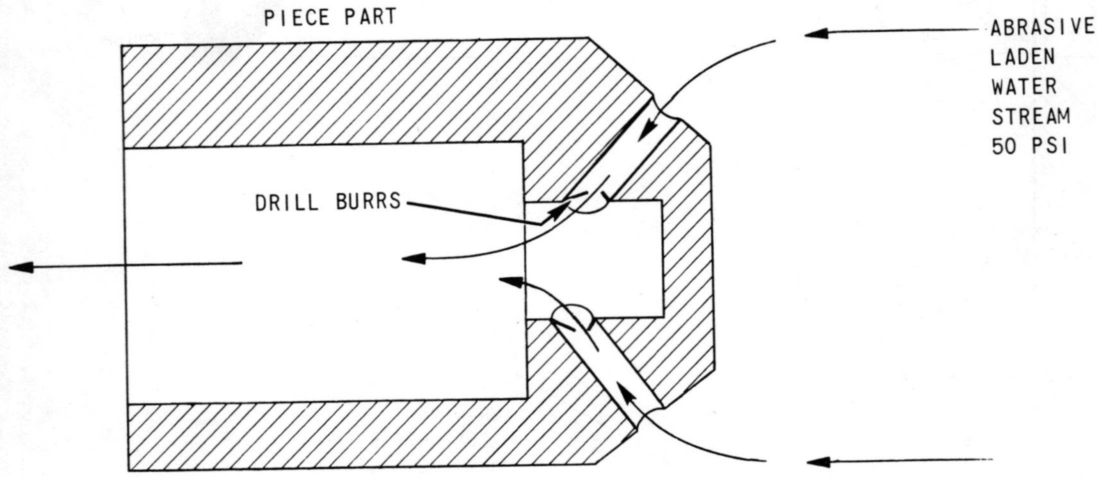

TABLE 7-8: LIQUID HONE DEBURRING

PROCESS ESSENTIALS
. Pressurized water
. Silicon carbide media

The silicon carbide particles hammer and abrade the burr

TYPICAL APPLICATIONS
. Thin burrs
. Steel parts

LIMITATIONS
. Burrs must be exposed to water flow
. Parts require cleaning
. Not suited to blind features

TYPICAL RESULTS
. 1 minute deburr time
. Maintains 0.005 max. radii

TABLE 7-9: POWER BRUSHING DEBURRING

PROCESS ESSENTIALS
- Wire bristles
- Tampico bristles + abrasive
- Nylon disks

Wire bristles cut the burr; filled brushes abrade the burr

TYPICAL LIMITATIONS
- Little control on radii produced at intersecting holes
- Tendency to push burrs in intersecting holes

TYPICAL APPLICATIONS
- Gears
- Complex features
- External features
- Also used to polish rough surfaces

TYPICAL RESULTS
- 0.005-0.030 radii
- 1-5 minute cycles

FIGURE 7-11: POWER SANDING DEBURRING

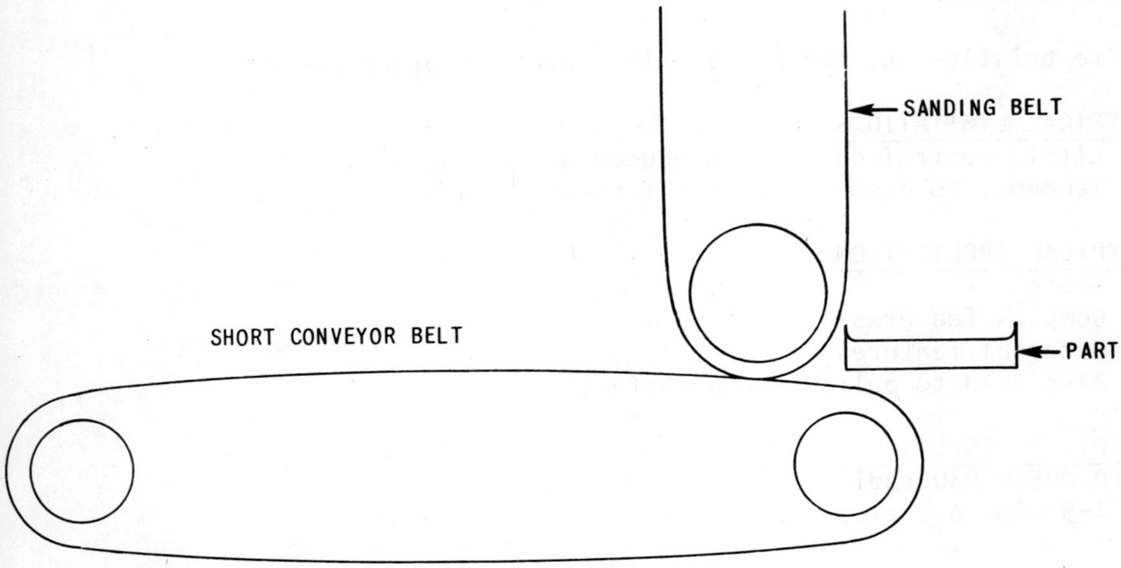

TABLE 7-10: POWER SANDING DEBURRING

PROCESS ESSENTIALS

. Continuous belt sandpaper

The burr is abraded off

TYPICAL LIMITATIONS

. Flat surfaces or straight edges

TYPICAL APPLICATIONS

. Punched parts
. Circuit boards
. Miniature parts
. Thin parts (0.007)
. Preliminary deburring

TYPICAL RESULTS

. 200 parts per hour

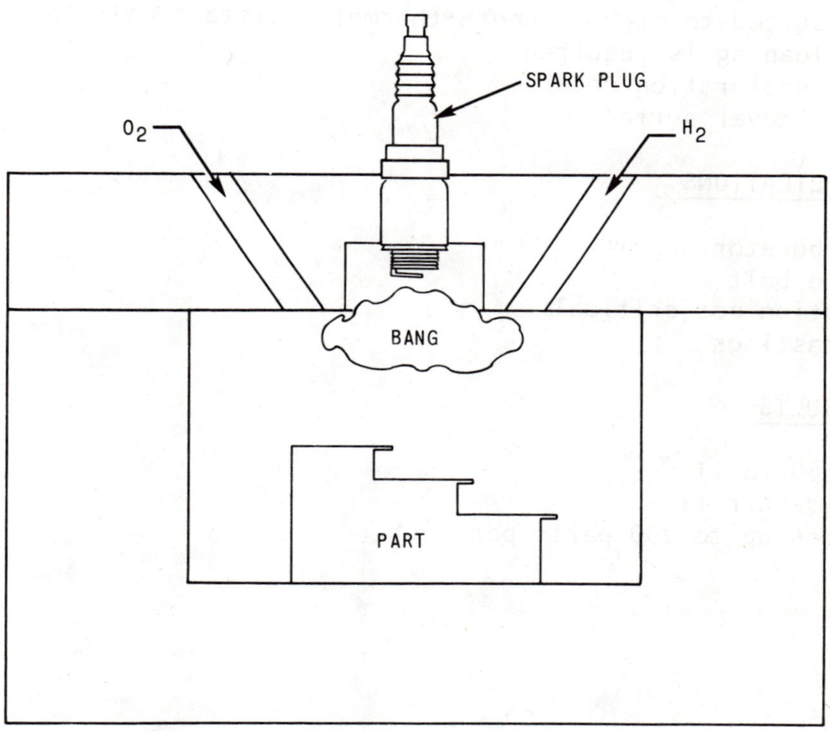

FIGURE 7-12: THERMAL ENERGY DEBURRING

TABLE 7-11: THERMAL ENERGY DEBURRING

PROCESS ESSENTIALS

- Hydrogen or natural gas and oxygen ignited
- Sealed chamber

The instantaneous high temperature flash front burns and/or vaporizes the burr

TYPICAL LIMITATIONS

- Not applicable to heavy burrs on parts with thin sections
- Surface pitting can occur
- Not well suited to high strength thermal resistant materials
- Surface cleaning is required
- Surface discoloration occurs
- Oxides may cover surface

TYPICAL APPLICATIONS

- Brass carburetor nozzles
- M-16 rifle bolt
- Burr location not critical
- Zinc diecastings

TYPICAL RESULTS

- 0.001-0.060 radii
- 4 second deburr time
- Cycle times up to 800 parts per hour

FIGURE 7-13: ULTRASONIC DEBURRING

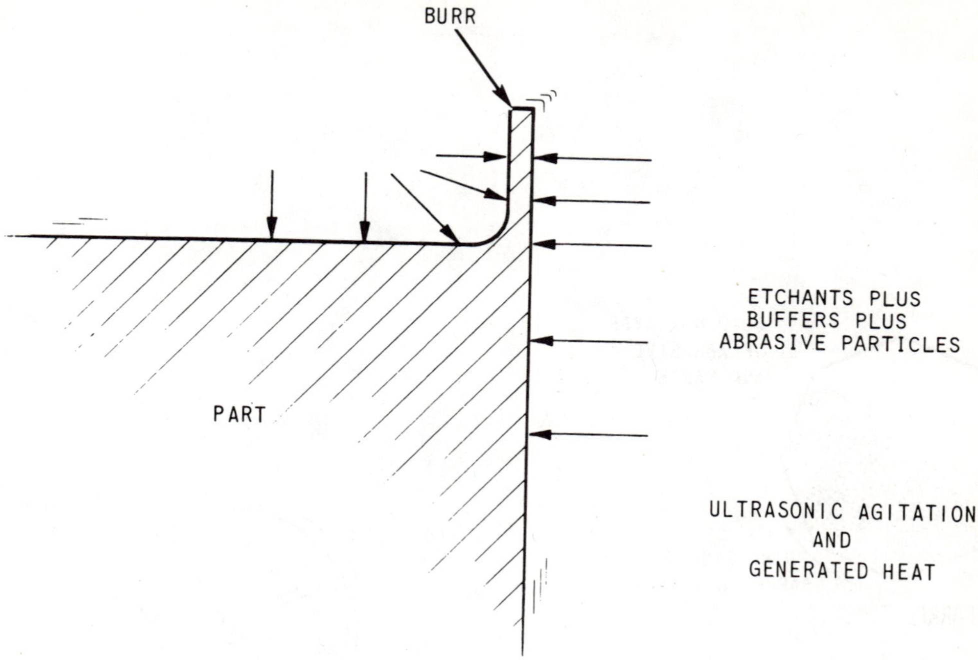

TABLE 7-12: ULTRASONIC DEBURRING

PROCESS ESSENTIALS

. Chemical etchant and/or abrasive slurry
. Tank
. Ultrasonic transducer

Cavitation, generated heat, abrasive bonbardment and etchant combine to remove burr

TYPICAL LIMITATIONS

. Will not remove thick burrs
. May change dimensions 0.0001-0.001
. Some pitting can occur
. Not generally applicable to stainless steel parts

TYPICAL APPLICATIONS

. Metal parts
    Copper
    Brass
    Zinc
    Aluminum
. High volume parts
. Thin burrs
. Inaccessible areas
. Small parts

TYPICAL RESULTS

. Process not widely used

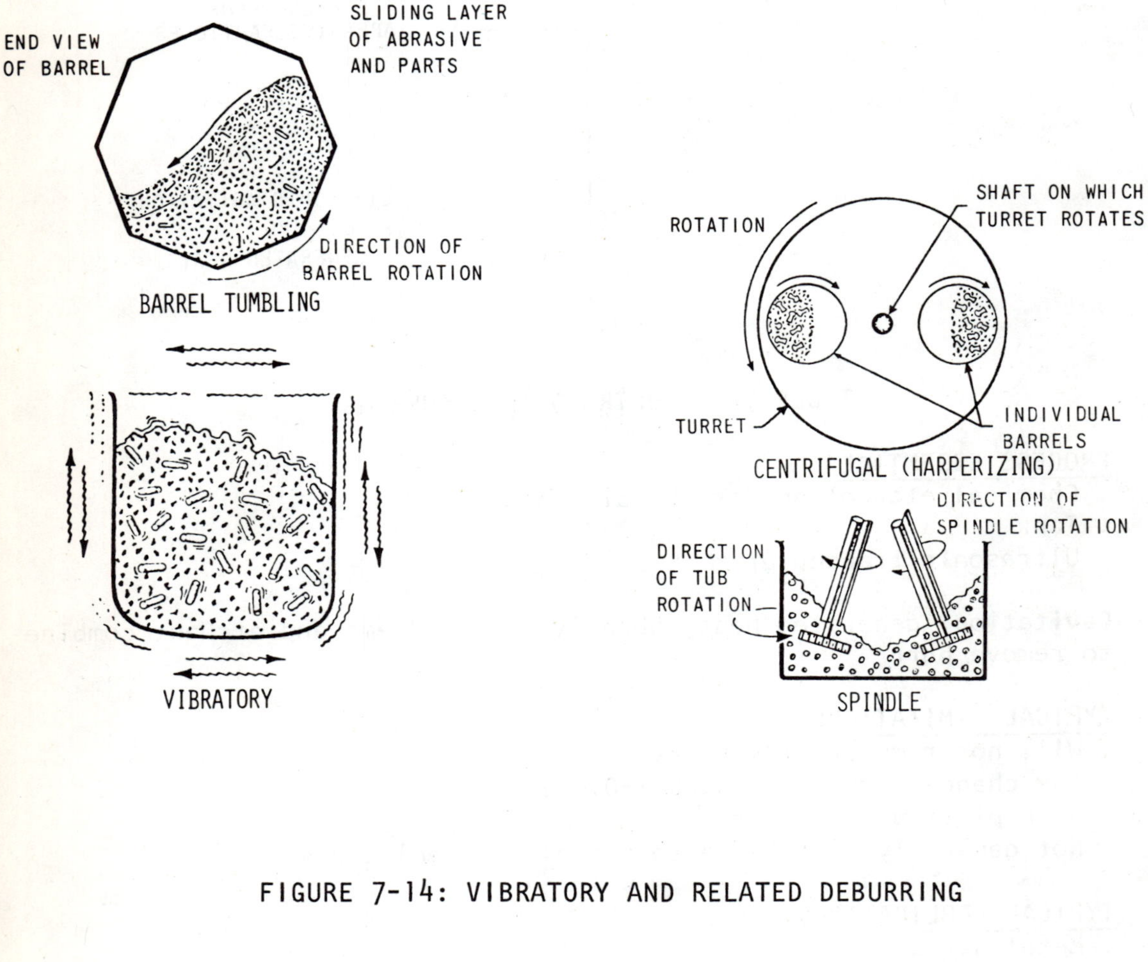

FIGURE 7-14: VIBRATORY AND RELATED DEBURRING

TABLE 7-13: VIBRATORY AND RELATED DEBURRING

PROCESS ESSENTIALS

. Vibrating tub
. Loose grain media (preforms)
. Abrasive compound
. Water
. Very flexible, easily altered process
. No tooling required

The abrasive action of the compound rubbing between the media and the burr abrades the burr. Brittle burrs are knocked off by the vibrating media.

TYPICAL LIMITATIONS

. Ineffective on extremely small parts with heavy burrs (centrifugal barrel, however, is very effective on small parts)
. Parts which will be plated or welded require special treatment
. Will not work on blind passageways or hard to reach areas
. Corners will be radiused up to 5 times as much as edges

TYPICAL APPLICATIONS

. Used for deburring and radiusing
. Used on any part where burr is exposed
. Also used for cleaning, descaling, improving finish, stress relieving, and mechanical plating
. Most effective on grinding burrs

TYPICAL RESULTS

. Depend on part size and geometry
. 0.003-0.015 edge radii
. Removes burrs up to 0.003 thick on miniature parts
. Can produce surface finishes of 20AA or better
. Typical run times are 2 hours per batch
. Maximum repeatability on small parts appears to be ± 0.0015 (varies with workpiece material)
. Lowest cost of all deburring methods

FIGURE 7-15: WATER JET DEBURRING

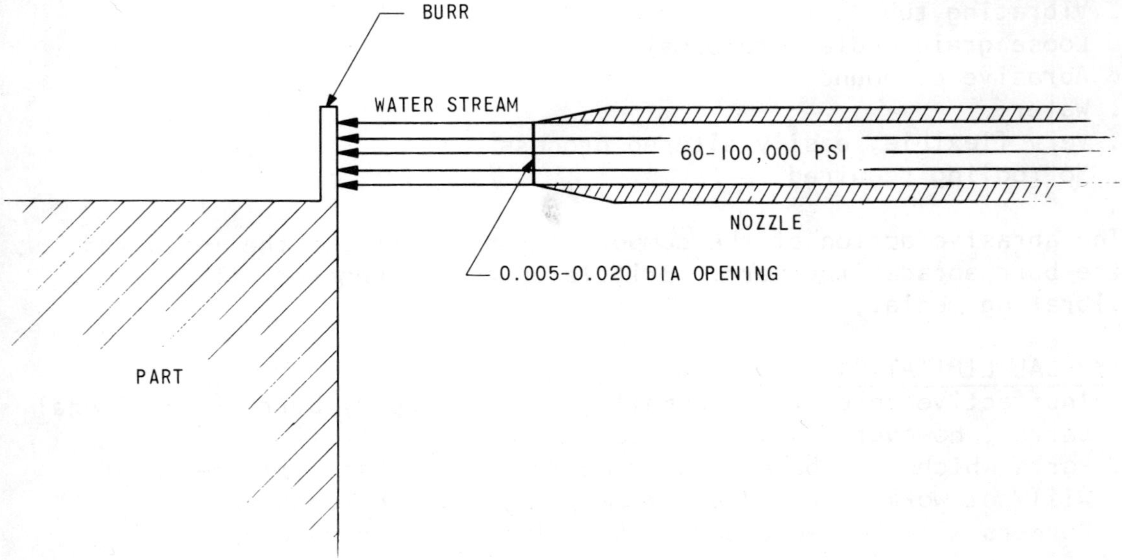

TABLE 7-14: WATER JET DEBURRING

PROCESS ESSENTIALS
. Water pressurized to 20,000 - 100,000 psi
. 0.005-0.020 diameter streams
. 0.35 gallons per minute

Combined momentum, narrow cutting edge, cavitation cut the burr off

TYPICAL APPLICATIONS
. Plastics
. Fabrics
. Wood
. Rubber
. Aluminum honeycomb
. Thin materials
. Deflashing
. Diecast aluminum
. Also removes dirt and chips

TYPICAL RESULTS
. Up to 600 IPM linear feedrate

TABLE 7-15: MISCELLANEOUS DEBURRING PROCESSES

Chemical Vibratory Deburring = Vibratory deburring + chemically reactive solutions instead of water and abrasive compounds

Electrochemical Vibratory Deburring = Vibratory deburring + electrolytic solution + spherical graphite media

TABLE 7-16: COMPARISON OF DEBURRING MATERIALS AND MECHANISMS

DEBURRING PROCESS	MATERIALS USED	DEBURRING MECHANISM	COMMENTS*
Abrasive Flow	Silicon Putty SiC, Al_2O_3	Erosion	Used on hard-to-reach burrs
Abrasive Jet	Al_2O_3	Abrasion	Used on hard metals
Barrel Tumbling	Al_2O_3, SiC Plastics	Abrasion and Fatigue	Widely used, typically least expensive of all processes
Brushing	Nylon Fibers, Al_2O_3, Steel Fibers	Abrasion or Cutting	Used on all burrs and accessible edges
Centrifugal Barrel Finishing	Al_2O_3	Abrasion	Used on small parts
Chemical	Buffered Acids	Chemical Attack	Used for thin burrs
Chemical Vibratory	$CuSO_4$, Al_2O_3	Chemical Reaction and Abrasion	Used for rapid stock removal
Chlorine Gas	Chlorine Gas	Chemical Attack	Laboratory method only
Electrochemical	Any Salt Solutions	Electrolyte Deplating	Used on hard-to-reach burrs
Electrochemical Vibratory	Caustic or Phosphoric Acid, Graphite	Electrolytic Deplating and Abrasion	Only one installation in United States
Electropolish	Buffered Acids	Electrolyte Action	Used for thin burrs

TABLE 7-16 CONT: COMPARISON OF DEBURRING MATERIALS AND MECHANISMS

DEBURRING PROCESS	MATERIALS USED	DEBURRING MECHANISM	COMMENTS*
Hand	Steel Knives, Al_2O_3, or SiC Points, Nylon Brushes	Cutting or Abrasion	Used for precision or on hard-to-reach areas
Liquid Hone	Water and SiC	Erosion	Used on thin, short burrs
Mechanized Mechanical	Cutters or Dies	Shearing or Cutting	
Sanding	Sandpaper, Al_2O_3, SiC	Abrasion	Used on flat parts and external edges
Spindle Finishing	Al_2O_3	Abrasion	Used on cylindrical parts like gears and turbines
Thermal Energy	H_2, O_2, Natural Gas	Melting, Shock	Used on thin burrs on zinc and some steel alloys
Torch or Flame	H_2 and O_2, Acetylene	Melting	Typically used only on large castings
Ultrasonic	Acids, Buffers, Al_2O_3	Chemical Attack	Used for very thin burrs
Vibratory	Al_2O_3, SiC Plastics	Abrasion, Fatigue	One of the most commonly used and least expensive of all processes
Water Jet	Water	Erosion and Cavitation	Used only on thin or brittle burrs

*Information provided represents typical results from steel piece parts (unless otherwise noted), based on a burr thickness of 50.8 μm (0.002 inch), a part volume of 16.387 cm^3 (1 $inch^3$) or less, and a batch size of 100 parts per lot.

TABLE 7-17: COMPARISON OF SIDE EFFECTS, CYCLE TIMES, AND REQUIRED SECONDARY OPERATIONS

DEBURRING PROCESS*	SIDE EFFECTS	CYCLE TIME PART (MIN.)	CYCLE TIME LOT (MIN.)	REQUIRED SECONDARY OPERATIONS
Abrasive Flow	Polishing	2		Air and Ultra-sonic Cleaning
Abrasive Jet	Matte Finish	1 to 5		Cleaning
Barrel Tumbling	Residual Stress Changes, Some Material Impregnation		120 to 480	Cleaning
Brushing	Polishing	1 to 5		Cleaning
Centrifugal Barrel Finishing	Produces Large Residual Compressive Stresses		5 to 20	Cleaning
Chemical	Polishing, Brightening		5 to 30	Precleaning
Chemical Vibratory			10 to 20	Cleaning
Chlorine Gas	Polishing			Cleaning
Electrochemical	Stray Etching	1		Film Removal
Electrochemical Vibratory			1 to 60	Cleaning
Electropolish	Possible pitting, streaking		1 to 10	Precleaning
Hand		1 to 20		None
Liquid Hone		1 to 5		Cleaning

TABLE 7-17 CONT: COMPARISON OF SIDE EFFECTS, CYCLE TIMES, AND REQUIRED SECONDARY OPERATIONS

DEBURRING PROCESS*	SIDE EFFECTS	CYCLE TIME PART (MIN.)	CYCLE TIME LOT (MIN.)	REQUIRED SECONDARY OPERATIONS
Mechanized Mechanical		1		Radiusing
Sanding		1 to 5		Cleaning
Spindle Finishing		1 to 5		Cleaning
Thermal Energy	Covers Part with Oxide Film		0.1 to 1	Cleaning
Torch or Flame				Cleaning
Ultrasonic	Polishing, Brightening		5 to 10	None
Vibratory	Residual Stress Changes Some Material Impregnation		120	Cleaning
Water Jet	None	1		None

*Information provided represents typical results from steel piece parts (unless otherwise noted), based on a burr thickness of 50.8 μm (0.002 inch), a part volume of 16.387 cm^3 (1 $inch^3$) or less, and a batch size of 100 parts per lot.

TABLE 7-18: COMPARISON OF LIFE OF MATERIALS AND SAFETY CONSIDERATIONS

DEBURRING PROCESS*	LIFE OF MATERIALS**	HEALTH AND SAFETY CONSIDERATIONS
Abrasive Flow	18.14 Kg each 1000 parts	None with adequate hydraulic press considerations
Abrasive Jet	1.81 Kg per hour	Control of airborne dust
Barrel Tumbling	90.72 Kg media or compound each 10,000 parts	None with proper chemicals
Brushing	1 brush each 10 to 1000 parts	Only a regard for rotating wheels
Centrifugal Barrel Finishing	90.72 Kg each 5000 parts	None
Chemical		Typical precautions for chemicals and fumes
Chemical Vibratory		None
Chlorine Gas		Chlorine may require strict safety procedures
Electrochemical	0.38 m^3 each 500 parts	Typical electrical, chemical, and gaseous precautions
Electrochemical Vibratory		
Electropolish		Typical electrical, chemical, and gaseous requirements
Hand		None
Liquid Hone		None

TABLE 7-18 CONT: COMPARISON OF LIFE OF MATERIALS AND SAFETY CONSIDERATIONS

DEBURRING PROCESS*	LIFE OF MATERIALS**	HEALTH AND SAFETY CONSIDERATIONS
Mechanized Mechanical		Precautions for keeping hands free of press and rotating cutters
Sanding		None
Spindle Finishing	90.72 Kg each 5000 parts	None
Thermal Energy		Typical control of combustible gases and consideration for controlled detonations
Torch or Flame		Combustible gases
Ultrasonic		Typical precautions for chemicals and fumes
Vibratory	90.72 Kg media or compound each 10,000 parts	None
Water Jet		None

*Information provided represents typical results from steel piece parts (unless otherwise noted), based on a burr thickness of 50.8 μm (0.002 inch), a part volume of 16.387 cm^3 (1 $inch^3$) or less, and a batch size of 100 parts per lot.
**1 Kg = 2.20 lb; 1 m^3 = 264.17 gallon

TABLE 7-19: COMPARISON OF TOOLING, EQUIPMENT, AND MATERIALS COSTS

DEBURRING PROCESS*	COST		
	TOOLING ($)	EQUIPMENT ($)	MATERIALS ($)
Abrasive Flow	100 to 500	20,000 to 30,000	10 to 30 per pound**
Abrasive Jet	0	2500 to 40,000	1.50 per pound
Barrel Tumbling	0	200 to 5000	0.80 per pound
Brushing	0 to 200	50 to 15,000	
Centrifugal Barrel Finishing	0	5000 to 40,000	0.80 per pound
Chemical	0	1500 to 20,000	
Chemical Vibratory	0	100 to 8000	
Chlorine Gas	100 to 300	Not available	
Electrochemical	150 to 500	8000 to 35,000	0.05 per gallon***
Electrochemical Vibratory	0	11,000	
Electropolish	200 to 500	1500 to 25,000	
Hand	5 to 50	0 to 100	
Liquid Hone	100 to 500	500 to 20,000	0.80 per pound
Mechanized Mechanical	5 to 500	50 to 10,000	

TABLE 7-19 CONT: COMPARISON OF TOOLING, EQUIPMENT, AND MATERIALS COSTS

DEBURRING PROCESS*	COST TOOLING ($)	EQUIPMENT ($)	MATERIALS ($)
Sanding	0	100 to 2000	
Spindle Finishing	0	7500 to 11,000	0.80 per pound
Thermal Energy	250 to 500	60,000 to 150,000	0.01 to 0.05 per part
Torch or Flame	0	150 to 1500	
Ultrasonic	0	1500 to 5000	
Vibratory	0	100 to 30,000	0.80 per pound
Water Jet	0 to 100	50,000 to 100,000	

*Information provided represents typical results from steel piece parts (unless otherwise noted), based on a burr thickness of 50.8 μm (0.002 inch), a part volume of 16.387 cm^3 (1 $inch^3$) or less, and a batch size of 100 parts per lot.
**1 lb = 0.45 Kg
*** 1 gallon = 3.785 dm^3

TABLE 7-20: COMPARISON OF APPLICABLE WORKPIECE MATERIALS AND PART GEOMETRY LIMITATIONS

DEBURRING PROCESS	APPLICABLE WORKPIECE MATERIALS	PART GEOMETRY LIMITATIONS	COMMENTS
Abrasive Flow	All Metals, Machined Plastics, Ceramics	Blind features will not be deburred	Used on hard-to-reach intersections for good finishes
Abrasive Jet	Metals, Plastics	Burr must be accessible to abrasive stream	
Barrel Tumbling	All Metals and Plastics	Edges must be readily exposed to deburr	Use appropriate media and compound if welding or plating follows deburring
Brushing	Metals, Plastics	Most miniature applications limited to external burrs	
Centrifugal Barrel Finishing	Metals, Plastics	Same as barrel tumbling	Same as for barrel tumbling
Chemical	Iron, Steels, Stainless Steels, Copper Alloys, Bronze, Aluminum, Brass		Parts must be clean prior to deburring
Chemical Vibratory	Steel Alloys, Zinc Alloys	Same as barrel tumbling	Same as for barrel tumbling
Chlorine Gas	Metals		Laboratory process only
Electrochemical	All Metals		Not well suited to aluminum because of oxide film deposit
Electrochemical Vibratory	Metals	Burr must be accessible to graphite spheres	Holes 3.175 mm in diameter are enlarged 5.08 μm in one 10-minute cycle*
Electropolish	Stainless Steels, High-Temperature Metals, Aluminum, Beryllium		

TABLE 7-20 CONT: COMPARISON OF APPLICABLE WORKPIECE MATERIALS AND PART GEOMETRY LIMITATIONS

DEBURRING PROCESS	APPLICABLE WORKPIECE MATERIALS	PART GEOMETRY LIMITATIONS	COMMENTS
Hand	All Materials	Burr must be accessible	Can be used to reduce burr size before commercial process
Liquid Hone	Metals, Plastics	Blind features will not be deburred	
Mechanized Mechanical	Metals, Plastics	Same as barrel tumbling	
Sanding	Metals	Usually limited to flat parts	Can be used primarily for deburring
Spindle Finishing	Metals, Plastics	Same as barrel tumbling	Same as for barrel tumbling
Thermal Energy	Metals, Plastics	Deburrs blind features** Burr area must be free from oil and water	Not well suited to heavy burrs on thin part sections
Torch or Flame	Metals, Plastics		
Ultrasonic	Copper Alloys, Brass Zinc, Aluminum	Some limitations on round parts	
Vibratory	All Metals and Plastics	Same as barrel tumbling	Same as for barrel tumbling
Water Jet	Molded Asbestos/Sisal, Aluminum Alloys	Limited to external burrs/ flash	Used by automotive industry

*1 mm = 0.3937 inch
**Provided that fixturing does not shield features from gases

TABLE 7-21: COMPARISONS OF BURR SIZES REMOVED AND EDGE RADII PRODUCED

DEBURRING PROCESS**	TYPICAL SIZES OF BURRS REMOVED (μm)	EDGE RADII TYPICALLY PRODUCED IN INDICATED MATERIAL		
		SST AND STEEL (μm)	BECU (μm)	AL (μm)
Abrasive Flow	5.02 to 76.2 thick	25.4 to 127		50.8 to 127
Abrasive Jet	12.7 to 127 thick	76.2 to 254		
Barrel Tumbling	2.54 to 76.2 thick	76.2 to 254	101.6 to 254	152.4 to 762
Brushing	5.08 to 76.2 thick	50.8 to 254	50.8 to 254	76.2 to 381
Centrifugal Barrel Finishing	5.08 to 76.2 thick	76.2 to 508	76.2 to 508	127 to 762
Chemical	2.54 to 12.7 thick			
Chemical Vibratory				
Chlorine Gas	5.08 to 101.6 thick	127 to 508		
Electrochemical	7.62 to 127 thick 25.4 to 508 long	50.8 to 254	50.8 to 254	152.4 to 762
Electrochemical Vibratory	50.8 to 127 thick	127 maximum		
Electropolish	2.54 to 25.4 thick	25.4 to 127	25.4 to 127	
Hand	5.08 to 76.2 thick	50.8 to 254	50.8 to 254	76.1 to 381
Liquid Hone	5.08 to 25.4 thick	127 maximum		
Mechanized Mechanical	0 to 254 thick	None	None	None
Sanding	5.08 to 76.2 thick			
Spindle Finishing	5.08 to 76.2 thick	25.4 to 508	76.2 to 508	127 to 762
Thermal Energy	5.08 to 50.8 thick	50.8 to 406.4		50.8 to 1524
Torch or Flame				
Ultrasonic	2.54 to 12.7 thick	25.4 to 76.2	25.4 to 76.2	

TABLE 7-21 CONT: COMPARISON OF BURR SIZES REMOVED AND EDGE RADII PRODUCED

DEBURRING PROCESS**	TYPICAL SIZES OF BURRS REMOVED (μm)	EDGE RADII TYPICALLY PRODUCED IN INDICATED MATERIAL		
		SST AND STEEL (μm)	BECU (μm)	AL (μm)
Vibratory*	2.54 to 76.2 thick	76.2 to 254	101.6 to 254	152.4 to 762
Water Jet	0 to 38.1 thick flash			

*In a 2-hour deburring cycle
**All values shown represent information applicable to small and miniature precision components;
1 μm = 39.37 microinches

TABLE 7-22: COMPARISON OF SIDE EFFECTS AND REPEATABILITY

DEBURRING PROCESS*	EFFECT ON ADJACENT SURFACES (μm)	EFFECT ON PART SIZE** (μm)	APPARENT BEST REPEATABILITY OF EDGE RADIUSING		
			SST (μm)	BECU (μm)	AL (μm)
Abrasive Flow	Polishes; 0 to 101.6 stock loss	None except near burrs	±25.4	±25.4	±50.8
Abrasive Jet	Matte Finish; 0 to 25.4 stock loss	None			
Barrel Tumbling	Surface finish and size change	0 to 12.7 size change	±38.1	±63.5	±63.5
Brushing	Polishes	None	±50.8	±50.8	±76.2
Centrifugal Barrel Finishing	Surface finish and size change	0 to 12.7 size change			
Chemical	1.27/minute etch rate on high carbon steel	1.27/minute etch rate on high carbon steel			
Chemical Vibratory		1.27 to 127 size change			
Chlorine Gas	10.16 to 254 stock loss	10.16 to 250 stock loss, each side			
Electrochemical	Slight etch; 0 to 254 local stock loss	Possible stray etching	±50.8		±101.6
Electrochemical Vibratory	Some etching possible	5.08 to 76.2 typical size change	±50.8		
Electropolish	Polishes; up to 12.7 stock loss per side	Up to 12.7 stock loss per side	±1.27		
Hand	None	None	±50.8	±50.8	±76.2
Liquid Hone		None			

TABLE 7-22 CONT: COMPARISON OF SIDE EFFECTS AND REPEATABILITY

DEBURRING PROCESS*	EFFECT ON ADJACENT SURFACES (μm)	EFFECT ON PART SIZE** (μm)	APPARENT BEST REPEATABILITY OF EDGE RADIUSING		
			SST (μm)	BECU (μm)	AL (μm)
Mechanized Mechanical	None	None			
Sanding	Surface finish changes	2.54 to 50.8 stock loss			
Spindle Finishing	Surface finish and size change	0 to 12.7 size change			
Thermal Energy	Surfaces are covered with an oxide	0 to 2.54 size change; surfaces are covered with an oxide	±76.2	±76.2	±127
Torch or Flame					
Ultrasonic					
Vibratory	None	0 to 12.7 size change	±38.1	±63.5	±63.5
Water Jet	Cleans	None			

*All values shown represent information applicable to small and miniature precision components; 1 μm = 39.37 microinches

**These values are conservative for hand-size or larger parts. Data in this column apply to all surfaces exposed to the deburring media and not just those adjacent to the burr-laden edge.

CHAPTER 8

VIBRATORY DEBURRING

Process Considerations

Although there are fewer than 13,000 vibratory machines in the U.S., they perform the bulk of all mechanized deburring. Their success is due to their ability to deburr an almost endless variety of part sizes and shapes in large batches in a short time. Machines are available which can deburr tons of workpieces per hour. Very low volume runs can be run in the very same equipment or smaller machines. Machines are available with essentially any degree of automation desired, from completely manual systems to fully automated, completely operator-free, in-line systems. In addition to the flexibility of the process, vibratory deburring also accomplishes other functions: cleaning, smoothing surfaces, brightening, inhibiting, drying, and transporting parts. These other benefits are obtained at little or no increase in cost above deburring cost.

Despite the many advances made in this field, vibratory deburring is still often considered an art rather than a science. The following paragraphs present some of the reasons for this and some of the considerations required to implement successful vibratory finishing.

Definitions

What is a burr? Webster defines a burr as a "roughness, especially on the edge of something made by drilling, turning, etc.; a rough or thin edge, ridge, or the like." We would add to this definition that burrs are "generally undesirable or have a generally undesirable degree of sharpness associated with them." Burrs are normally developed in the fabrication or machining of the parts.

Deburring, then, is the removal of burrs. But many parts "deburred" do not have the requirement of complete burr removal. Merely a reduction in sharpness or a reduction in the height of the projection often is sufficient so that mating parts can move freely without interference or so that personnel will not cut themselves.

Deburring, therefore, can and should be categorized by degrees. A very light deburring can result in the removal of sharpness only by peening or a light cut. A slightly more aggressive deburring could be called edgebreak. Complete removal of the burr may be required. Finally a measurable radius is sometimes required on the part. Because of our broad definition of deburring, the removal of flash from diecast metal parts or injection molded plastic or molded ceramic parts can also be considered deburring. Categories for parts such as these include also the removal of flash or the removal of parting lines.

As indicated in Chapter 2, it is economically unwise to require a radius on a part where only a light cut or sharpness removal is adequate. The relative degree of deburring, therefore, is dependent entirely on the requirements of the part by its manufacturer. When well defined, optimum economy results.

Manufacturers processing vast tonnages of many parts who currently utilize vibratory and barrel finishing do not specify a radius for deburring on a great many of their parts because it is not necessary. Drawings for these parts may only specify "deburr" which, as we have seen, covers a range of tolerance. Preferable to this would be "remove sharpness," "edge-break," "complete removal," or "radius _____ inches." These definitions would more clearly indicate the requirements for these parts and thereby help their manufacturing arm develop the optimum procedures to accomplish this end.

And many parts do carry specifications regarding deburring or radius. Many of these specifications are functional in nature, defining sharpness by what the part cannot cut, for example.

Setting Up the Process

With an understanding of the part requirements, how do those in the vibratory and barrel finishing industry reduce these requirements into a practical application? It is conceivable that one could determine the hardness and toughness parameters of the base material as well as the volume of burr to be removed and match these with a media of given abrasiveness capable of performing this task in a desirable time period. From these determinations then, time and cost estimates could be developed.

Essentially, this is our approach but on a much more practical scale. First accessibility of the media to the burr area is of paramount importance. Next potential lodging problems will often influence the shape and/or size of the media and, in certain cases, will determine the type of media required for the given task. Conical shaped, agressive cutting ceramic media is not available, nor is cylindrically shaped resin bonded media. If one of these shapes is required due to accessibility and/or lodging, the optimum media type normally selected for cutting cannot be used. By the same token, many applications where accessibility in small places is required eliminate use of any cutting grade of media and hardened steel shapes must be used. As opposed to grinding wheels or similar metal removal techniques, lodging and separation are critical considerations. Rather than reduce the scientific thought process, they materially compound its difficulty.

From a production standpoint, in bulk handling on volume parts, abrasive wear, especially on heavy parts, must be watched very carefully. Worn abrasive can cause enormous time loss not only in not accomplishing the job for which it was intended because of its decreased size or change in form, but also in becoming lodged in the work part which could be troublesome to remove. Unless detected, this can cause a malfunction in assembly. The selection of the proper form in the right material to remove the burr and create a desired radius is something that requires proper thought. Provided there is no past history on a particular job and one has a variety of abrasives at his disposal, he sould try a few sample runs before committing the lot. The selection of the proper compound and water level can offer some very important variations. The numerous compounds and abrasives that are on the market today make for a very interesting study. There are some for instance that provide a good cut or metal removal for parts with a stubborn burr where the tolerance is open. This type is available for almost every material, however one must be careful to select ones which prevent excessive burr roll-over, edge radius, or discoloration.

Water level in barrels and vibratory finishing plays a major role in the deburring action. High water levels retard the action. This is a useful method to control excessive cutting when burrs are small and tolerances are close. In all cases, it is essential to maintain an accurate record or log of conditions used in order to obtain repeatable results.

Media selection is dependent also on the type of equipment used or needed. The rotary barrel is low in initial cost, low in supplies cost, high in labor cost, slow in process time and messy. It is also the outstanding tool for generation of radii on stampings. The tub vibrator is preferred for large or very long parts in limited production volumes. Round vibrators, especially those with simple, integral separating systems, are preferred for most other applications because of their ease of use and low operational costs.

Once the media and the machinery have been selected, the equipment must be set up and adjusted according to the variables it contains. This might be by weight settings which control the amount of amplitude on the tub as well as feed rate in the case of round machines. Feed in round machines containing integral separation of devices should be adjusted at the same time for separation speed. Some vibratory equipment contains variable frequency. Compound and water flow rates, as well as compound type, must also be adjusted in order to perform the desirable tasks at minimum costs. The volume ratio of media to parts, see Table 8-1, must be determined based on the equipment, its settings, and the amount of contact one part can have with another during the finishing cycle. When no part on part contact can be tolerated, a compartmented machine must be selected on the basis of the production volumes required. Media to part ratios will vary depending on the equipment type. For example, tub type vibrators have a pronounced end effect, and segregation of parts and media will occur in these machines because of their poor mixing capabilities. Parts will tend to collect at either end of these machines with a much lower part density in the center. Round machines on the other hand can normally process parts at higher parts loading or lower media to parts ratios.

Separation of media and parts must be considered because any time parts are mixed with media, they must subsequently be separated from it. While mesh screens are standard; tie rod screens, inverse separation, magnetic separation, tie rod screens with steps to turn over parts, tie rod and screen combinations, auxiliary screens with independent amplitude control and many others are used to solve a great variety of separation problems.

The tremendous number of variables illustrated very briefly above only serves to indicate the reason that a "scientific" approach has not yet been implimented throughout industry. The many variables associated with vibratory finishing are well known to the manufacturers of this equipment, and computation of these variables at this time is difficult or impossible for industry as a whole. Such approaches are feasible and are in use within specific companies and industries. For example, media lodging is a relatively simple determination under normal considerations, but in vibratory equipment two or three pieces of media can stick or lodge in a hole or recess. If it can happen, it <u>will</u> happen in vibratory equipment. Experience with various shapes, therefore, is essential to minimize these problems. And even the best experience must be verified by actual tests!

Other Influences

Thus far, the entire consideration has been devoted to deburring. While deburring may be the primary consideration in processing many parts, it is certainly not in all of them. Surface finishing can be just as important to many manufacturers of parts. Washing or cleaning is a growing requirement for a tremendous number of applications where solvent degreasers cannot be used. Surface luster, whether a dull matte or whether a bright burnished surface is required, is controlled by media and compound type. Descaling of ferrous metals or removal of oxides from copper alloys, silver, zinc, and aluminum is critical to many applications. Finally, corrosion inhibition--any metal part can corrode. The degree of inhibition required depends on the next step in the part's manufacture, whether it is painted or plated or used as it is, and its storage time prior to use.

Surface reflectivity and surface roughness can be measured and therefore defined when necessary. Prepaint finishes depend on the leveling ability of the paint used. Surface roughness required to hold an adherent lubricating oil film is more difficult to define but practical. Of the above, surface luster is probably the most difficult to predict. Individual decisions in this area are frequently emotional because of appeal or lack of appeal to those responsible for these decisions.

The vibratory finishing process is inherently stupid. It cannot think. It cannot do more work in one area than it can in another. As a result, the uniformity of the process, month after month, is outstanding when given minimal care. For this reason, a greater number of users is using the process as a quality control tool. They realize that the biggest single variable in the entire process is in the parts themselves! Stamping dies and drill bits get dull after long use. Diecasting dies lose their fit and surface deterioration progresses slowly and steadily. The burr conditions slowly change. They get worse. When the parts are not satisfactorily deburred or finished in a given cycle, a longer cycle must be used which increases costs and reduces equipment productivity. It is time to sharpen the tools which make the parts. This variation, once understood, can and should be used to great advantage. A simple, effective means of quality control is thereby easily established.

Parts Processing by Vendors

Therefore, in order to serve potential customers better, parts are normally processed in two or three types of media for various lengths of time and in one or more different compounds. The user then has the opportunity to select that process which best fits his needs from a deburring standpoint, as well as from appearance and cost standpoints. The educational process involved shown him more vividly than words ever could exactly what these variables can do to his parts. He can make his selection based on facts such as part appearance, compound consumption rate, media depreciation rate, and time cycles, and thereby determine with excellent accuracy parts costs on a production basis.

Computations

Once the media type and time cycles have been determined, it is necessary to determine the size of equipment which will handle a given production volume. Capacity calculations are easily determined by use of media to part ratios (Table 8-1), part dimensions, time cycles, and accurate volumetric capacities of vibratory equipment. Nomographs like that in Figure 8-1 are available which readily determine machine size based on the volume of parts to be processed per load and the media:parts ratio selected. With these data and a final determination on equipment size, production rates can be determined, accurate parts loading can be specified, media depreciation rate and compound consumption rate can be developed and, as a result, highly accurate, guaranteed costs are predicted. In the ultimate analysis, the ease of parts handling and the cost of a particular finishing process are the only matters of extreme importance to the potential user of this equipment. This system works. It has worked for years, and it will continue to be used because of its accuracy and reliability. It has to work well because it is usually guaranteed!

Science is defined as knowledge amassed, tested, coordinated, and systematized. This approach has been amassed and tested, often the hard way. It is now coordinated and systematized. This is the science of vibratory finishing.

Design for Deburring

This discussion should have made the reader aware of potential problems with media lodging, media access to burrs, media-parts separation, and parts uniformity as a function of deburring economy. The farsighted engineer has already concluded that part design can influence deburring and finishing costs just as it can for plating, painting, and assembly. Review of part design by those knowledgeable in deburring can cause relocation of webs, hole size changes, etc., which have no adverse effect on the function of the part. Finishing cost reductions possible by such foresight can easily be 50 percent or more. If a part will be manufactured with intolerable burrs, they must be removed. Plan for it.

TABLE 8-1: MEDIA TO PART RATIOS

Media-to-Part Ratio by Volume	Normal Commercial Application
0:1	No media. Part-on-part. Used for beating off burrs. No media for cutting.
1:1	Equal volumes of media and parts. Forgings, sand castings. Crude, very rough surfaces
2:1	More gentle, more separation, still severe part-on-part damage.
3:1	About minimum for nonferrous parts. Considerable part-on-part contact. Fair to good for ferrous metals.
4:1	Probably "average" conditions for nonferrous parts. Fair to good surfaces. Good for ferrous metals.
5:1	Good for nonferrous metals. Minimal part-to-part contact.
6:1	Very good for nonferrous parts. Common for preplate work on zinc with plastic media.
8:1	For higher quality preplate finishes.
10:1 to 15:1	Even better. Used for very irregular shaped parts.
No contact	Absolutely no part-on-part contact. One part per machine or compartment. Use doughnut shaped machine or fixturing.

SPIRATRON MODEL NO.	CHANNEL DIAMETER IN.	CM.
100	27.5	70
70	23.5	60
60	21.5	55
40	19.5	50
20	16.5	42
12	13.5	34
4	8.8	22
65B	27.5	70
50B	17.5	44
30B	14.0	36
20B	12.0	30
7B	8.8	22
4B	8.8	22
60D	21.5	55
20D	17.8	45
12D	14.5	37
4D	9.8	25

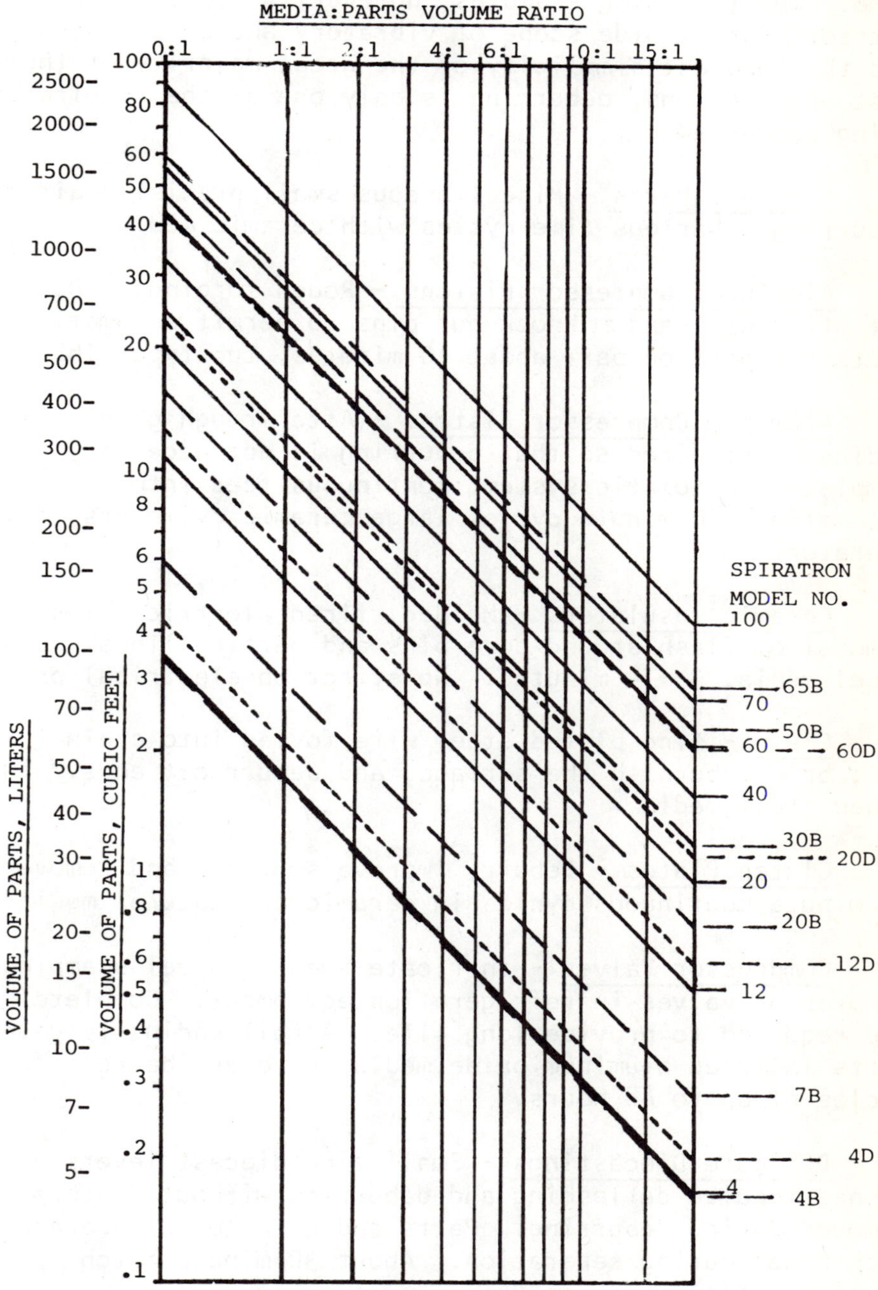

DIRECTIONS: Determine volume of parts to be run per cycle. Read across to selected ratio, then up to first machine of type required.

NOTE: Capacities given assume batch operation, gate down process. Add 15% to capacity for continuous process, gate always up.

FIGURE 8-1: PART LOAD VERSUS RATIO

Deburring Applications

In any discussion of deburring, the reader normally associates with the parts involved in his plant where burr problems exist and are troublesome. The following examples, however, are presented to illustrate the extraordinarily wide scope of vibratory and barrel deburring applications and the inherent simplicity of the great majority of them. Note that in most applications, deburring is only one of the requirements for the parts being processed.

Aircraft Parts - Miscellaneous small precision aircraft parts require deburring. Various time cycles with ceramic media.

Aluminum Compressor Pistons - Rough forgings. Deflash at parting line, cut off areas, and at knock out pins to permit automatic feed to grinder. Parts run part on part about 10 minutes, tub type vibrator.

Aluminum Compressor Pistons - After rough grind on outside diameter, radius is required so that deburring is not necessary after final grind. Completely automatic system, continuous feed into and from Spiratron with Rotomation, 30 minute cycle, large ceramic cylinders, fully automatic, no operator.

Ceramic Insulators - Molded, fired electrical insulators require removal of flash around core pins and parting lines. Continuous process in steel media, 3 - 4 minutes. No effect on electrical properties.

Chain - Zinc plated steel wire formed into chain. Remove lubricant oil, bright burnish the surface, and deburr cut ends. 20 minutes in hardened steel media.

Clutch Plates - Deburr, improve surfaces and remove organic soils. 30 minute continuous cycles in ceramic and natural media.

Compressor Valves - Intricate Swedish steel stampings are used as compressor valves in refrigeration equipment. Completely notch free edges are required to provide long life. A full radius is developed on these parts in fused aluminum oxide media in rotary barrels for relatively long cycles of up to 48 hours.

Delicate Diecastings - Small zinc diecast levers used in dial telephones require deflashing and deburring without distortion. Gate to be removed during deburring. Parts and gate to be separated from media and each other during separation. About 30 minute batch operation. Small ceramic media.

Diesel Engine Connecting Rods - Remove heavy burrs on connecting rods and caps in 30 to 40 minute continuous time cycles. Clean parts of machining oils and radius to increase strength of the connecting rod arm, both before and after assembly. Very fast cutting ceramic media.

Eyeglass Frames - Nickel silver formed and stamped frames. Require preplate finish and complete removal of burrs. Plastic media, 4 hours.

Fountain Pen Clips - Preplate finish and complete burr removal required on stamped pen and pencil clips. Plastic media, 4 hours.

Garden Tractor Parts - Machined, stamped, welded, and formed steel, cast iron, and bronze parts are cleaned and deburred for assembly. 10 to 12 minute continuous cycle, hardened steel media.

Magnesium Diecastings - Chain saw housings require deburring and pre-paint finish. 10 minute continuous cycle in ceramic media or about 20 minutes in plastic media.

Phenolic Plastics - Relatively brittle phenolic plastics, wood flour filled, require deflashing without impairment of surface luster. Iron handles, electrical insulators, etc. Steel media, continuous cycles, about 6 to 9 minutes. Plastic insulators are run dry with exhaust to remove dust and cool mass.

Stainless Steel Bezels - Deburr and burnish stainless steel trim for instruments and automotive applications. Hardened steel media. Cycles from 10 to 30 minutes. Batch and continuous.

Surgical Instruments - Polish and radius stainless steel surgical instruments. Complete radius required. Fast cutting media plus polishing. Long cycles.

Toy Stampings - Prepainted and preplated steel strip is stamped into components for toys. Parts require deburring to prevent cutting of children. Plated or painted finishes cannot be damaged. Rust inhibition necessary on exposed edges. Deburr 7 minutes in steel burnishing media, continuous basis. Automatic rinsing and continuous drying. Ultra low toxicity required for all compounds.

Transmission Gears - Large gears require extensive cleaning and deburring. Continuous 25 minute cycles with aggressive cutting natural material.

Unless indicated otherwise, "Spiratron" round vibratory finishing machines were used in the above examples. The applications presented were selected solely for the purpose of indicating some of the capabilities of vibratory and barrel finishing. These applications are commercial.

Scope of the Industry

Diverse applications indicate only in part the current scope of a growing and highly dynamic industry. The rotary barrel as a machine tool began its serious development before the turn of the century. While less dramatic, the changes in the rotary barrel since its early days have been great. The octagonal or eight sided barrel has become the industry standard because of its ability to lift and rotate the mass. The vibratory barrel and centrifugal barrel were developed to speed up the finishing process. Some degrees of automation have been developed to reduce labor, simplify separation of media and parts, and to increase the ease of introducing fixtures in large rotary barrels. Dual action machines wherein fixtured parts rotate in a direction opposite to the media mass are also available.

The vibratory finishing industry began in 1957 with the first tub type vibrator introduced commercially in this country and in Germany. The first tubs merely shook. Parts and media were separated by hand. Changes in this equipment were the introduction of tilting devices to dump out media and parts onto a screen, the addition of a door in the end of the vibrator for discharge of media and parts onto a screen deck, the addition of drains to permit soiled solution to flow through the vibrating chamber, the addition of media return conveyors and hoppers to develop semiautomatic production, the addition of dividers to separate large parts, flow-through solution systems, automatic tilt mechanisms to facilitate emptying out parts and media, long continuous single tubs up to 20 feet (6 meters) in length for continuous processing, and long connected multimodular machines with unlimited length capabilities for the processing of parts 30 feet (9 meters) or more in length. Variable speed, direct and indirect drive systems, two or more eccentric weight assemblies, and pressurized devices were also developed to increase the capabilities of the tub vibrator.

The round or doughnut shaped vibratory machines were developed about 15 years ago. These machines offer simple weight adjustments and the potential for greatly simplified media-parts separation. Integral screen decks, automatic or semiautomatic gate operation, flow-through solution systems, underscreen feed chutes, noise covers, automatic weight shifters for control of continuous processing time, and simple or complex screen decks offer improved mixing and easier parts handling techniques when compared to tub vibrators. These units come in small and large sizes--to 100 cubic feet (2800 liters) in size. Round vibratory steel ball burnishers carry up to 18,000 pounds (8160 kilograms) of steel media and exceed 12 feet (3.7 meters) in overall diameter.

The lubrication of bearings, especially in larger vibratory equipment, has gone from grease packed to grease lubricated to oil mist to automatic lubrication systems, greatly extending the life of these vital components. Lining materials have changed from natural and synthetic rubbers to polyurethane elastomers, a change essential to increase durability in the area of intense wear. Life of these linings is now measured in years.

Natural stone media has almost been replaced by, in turn, synthetic random shaped media, preformed ceramic media of a wide range of abrasiveness, preformed resin bonded media, hardened and stress relieved steel media, and a whole host of other available products. The range of shape and size, size control, degrees of cutting ability and durability have been a great asset in promoting growth of vibratory finishing. Compound developments with improved cleaning, inhibition, oxide removal, and mass control characteristics further increase the versatility of these systems. Compounds deep media clean and clean media produces bright clean parts, light in color.

Since 1950 over 100 patents have been issued in the field of vibratory finishing worldwide, excluding duplication of foreign filings. The growth of this industry is predicated on these technological advances to improve and simplify as well as to increase the scope of vibratory and barrel finishing. Several companies have licensees worldwide for the manufacture of equipment and materials under their patents. As a result, worldwide sales of vibratory finishing machines since 1960 through 1974 are over $150 million. And this does not include supplies or spare parts. In installed capacity, vibratory equipment is approaching one-quarter million cubic feet (7 million liters).

Research in Vibratory Deburring

Although vibratory deburring has been a widely used deburring process for 20 years, no serious attempt has apparently been made to characterize its capabilities in a quantitative manner, even for simple shapes. While some researchers, such as Matsunaga and Haguida[1] and a number of Soviet authors, have provided some excellent quantitative information, none have described the burr removal characteristics of the process.

In trying to numerically define the capabilities of a process such as vibratory deburring even for a limited range of parts, several factors must be considered. From a strictly capabilities standpoint, these factors are significant:

- Size of burr which can be removed without adversely affecting surface finish, dimensions, or edge radius;
- Workpiece material properties; and
- Part geometry and burr location.

Notice that two major factors in vibratory finishing are absent in the above list: economics and processing parameters.

In selecting an appropriate deburring process, one has to know three sets of facts.

- Will the process meet the technical requirements shown in the previous list?
- What process parameters are required to produce the results?
- What are the economics of the process?

For example, in conventional vibratory finishing, the very nature of the process makes it impossible to remove a 0.005 inch thick (127.0 μm) burr and maintain size within 0.000010 inch (0.25 μm). If it were possible then one would want to know the details of how it could be done, and knowing the details it would be possible the calculate the economics. Therefore the first question that has to be answered is, "Will the process meet the technical requirements?"

In the following paragraphs, some empirical data is presented on the process capabilities of vibratory deburring for three sets of conditions. In addition, an approach is presented for classifying the general capabilities of the process.

Edge Radiusing Capabilities

In 1966, Robbins[2] published 120 curves showing the edge radius produced on 0.5 inch (12.7 mm) cubes. The cubes, which initially had sharp edges (no burrs and no radius above 0.001 inch (25.4 µm)), were vibratory finished in four different machines using four different media. Four workpiece materials were used. Figures 8-2 through 8-5 illustrate the results obtained on two materials using two media. Similar curves, using plastic preformed media, were recently reported by the author.[3] Two of these curves are shown in Figures 8-6 and 8-7.*

As indicated in Figures 8-2 through 8-7, the larger media produced faster radiusing than the smaller media. In addition, it is impossible to achieve a radius larger than 0.004 inch (101.6 µm) in stainless steel using the N7 nuggets in any reasonable time frame. Using the large plastic cones however a 0.006 inch (152.4 µm) radius was produced in only two hours. The aluminum, which is a much softer material, radiused roughly twice as fast as the stainless steel.

Two process limitations were noted from this test. First, using small media, it is not possible to generate large radii.** This is significant if small parts are involved, because large media cannot be effectively used on many small parts. The second limitation (not shown in the figures) is that the repeatability of the radiusing was only ± 0.001 inch (±25.4 µm). Thus, while the average radius measured followed the curves shown, actual measurements would vary ± 0.001 inch from this value. This is for initially burr free parts. In additional tests involving the removal of small milling burrs using the 9/16 inch (14.35 mm) plastic cones, it was noted that radii varied within ± 0.0022 inch (55.9 µm) on aluminum specimens and ± 0.0012 inch (30.5 µm) on the stainless steel specimens.[3]

While this is not a serious limitation in most applications, it is on small precision parts requiring ± 0.001 inch tolerances on edge radii.

Size, Weight, and Surface Finish Changes

In the second test, weight losses and size and surface finish changes were noted over a four hour cycle. Cylindrical specimens, each 0.490 inch (12.45 mm) in diameter and 0.5 inch (12.7 mm) long, were used in this test. Ten specimens of each material were weighed to the nearest milligram and measured to the mearest 0.000020 inch (0.51 µm). Three groups of these specimens were used to compare the results of three media. To assure initial uniformity, the parts were centerless ground to a tolerance of

*Details of the tests and equipment involved are described in the appendix of this chapter.

**Throughout this discussion it is assumed that a conventional vibratory process is used. It is possible by fixturing parts, masking surfaces, and similar approaches to obtain different results. These techniques, however, are not part of what most authorities consider normal approaches.

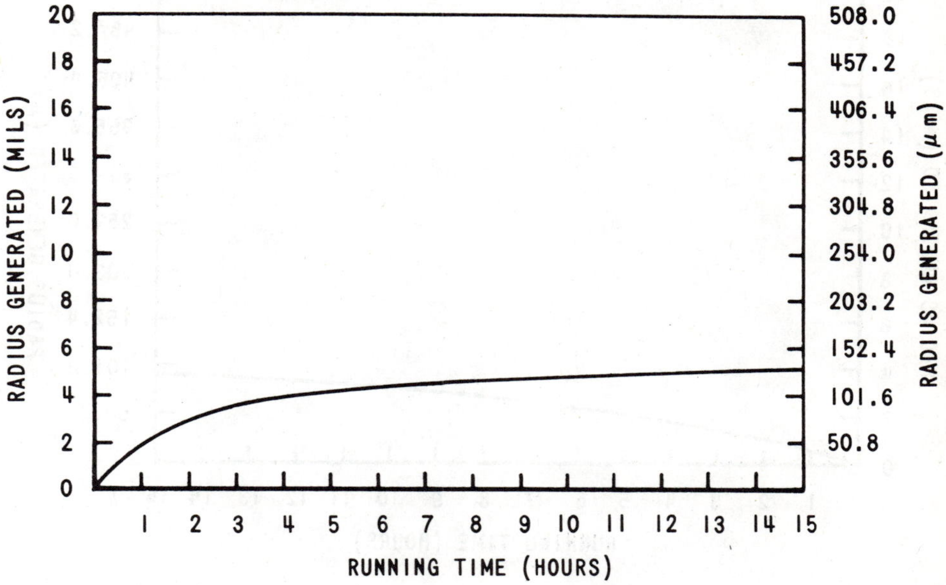

FIGURE 8-2: EFFECT OF OPERATING TIME ON EDGE RADIUS FOR 3/16 INCH (2.381 MM) TRIANGLES ON 303SE STAINLESS STEEL

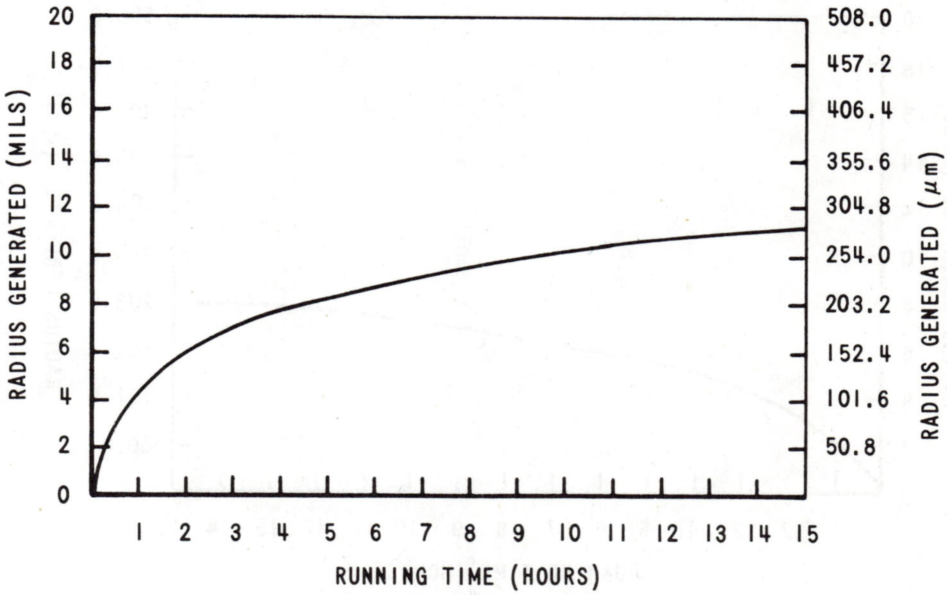

FIGURE 8-3: EFFECT OF OPERATING TIME ON EDGE RADIUS FOR 3/16 INCH (2.381 MM) TRIANGLES ON 6061-T6 ALUMINUM

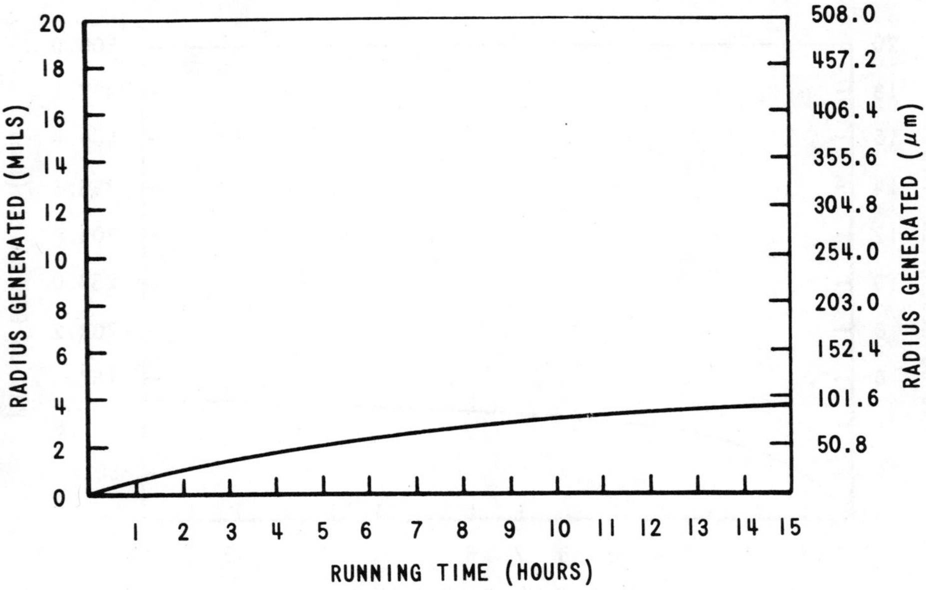

FIGURE 8-4: EFFECT OF OPERATING TIME ON EDGE RADIUS FOR NUMBER 7 CHIP ON 303SE STAINLESS STEEL

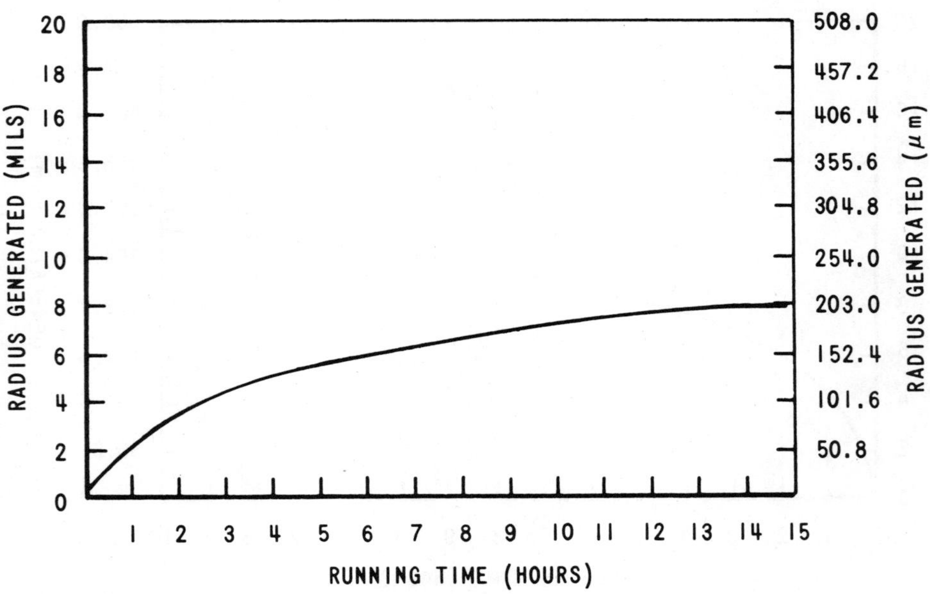

FIGURE 8-5: EFFECT OF OPERATING TIME ON EDGE RADIUS FOR NUMBER 7 CHIP ON 6061-T6 ALUMINUM

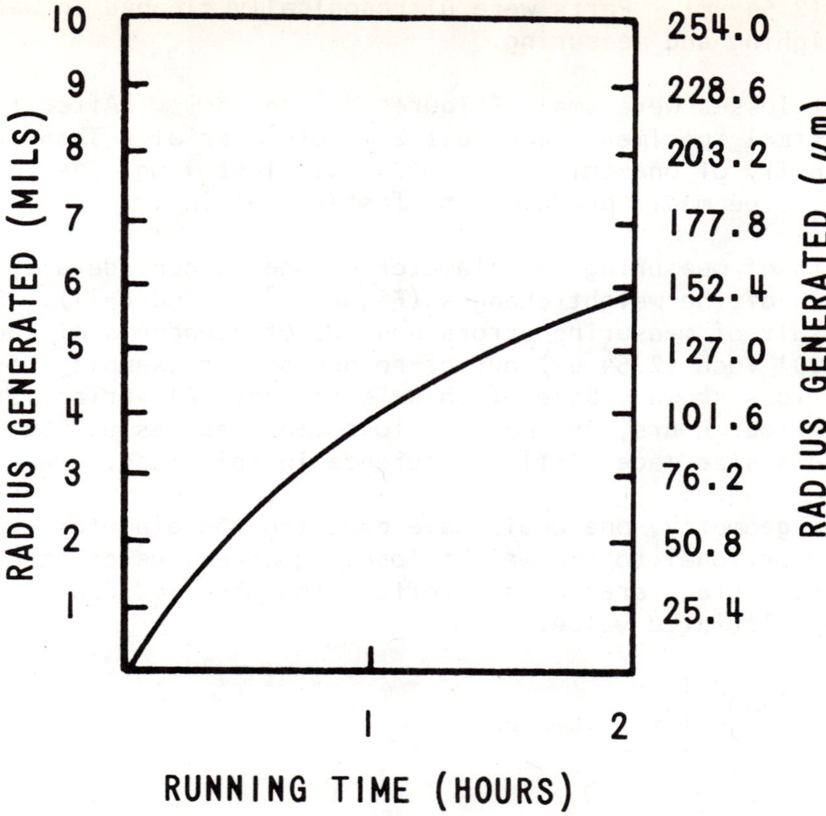

FIGURE 8-6: EFFECT OF OPERATING TIME ON EDGE RADIUS FOR 9/16 INCH (14.29 MM) PLASTIC CONES ON 303SE STAINLESS STEEL

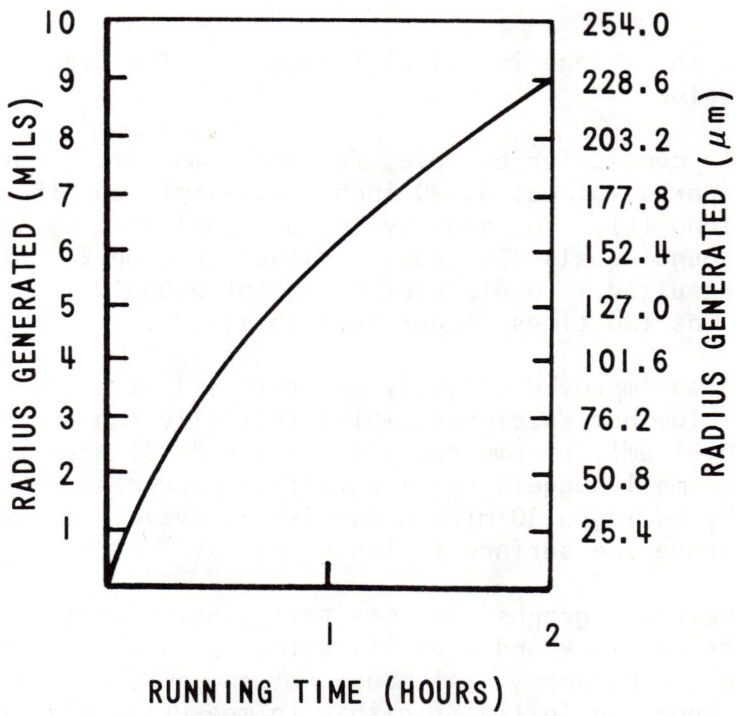

FIGURE 8-7: EFFECT OF OPERATING TIME ON EDGE RADIUS FOR 9/16 INCH (14.29 MM) PLASTIC CONES ON 6061-T6 ALUMINUM

± 0.0001 inch (2.54 μm). Parts were ultrasonically cleaned in warm detergent before weighing and measuring.

The weight losses were small (Figures 8-8 and 8-9). After three hours the stainless steel specimens each lost 2 mg of material. The aluminum, which has a density of one-third that of steel, lost 1 mg. As in edge radiusing, the large media produced the fastest cutting.

The results of measuring the diameter change is considerably different from the results of the weight changes (Figures 8-10 and 8-11). Some of this is the result of measuring errors and out-of-roundness of the workpieces. A 0.0001 inch (2.54 μm) out-of-roundness, for example, would be half the total loss shown. Some of this is the natural variability of the process. After four hours, the maximum loss measured was 0.0002 inch (5.08 μm). Media size made little difference in this case.

Because of geometry, one would have expected the diameter change to be directly proportional to the weight loss. However, using the following equation for the surface area of a material, the observed change does not agree with the calculated value.

$$\Delta W = \left[\frac{\pi D^2}{2}\right] + \pi DL \quad \rho\Delta t \tag{1}$$

Thus,

$$\Delta t = \frac{\Delta W}{\rho\pi D(D/2+L)} \tag{2}$$

Note that Δt is the change in radial thickness. The diameter change will be twice this value.

The 9/16 plastic cones, for example, removed 8 mg after four hours. The diameter of the parts (D) was 0.490 inch (12.45 mm), and they were 0.5 inch (12.7 mm) long (L). The density (ρ) of steel is 0.29 lb/in^3 (8024 Kg/m^3). One pound equals 454 grams. Using Equation 2, this weight change should have resulted in a diameter change of 0.00010 inch (2.54 μm). The measured change was two times larger than this.

The surface finish improved slightly on stainless steel specimens (Figure 8-12). The aluminum specimens, which initially had a finish of 23 microinches AA (0.61 μm), became rougher (Figure 8-13) when the larger media was used. The small nuggets reduced surface roughness to 18 microinches (0.46 μm). By adding a 30 minute burnishing cycle, it would have been possible to improve the surface finish slightly.

In assessing these six graphs, one can tell that vibratory finishing removes small amounts of stock and that its effect on surface finish depends on the workpiece hardness. Although not shown here, surface finish improvement depends upon the initial finish. In measuring diameter changes, it appears that there is a variation of at least ± 0.000025 inch (0.635 μm)

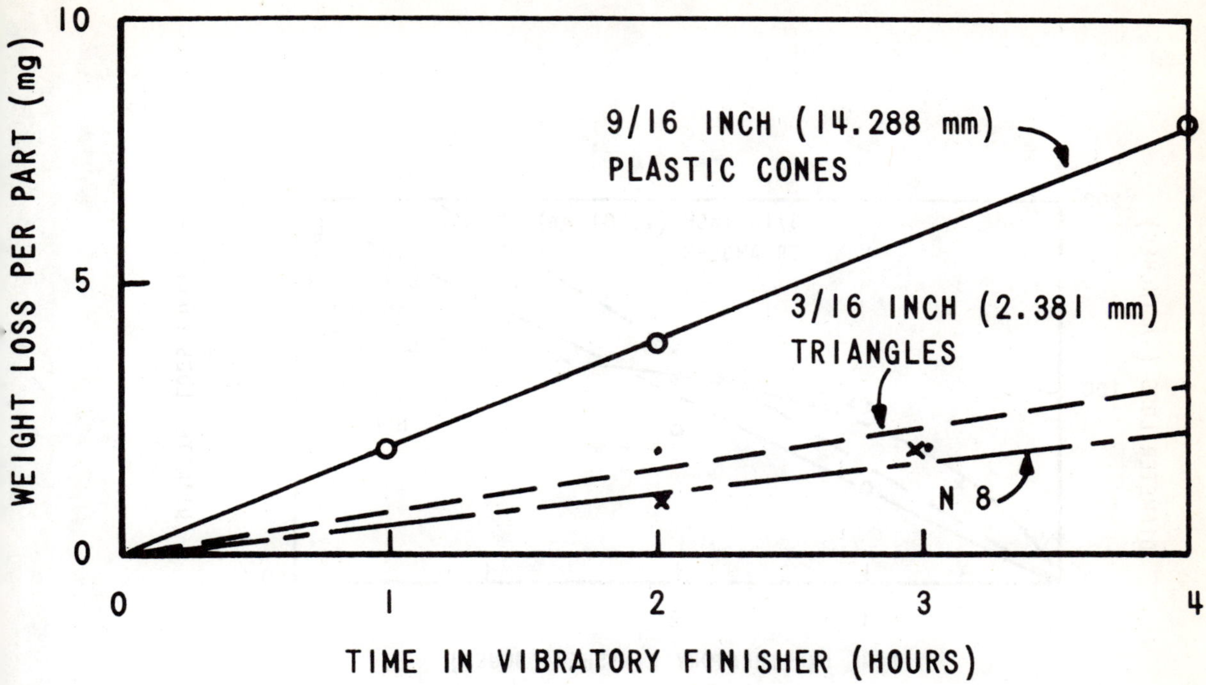

FIGURE 8-8: EFFECT OF MEDIA AND TIME ON WEIGHT LOSS FOR 303SE STAINLESS STEEL (R_C20)

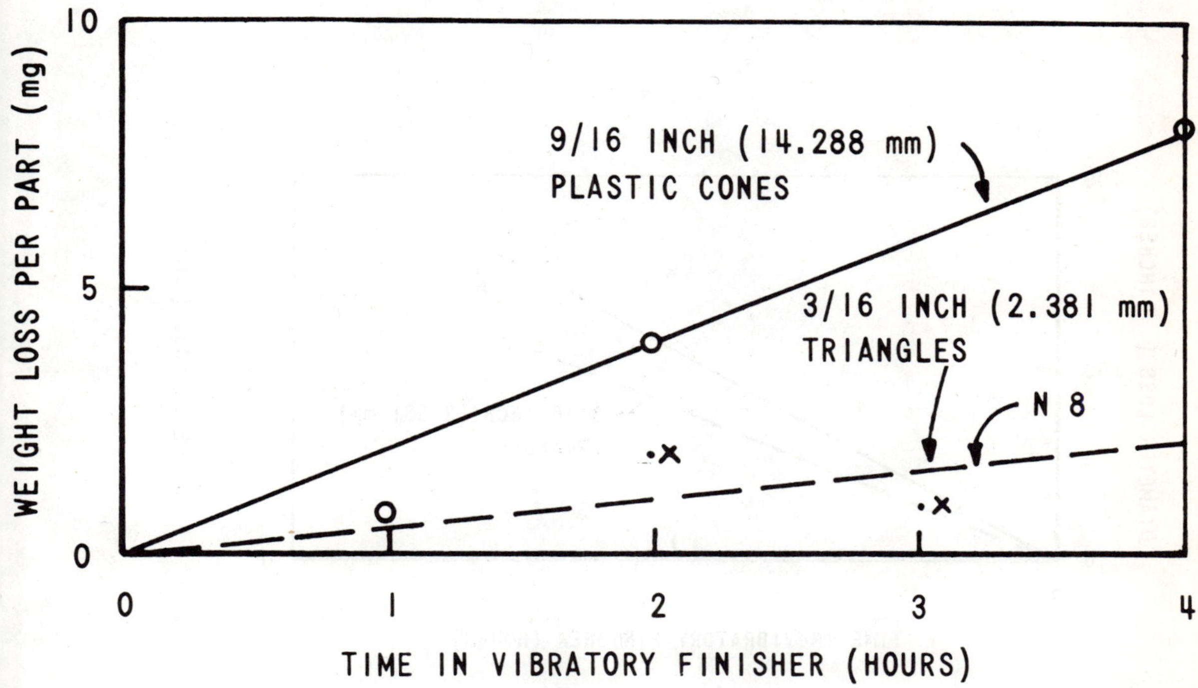

FIGURE 8-9: EFFECT OF MEDIA AND TIME ON WEIGHT LOSS FOR 6061-T6 ALUMINUM (R_B61)

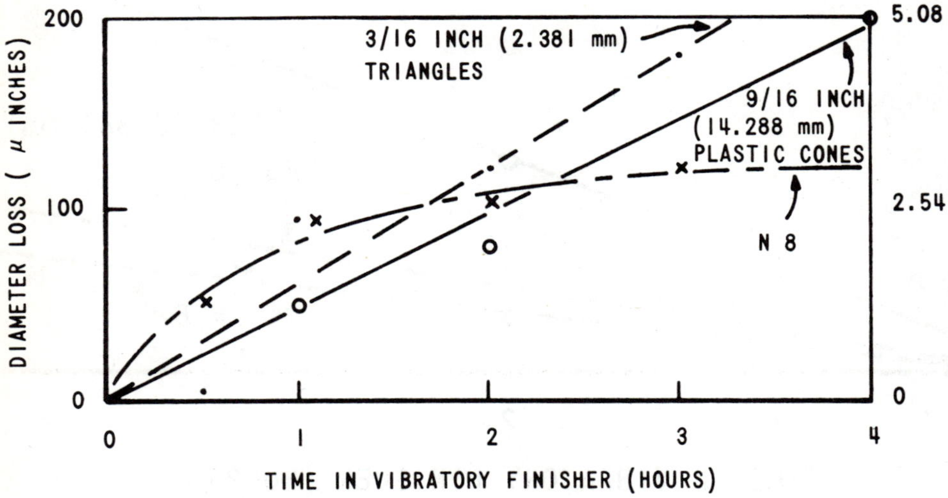

FIGURE 8-10: EFFECT OF MEDIA AND TIME ON CHANGE IN DIAMETER FOR 303SE STAINLESS STEEL (R_C20)

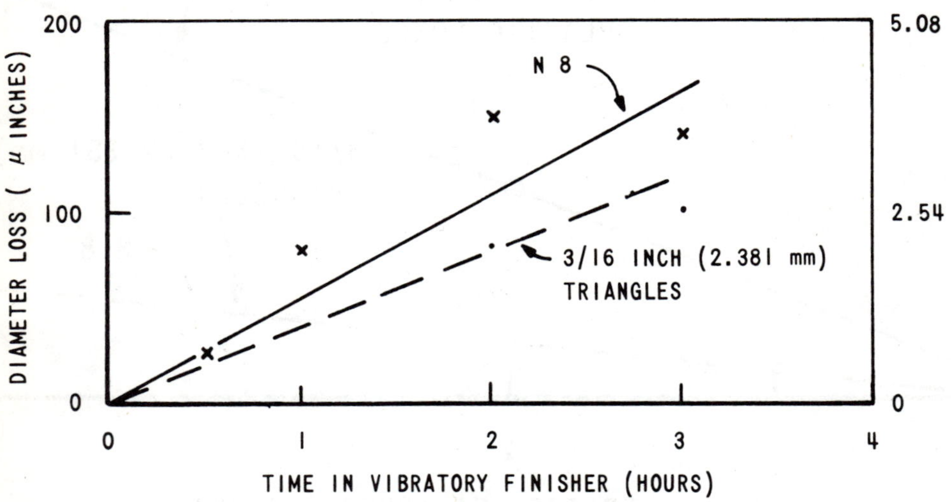

FIGURE 8-11: EFFECT OF MEDIA AND TIME ON CHANGE IN DIAMETER FOR 6061-T6 ALUMINUM (R_B61)

because of process variability, measurement error, or other sources. While the stock losses are extremely small, it is obvious that on ultraprecision parts with tolerances of $\pm$ 0.0001 inch (2.54 µm) one would probably not want to use this process to remove large burrs. In addition, using weight changes to predict dimensional changes may result in noticeable errors.

Burr Removal Time as a Function of Burr Size

Simulated burrs were used to determine how burr size affects burr removal time. Steel shim stock of various thicknesses were spot welded to 0.5 inch (12.7 mm) cubes of 303Se stainless steel. These simulated burrs projected above the top surface by varying amounts (Figure 8-14). After measuring burr height, the parts were vibratory finished in 3/16 triangles and 1A-1 abrasive compound. Changes in burr height were then recorded each hour for four hours.

As seen in Figure 8-15, the 0.001 inch (25.4 µm) thick simulated burrs wore much quicker than the 0.005 inch (127.0 µm) thick burrs. A regression analysis of the data indicates that the reduction in burr height (dH) can be expressed by

$$dH = (41.97 - 23.77b_t + 15.64t - 6.15b_t t + 4.34b_t^2 + 2.82t^2) \times 10^{-4} \qquad (3)$$

where

b_t = burr thickness in 0.001 inch units,
t = time in vibratory finisher (hours), and
dH = change in burr height in inches.

Differentiating with respect to time provides an equation for the rate at which burr height decreases

$$\frac{dH}{dt} = (15.64 - 6.15b_t + 5.64t) \times 10^{-4} \qquad (4)$$

Figure 8-16 illustrates how a burr initially 0.005 inch high is affected by burr thickness. Figure 8-17 is an extrapolation of Figures 8-2 and 8-16 to illustrate how the final edge radius would be affected by burr thickness. In practice, the transition from a small burr to a radius would probably be a sharp discontinuity since the radius actually begins forming on the burr (Figure 8-18).

In viewing Figure 8-16, it is clear that burrs 0.005 inch (127.0 µm) high and thinner than 0.001 inch (25.4 µm) are worn away very quickly. Burrs which are 0.003 inch (76.2 µm) thick are not entirely worn away in four hours when the 3/16 inch (4.75 mm) triangles are used. Little of the 0.005 inch thick burrs was removed after four hours.

As indicated in Figures 8-2 and 8-17, when no burr is present, a 0.004 inch (101.6 µm) radius is produced after four hours. When a 0.001

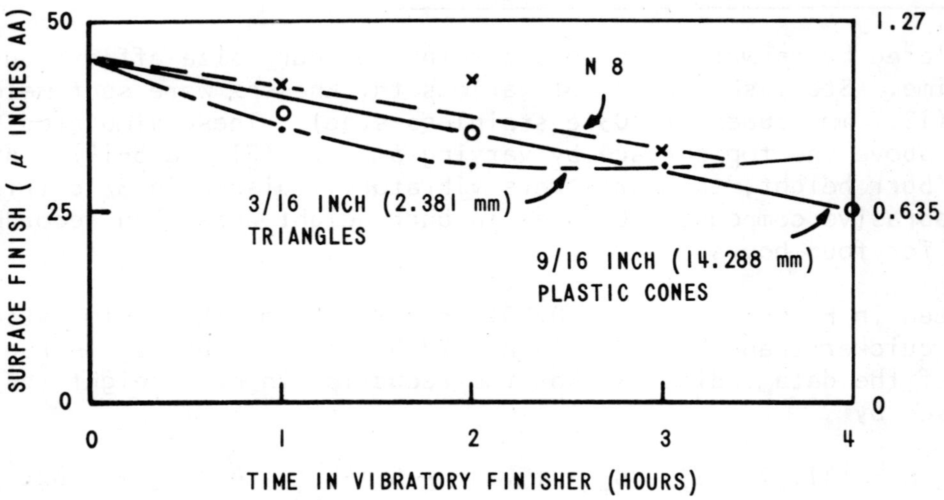

FIGURE 8-12: EFFECT OF TIME AND MEDIA ON SURFACE FINISH FOR 303SE STAINLESS STEEL (R_C20)

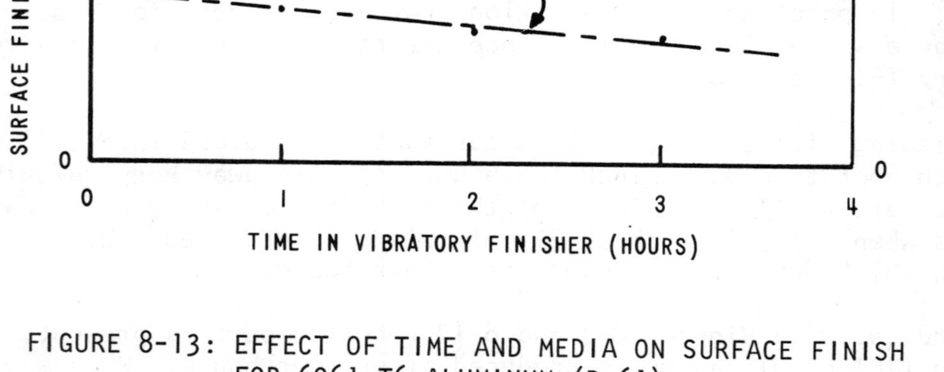

FIGURE 8-13: EFFECT OF TIME AND MEDIA ON SURFACE FINISH FOR 6061-T6 ALUMINUM (R_B61)

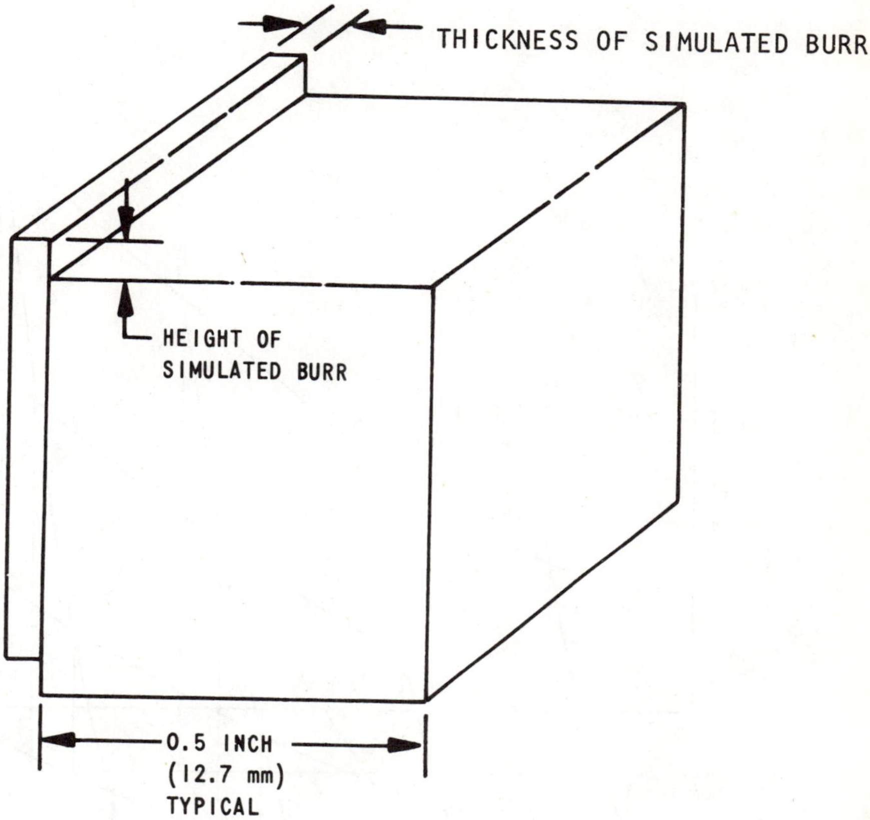

FIGURE 8-14: GEOMETRY OF SIMULATED BURR SAMPLE

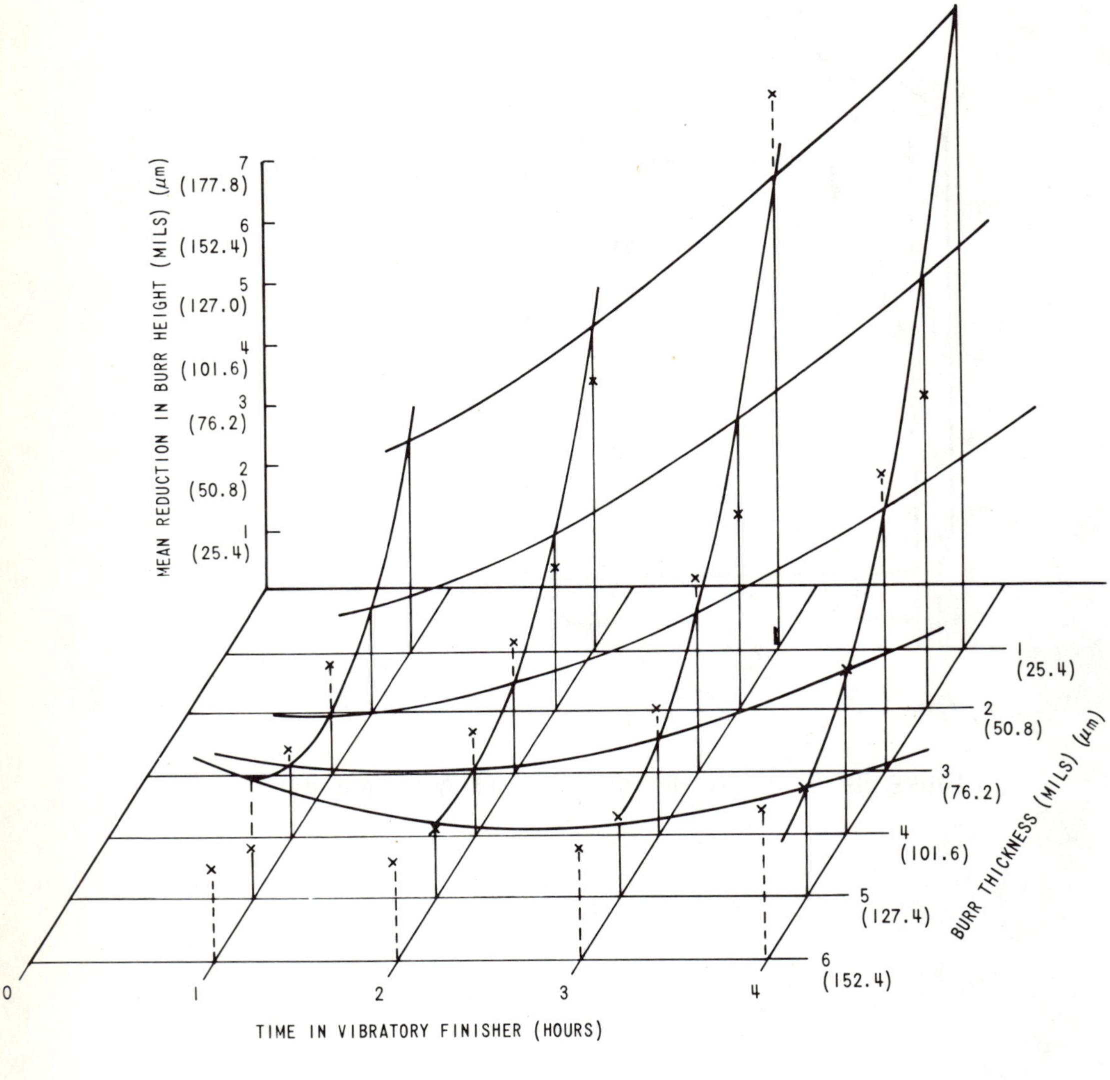

FIGURE 8-15: EFFECT OF BURR THICKNESS AND TIME ON BURR HEIGHT REDUCTION FOR SIMULATED BURRS

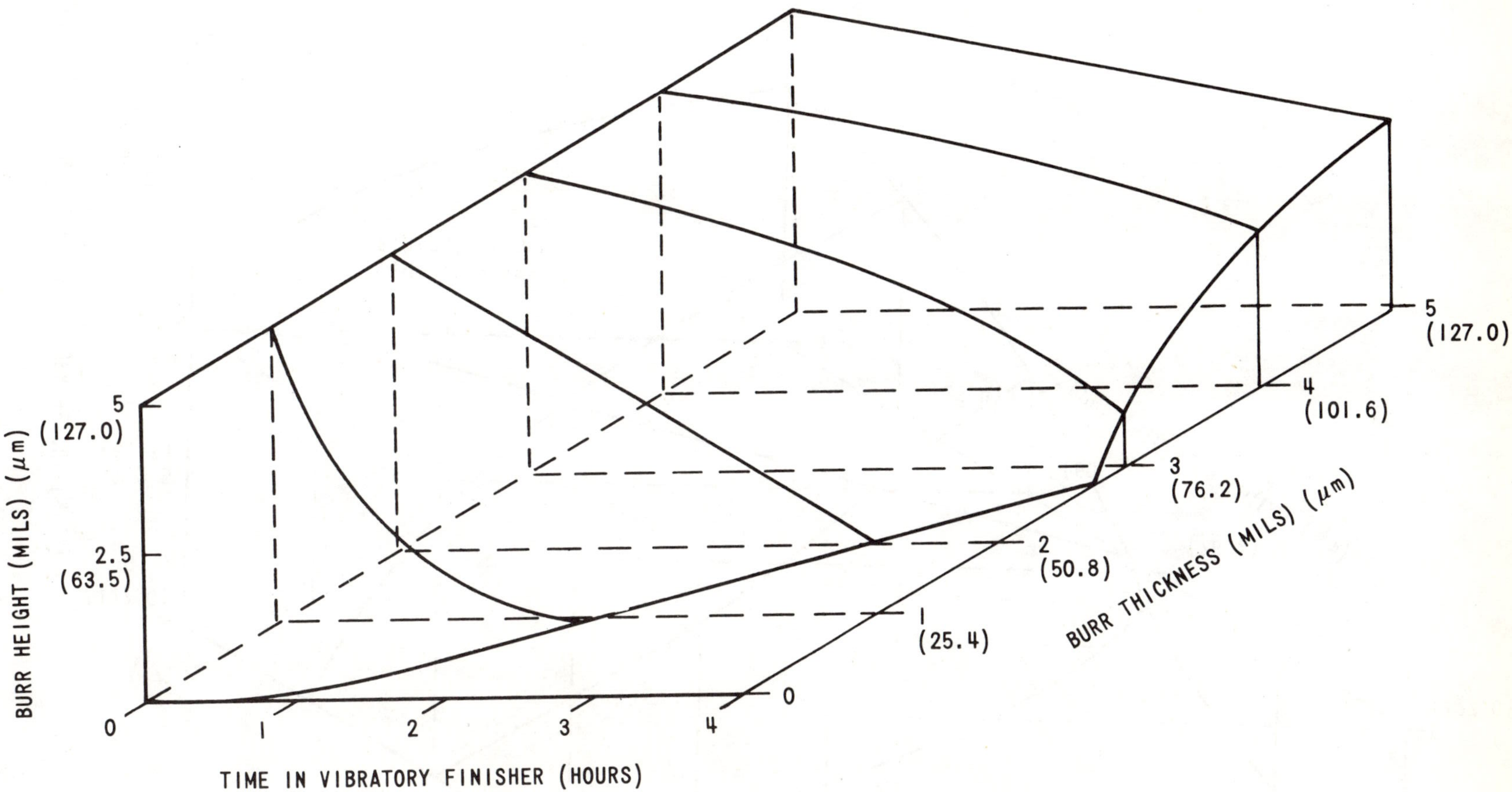

FIGURE 8-16: EFFECT OF BURR THICKNESS AND TIME ON BURR HEIGHT FOR SIMULATED BURRS

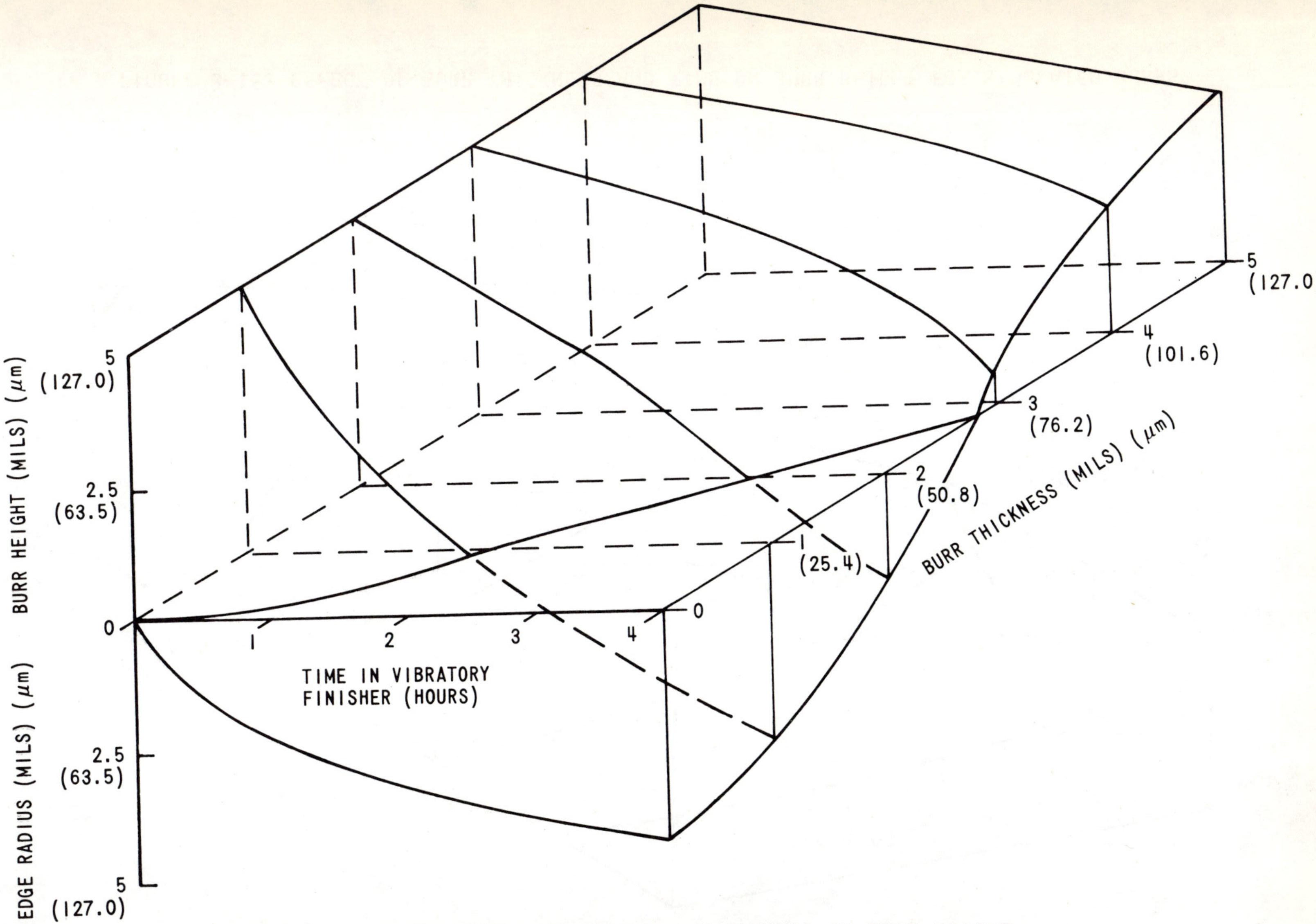

FIGURE 8-17: EFFECT OF TIME AND BURR THICKNESS ON BURR HEIGHT AND EDGE RADIUS FOR SIMULATED BURRS

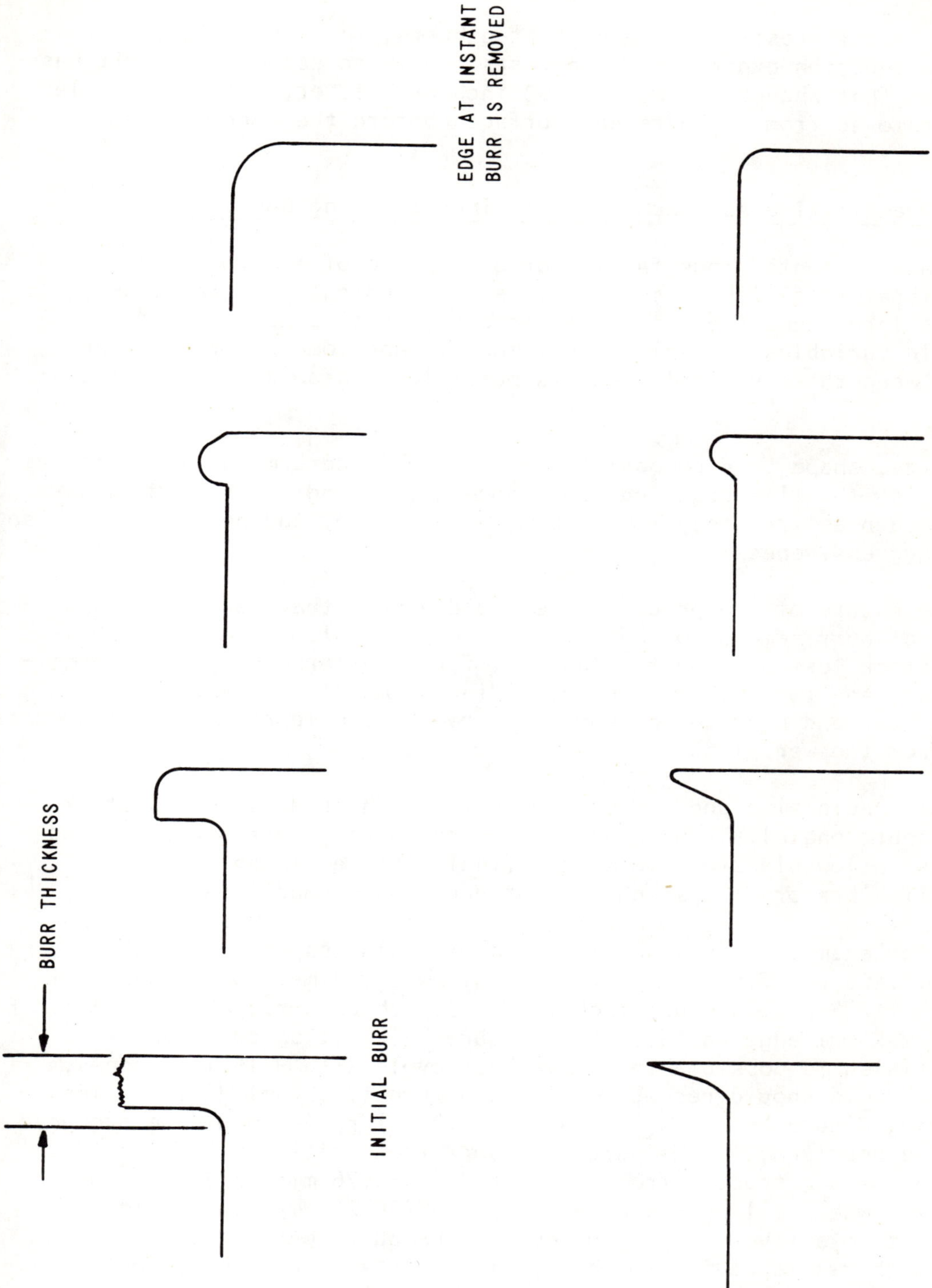

FIGURE 8-18: RADIUS ON BURR AND ON EDGE

inch thick burr, 0.005 inch high is present the burr is gone after 1.8 hours. After four hours, a 0.003 inch radius exists on the edges. After four hours, a burr initially 0.0027 inch (63.6 μm) thick and 0.005 inch high would be removed, but no measurable radius would occur.

Although this test utilized simulated burrs, the results are consistent with production experience. In tests involving actual burrs, it has been noticed that when burrs were 0.003 inch or thicker, some measurable stock was removed from the external surfaces before the burr was removed.

A General Quantitative Approach for Defining Deburring Capabilities

The data presented thus far is for only three of an almost unlimited range of vibratory finishing combinations. As indicated in Table 8-2, there are a large number of variables in addition to a near limitless range within variables. Despite the immensity and complexity of interactions between these variables, it is possible to draw some general conclusions.

The size, shape, and composition of the media determine its agressiveness (that is, the time required to produce given conditions). Machine parameters such as frequency, amplitude, hopper size, and hopper shape also influence aggressiveness.

By the nature of the process, one would assume that reaching a specified level of burr removal or edge radiusing would also result in a given amount of stock loss on a part. That is, for an external edge in a specific material, any two combinations of media and machine parameters which remove the burr and produce the same edge radius should remove equal amounts of stock from the workpiece.

If only media size and composition affect edge radiusing and stock loss disproportionately*, then one only needs to know burr size, desired edge radius, allowable stock loss, and final surface finish limits to determine if vibratory finishing will produce these conditions.

Take for example a rectangular piece of stainless steel having a 16 microinch (0.406 μm) finish requirement, allowable dimensional change of 0.0005 inch (12.7 μm), a 0.005 inch (127.0 μm) thick burr, and a 0.010 inch (254.0 μm) maximum edge radius. From Figure 8-12, it appears that a 16 microinch finish is possible in a five hour cycle with 9/16 inch cones. A five hour cycle should result in a 0.0003 inch (7.62 μm) diameter loss. By projecting Figure 8-6, a 0.008 inch (203.2 μm) radius would be produced after a five hour run, if the part were burr free initially. Since Figure 8-16 was constructed from data for 3/16 (4.76 mm) triangles, one can only estimate what will happen with 9/16 inch (14.29 mm) cones. It is possible that in a five to ten hour cycle, the burr would be entirely removed and that some radius (probably smaller than 0.010 inch) would be

*Knowing the final edge radius is not sufficient to predict size changes. As evident in the previous figures, media size greatly affected radii but only slightly influenced stock loss.

Compound	Media	Machine	Workpiece	Loading
Composition	Size	Frequency	Material	Amount of Media
Grit Size	Shape	Amplitude	Burr Thickness	Amount of Abrasive
Grit Material	Material	Basic Design*	Burr Length	Amount of Water
Emulsifying Agent	Coarseness of Grit	Shape	Burr Toughness	Volume of Parts
		Capacity	Geometry	Weight of Parts
	Composition		Burr Location	
	Weight		Size	
			Weight	

*Some machines have "dead" zones in which little deburring action occurs.

TABLE 8-2: VARIABLES SIGNIFICANT IN VIBRATORY DEBURRING

produced. Therefore, vibratory finishing is a viable candidate for this part.

By evaluating three or four examples like this, one will be able to construct tables such as that shown in Table 8-3. Since surface finish is more affected by media used than by material removed from the workpiece, tables like Table 8-4 have to be constructed for various levels of surface finishes (that is for 8, 16, 32, and 64 microinch (0.2, 0.4, 0.8, and 1.6 µm)). Because material hardness affects all measured variables, tables have to be constructed for at least two levels of hardness. Aluminum and steel represent two logical choices.

In constructing and using such tables, one needs to consider several factors. The curves shown in Figures 8-2 through 8-17 represent average trends. Half of the parts will have slightly smaller values and half will have larger values than shown. If tables are prepared from such curves, it should be noted that the tables are for average effects. On precision parts it may be necessary to construct tables in which 95 percent of the results will be smaller than the value shown.

Since deburring of holes and other cutouts is directly influenced by media shape and size as well as aggressiveness, tables constructed for external edges will not directly apply to these features. One also must assume that larger media will not be used on miniscule parts because little action will occur in this case.

From Figures 8-2, 8-3, 8-4, 8-9, and 8-10, it is apparent that large media produced large radii but was not significantly more aggressive in removing stock from the workpiece. By the nature of the process, it is logical to assume that the machine parameters should affect both stock removal and edge radiusing proportionately, whereas media size affects one more than the other. If this assumption is true, then one can ignore machine parameters in preparing tables of physical process capabilities. Thus, the machine parameters only indicate the length of time required to reach the desired capabilities.

To prepare process capability tables for cases where minimum edge radiusing and stock loss are critical, one will find it sufficient to take table data from three or four of the smaller media sizes. In the aerospace and die casting industries, large radii and/or stock losses are desirable; therefore data for the large media sizes would be beneficial. These two extremes then would define the quantitative capabilities of vibratory deburring.

The question frequently arises, "Is it worth all this effort to quantitize capabilities?" The answer, of course, depends upon who is using the data. The novice in vibratory deburring needs material like this if he is to be effective immediately. The departmental manager needs easy-to-read facts to make his decisions on equipment needs and general processing capabilities. The engineer responsible for high precision, job shop runs needs the data. Being wrong means unacceptable delays and scrap charges. Even the individuals who do not fall into these categories would use the information if it were available because there is always that one part for

TABLE 8-3: ESTIMATED CAPABILITIES OF VIBRATORY FINISHING FOR STEEL ALLOYS WITH SURFACE FINISH REQUIREMENTS OF 32 MICROINCH (0.813 μm) OR BETTER AND PRECISION LOW LOSS CONDITIONS

Burr Size Thickness x Length (Inch x Inch)*	Allowable Thickness or Diametral Loss** (inch) 0.0001	0.0005	0.0010	0.0050
Maximum Edge Radius (Inch) (μm)				
0.0001 x any length				
0.002 (50.8)	x	x	x	x
0.005 (127.0)	x	x	x	x
0.010 (254.0)	x	x	x	x
0.015 (381.0)	x	x	x	x
0.0005 x 0.0005				
0.002 (50.8)	x	x	x	x
0.005 (127.0)	x	x	x	x
0.010 (254.0)	x	x	x	x
0.015 (381.0)	x	x	x	x
0.003 x 0.003				
0.002 (50.8)	--	--	--	--
0.005 (127.0)	--	x	x	x
0.010 (254.0)	--	x	x	x
0.015 (381.0)	--	x	x	x
0.005 x 0.005				
0.002 (50.8)	--	--	--	--
0.005 (127.0)	--	--	x	x
0.010 (254.0)	--	--	x	x
0.015 (381.0)	--	--	x	x

*0.001 inch = 25.4 μm

**x indicates process can maintain the indicated size loss and edge radius while removing burrs of the size shown. A dash (--) indicates the process is not capable of maintaining both of these features while removing the indicated burr size.

which they need to know immediately whether a process will work. Unfortunately, this information has not been available to date.

Once tables like Table 8-3 are available, the next need will be for graphs like those shown in Figures 8-2 and 8-17 or for their corresponding formulas. Once it is known that a process can maintain the desired capabilities, the next question is, "What parameters are required to produce these conditions?" Fortunately whoever prepares the tables will have had to construct curves such as those indicated. Thus, if the tables are available the associated graphs will also be available.

Because of the large number of variables involved, vibratory finishing is a complex process. Tables alone will not completely describe a process yet they are needed to provide insight. A large amount of additional research is required to make vibratory finishing a fully predictable process. Progress has been slow, but it is accelerating and it must continue to do so if deburring productivity is to increase.

References

[1] M. Matsunaga and Y. Haguida, "Vibratory Finishing--Fundamental Research," Metal Finishing, Vol. 63, No. 9, September, 1965, pp. 52-57; No. 10, October, 1965, pp. 88-94, 97.

[2] G. V. Robbins, Radius Generation in Vibratory Deburring, Bendix Corporation (Kansas City, Missouri), September, 1966 (Available from NTIS under accession Number N67-17437).

[3] L. K. Gillespie, Vibratory Deburring, Bendix Corporation (Kansas City, Missouri), Report BDX-613-735, April, 1973 (Available from NTIS).

[4] John B. Kittredge, "Vibratory Deburring--The Dynamic Science," SME Paper MR75-487, 1975.

Appendix

Factors Common to All Tests

A 0.33 cubic foot (0.00944 m^3) vibraslide Model 33 Vibratub was used in these tests. This machine has an amplitude of vibration of 0.25 inch (6.35 mm) and a variable frequency. The frequency was not monitored, but it was somewhere near the middle of its 12 to 29 cps range. In each case the vibrating tub and media were rinsed after parts were removed from the machine for measurement. New compound was added each time the test was resumed. Carborundum's* 1A-1 abrasive compound was used in all tests. One-half pound (227 g) was inserted for each start or restart. Workpiece materials included 303Se stainless steel (R_C 20) and 6061-T6 aluminum (R_B 61).

*Carborundum Corporation, Niagra Falls, New York

TABLE 8-4: EXAMPLE INDEX TO TABLES ON VIBRATORY DEBURRING CAPABILITIES (SEE INDICATED TABLE NUMBER TO DETERMINE PROCESS CAPABILITY)

	Material							
	Steel				Aluminum			
	Surface Finish (Microinch)				Surface Finish (Microinch)			
	8	16	32	64	8	16	32	64
Case								
Precision Low Loss	1	2	3	4	5	6	7	8
Large Radii Large Size Change	9	10	11	12	13	14	15	16

TABLE 8A-1: MEASURED MEAN CHANGE IN BURR HEIGHT

Mean Reduction in Burr Height (Inch)

	Time in Vibratory Finisher (Hours)				
Burr Thickness (Inch) (μm)	X_{1}*	1	2	3	4
0.001 (25.4)	1	0.0035	0.0043	0.0089	0.0121
0.002 (50.8)	2	0.0020	0.0022	0.0031	0.0050
0.003 (76.2)	3	0.0018	0.0021	0.0031	0.0047
0.004 (101.6)	4	0.0014	0.0018	0.0019	0.0025
0.005 (127.0)	5	0.0008	0.0011	0.0013	0.0018
0.006 (152.4)**	6	0.0015	0.0015	0.0017	0.0022

*Value substituted into Equations 3 and 4

**Omitted from regression analysis

Almco's** Supercut 9/16 inch (14.35 mm) plastic cones, Carborundum's numbers 7 and 8 random shaped, fused aluminum oxide nuggets, and Fortune Industry's*** 3/16 by 3/16 by 1/8 (4.76 by 4.76 by 3.18 mm) AX-90 ceramic bonded preformed triangles were used in the reported studies.

Simulated Burr Study

The burrs in this study were of controlled thickness, made by spot welding shim stock to steel blocks. The material was in increments of 0.001 inch thickness, from 0.001 to 0.006 inch for steel shim stock. There were six specimens for each thickness with variable initial height or extension above the surface of the steel blocks. The heights at the mid-point were gaged and recorded at the start (t = 0) and at one hour intervals up to four hours. The data was converted to change in height by subtracting height at the end of each time period from the initial height. This gave the accumulated loss in "t" hours. The means (shown in Table 8A-1) for the eight specimens were used for each thickness after each time period.

It will be noted the mean material removed at each time period for the 0.006 inch burr thickness was more than for the 0.005 inch thickness which is contrary to prediction and also showed none or little material removal after the first hour. The data for the 0.006 inch thickness was omitted from the regression analysis as it obviously was not consistent.

**Almco Division of Queen Seeley Thermos, Albert Lea, Minnesota
***Fortune Industries, Chelsea, Michigan

CHAPTER 9

SPINDLE FINISHING

One of the often overlooked deburring processes is spindle finishing. While there are many "offshoots" or refinements of this type of equipment, this chapter will cover only the basic spindle finishing concepts.

Processing Principle

The spindle finishing concept, sometimes referred to as a "slurry machine" or "flowing abrasive" machine, is not new to the industry. It involves principles used in varying finishing applications for many years but considered a specialty process. Spindle finishing is based on the principle of an abrasive flowing mass resulting with the effect of a "form fitting grinding wheel"--allowing controlled, precision, deburring and surface improvements.

Equipment generally includes a circular rotating tub holding loose abrasive media and a rotary spindle to which the part is held or fixtured. (Figure 9-1)

The principle of this process is to deburr or finish the workpiece mounted on the spindle by immersing it into a rapidly moving abrasive slurry. The velocity of the abrasive is generated by rotating the tub at a speed suitable for the specific process requirement. The abrasive particles swiftly flow over rough edges and through holes in the workpiece removing sharp edges and improving surface finishes. (Figure 9-2)

The holding fixtures are designed to hold the workpiece in such a manner that the areas to be treated are exposed. The tub is rotated in one direction, while the spindle or spindles normally turn in the opposite direction. As the tub is rotated the media is forced to the outside by centrifugal force. The layer of abrasive formed as a result of this action is dense and firm and therefore provides agressive action on the workpieces immersed in it. Fixtures usually consist of chucking devices which are spring loaded for clamping and released by an air cylinder--making the machine operation extremely simple.

Because parts are fixtured, there is no danger of impingement, nicks or scratches from parts colliding with one another. Since fixturing is needed these machines are most effective for mass production operations on a single or similar range of parts.

Generally the spindle is adjusted to place the workpiece as close as possible to both the outer periphery and the bottom of the rotating tub for maximum effect. There are two reasons for this--first, the abrasive velocity is proportional to the radius from the tub's center, and second,

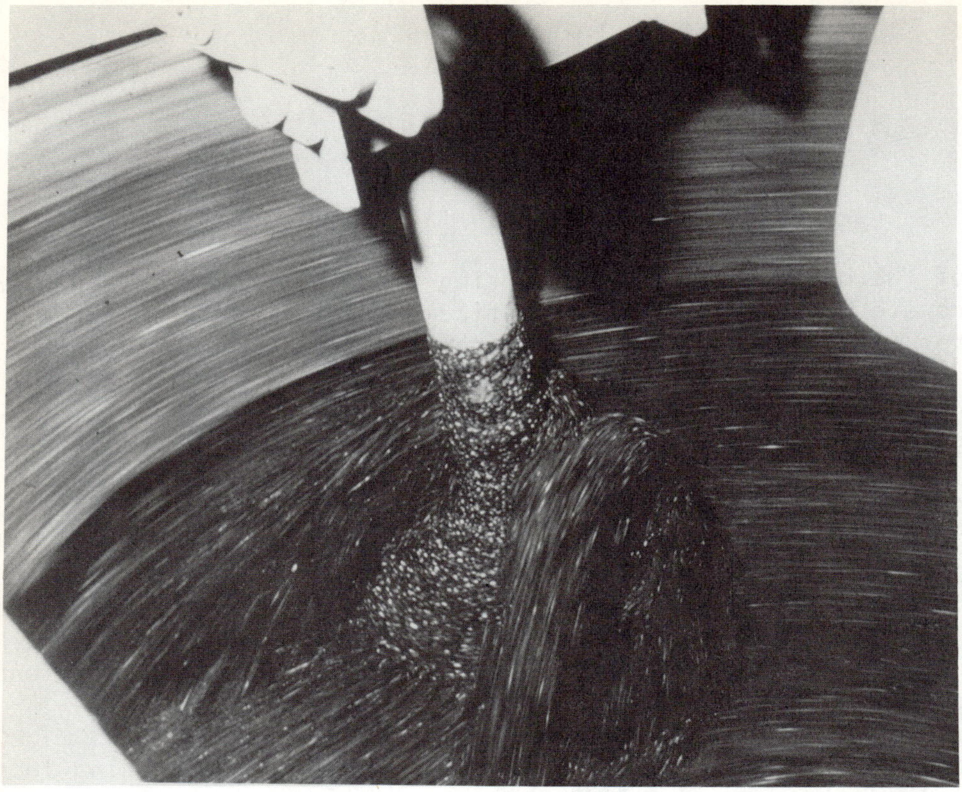

FIGURE 9-1: TWO SPINDLE FINISHING CONCEPT

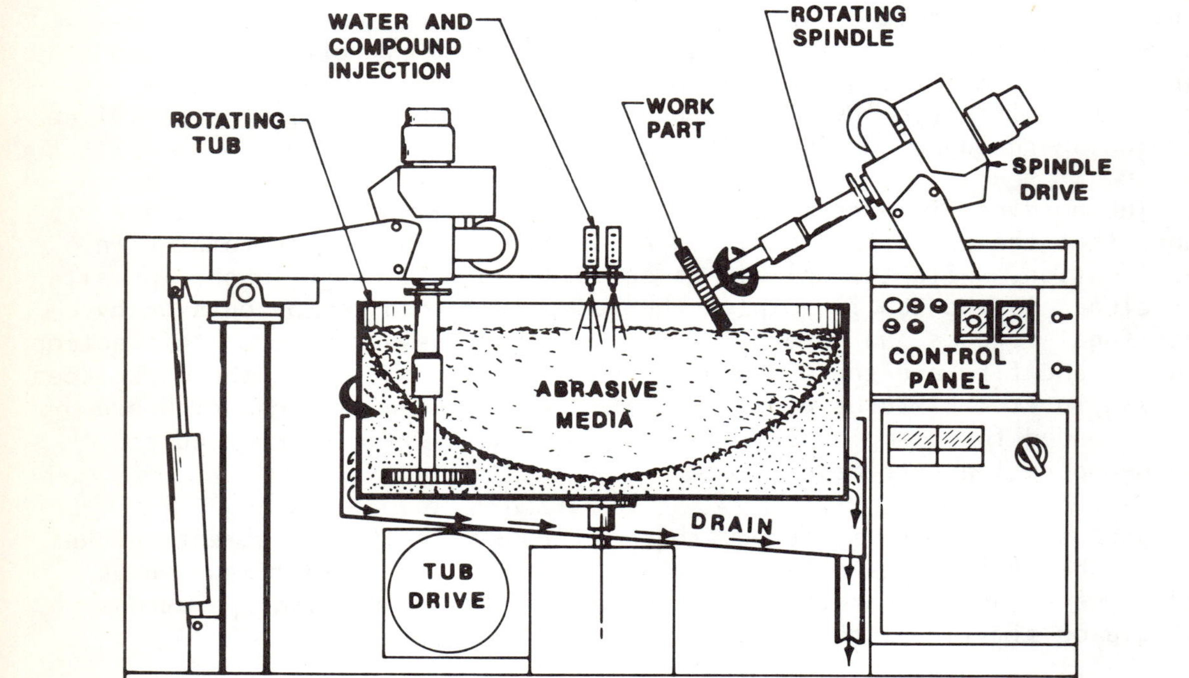

FIGURE 9-2: SPINDLE FINISHING IN MOTION

there is advantageous pressure effect created by compressing the moving abrasive between the workpiece and the tub bottom and side wall.

Processing cycles for spindle finishing are comparatively short. In many applications, a cycle of 5 to 20 seconds is sufficient although cycles up to five minutes and longer are not uncommon--determined by the complexity of the parts, which can range from engine valves to complex gears, while surface finishes range normally between 10 and 5 RMS, and with edge radii consistently produced down to 1 thousandth of an inch.

Because of repetitive loading and unloading requirements due to the relatively short time cycles full time operators are required, however, in some specific and more refined operations, mechanical loading arms and industrial robots can be incorporated to reduce operator time.

Equipment and Operation

Spindle finishing equipment is available in a variety of configurations and with design features unique towards specialized process requirements. However, the majority of equipment manufactured to take advantage of the flowing abrasive mass principle consists basically of a structural steel frame and housing which support the rotating processing tub and fixture holding spindles.

Depending on the requirement, spindle machines may commonly be designed with from one to twelve spindles, with the single or two spindle machine at this time being the most popular because of greater versatility.

The rotating tub is motor driven and usually by a variable speed drive to allow flexibility of operation.

The spindles are also motor driven either by constant or variable speed drives to rotate the part in the abrasive mass to equally expose all surfaces for finishing. Spindles are generally pivoted or lowered into the finishing mass by means of air cylinders for processing and raised to a convenient position for loading and unloading. The majority of spindle machines used for high production operations also include provisions for air operated fixtures for quick loading and unloading of parts.

Machines designed for wet processing incorporate a rotating tub having perforated sections on the bottom or side wall to allow water based chemical compounds, that are injected automatically and continually, to flush away any metal chips, abrasive "fines" or sludge for maintenance of efficient cutting action and continuously clean parts and a contamination-free mass. Discharged liquids flow into a collection ring positioned around the tub and further to a common drain fitting where "spent" solutions may be channeled to drains, settling tanks, or filtration/recirculation systems for reentry into the processing tub.

Accurate control of processing cycles is maintained by electric timers, process tub speed by direct reading tachometers showing surface feet per minute, spindle speed by tachometers showing revolutions per minute, and liquid flow meters accurately measuring the amount of water and/or liquid compounds introduced into the process.

Established operating procedures are effectively controlled with repeatable cycle accuracy through these control features.

Operator controls are conveniently located near the operator's station with proper emergency shut-off generally included for maximum safety. Electrical components are housed in an electrical housing usually mounted on the machine for convenience and reduced floor space.

To allow versatility in processing a relatively wide range of parts and configurations, most spindle equipment includes "adjustment features" in the spindle design including a provision to adjust the spindle up or down, to adjust the spindle towards or away from the tub wall, to pivot or angle the spindle head to allow proper positioning of part for maximum exposure to flowing mass.

At the end of the predetermined process cycle, the spindle is automatically raised out of the processing tub where finished parts can be manually or automatically sprayed with water or with a cleaning solution to flush away any loose abrasive particles and residue from the abrasive mass (Figure 9-3).

With the process completed, the operator can remove the finished parts which, in most production installations, is accomplished by pressing a foot switch which actuates the fixture collet to release the finished parts. Upon removal of finished parts, the next process cycle begins by loading unfinished parts on the fixture and repeating the process cycle.

Processing Materials

Processing media includes a wide variety of abrasive materials and chemical compounds. The most popular media is fine grit aluminum oxides because it can effectively accomplish common deburring and edge radius operations with controlled accuracy on precision metal parts. Subject to part configuration and degree of cut or finish requirements, larger size random shaped aluminum oxide or preshaped ceramic based media is used for more aggressive finishing action. Sometimes a large sized media is mixed with a smaller sized media which can allow aggressive action on the outer edges and surface of a part and still deburr or improve the surface finish of small, restricted areas and holes by allowing the smaller medis to reach areas not effectively reached by the larger media.

Where fine, low micro finishes are required, plastic based media with low abrasive content is used; and where high luster and polishing effects are desired, dry media such as maize or walnut shell compositions along with steel balls for "burnishing" is highly effective on parts such as aluminum die castings and related soft metals.

Along with a wide range of abrasive media utilized in spindle finishing, there is also a wide range of chemical compounds available to enhance

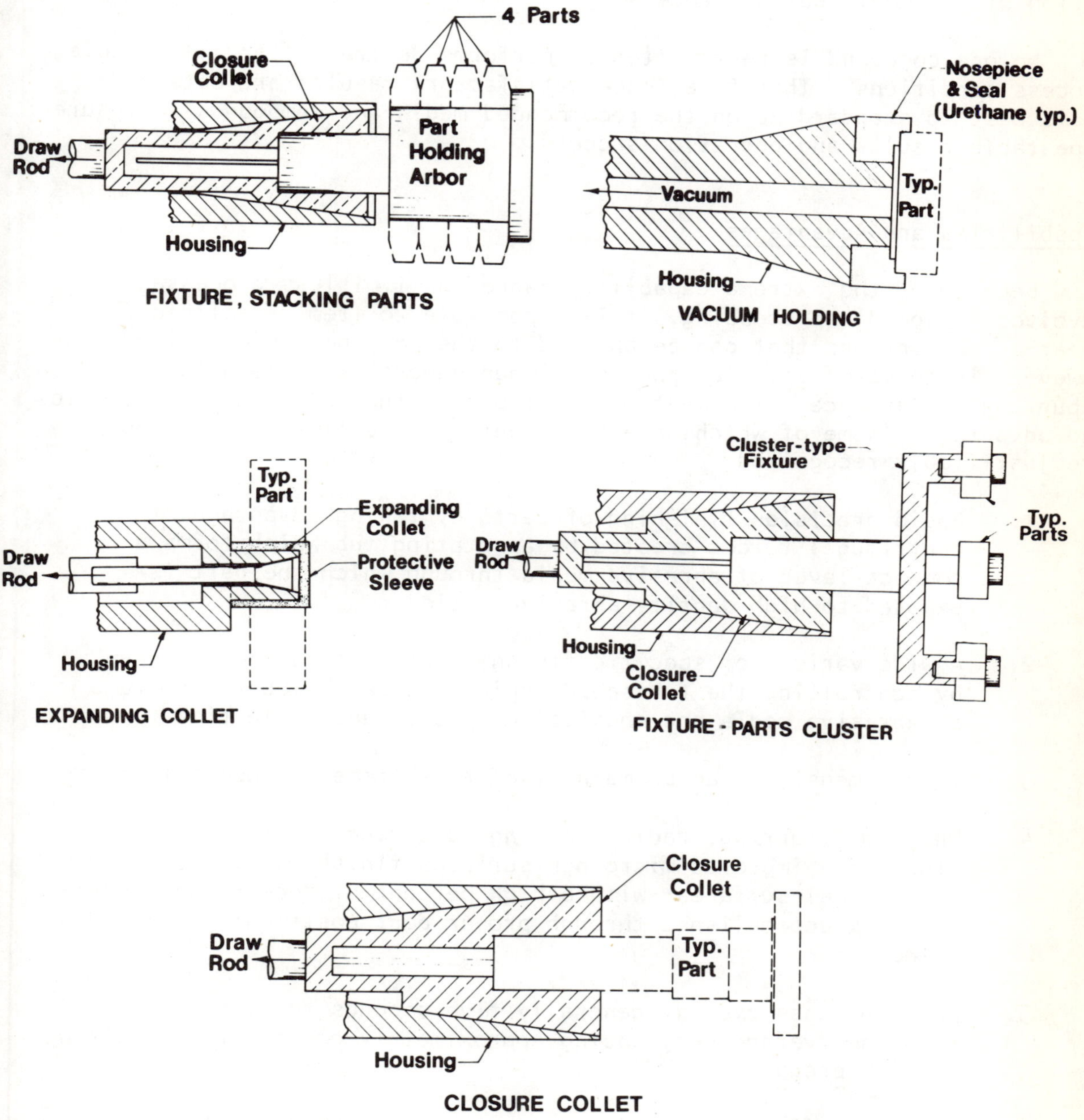

FIGURE 9-3: SPINDLE FINISHING FIXTURING TECHNIQUES

or complement the finishing process. Compounds can be in powder or liquid form, with liquids being more popular in wet spindle processing because they can automatically be injected through controlled injection systems.

Generally, specific compounds are required for each different application, however, there are standard compounds formulated to be compatible with a wide variety of processes to keep the parts and media clean, provide lubrication, provide protection against rust, and to aid in the abrasive action of the media during processing.

No one compound is best suited or performs desirably for all spindle process conditions. Therefore, once satisfactory results are obtained, it is best to standardize on the recommended media and compounds to insure repeatable results and operation success.

Capabilities and Advantages

Because of the extreme capability range and limitless processes involved in spindle processing, it is impossible to itemize defined or clear-cut advantages that can be applied to the many possible applications. However, in terms of spindle processing when generally applied to precision deburring and surface refinement of metal parts, the following capabilities and advantages (some of which have been mentioned earlier in this book) are justifiably recognized.

1. Rapid precision finishing of parts by taking advantage of centrifugal forces created by a rotating tub which forms a compact layer of abrasive media through which the part is exposed to high density abrasive action.

2. A wide variety of specific finishing results can be obtained by controlling the tub speed, spindle speed, amount and grade of abrasive media and chemical compound, and cycle time.

3. Part dimensions can be maintained to extremely close tolerances.

4. Uniform deburring, radius forming, blending of machine or grind lines, and improved microinch surfaces finish can be accomplished on external surfaces--with a high percent of recessed or internal work accomplished through proper part positioning and media selection.

5. Short process cycles, generally five seconds to five minutes, with the average cycle being approximately one minute--resulting in high production.

6. Parts are fixtured resulting in no damage due to part on part impingement.

7. Eliminates rejections due to discrepancies attributable to variations in manual performance.

8. Can effectively reduce direct and indirect labor costs.

9. Process repeatability--under controlled process conditions with operation established for tub speed, part position, media and compound, etc.

10. Production requirements or operational standards are easily maintained once operation is established.

11. Self-contained machines requiring a relatively small amount of floor space and can be used in production line sequence planning.

12. Installation is relatively inexpensive and does not require special facilities. Services include only normal plant water, air, electrical and drain facilities.

13. Low operational noise level, generally in the range of approximately 80-85 decibels (dBA).

14. Minimal media wear which in normal operations amounts to a media loss of approximately one to two percent of the tub's volume over an eight hour shift.

15. One operator can run two machines subject to number of spindles and time cycle.

16. Process applications include: rough deburring, internal or recessed burr removal, descaling, ball burnishing, dry polishing or scuff sanding, micro finish improvement and stress relieving.

17. Automatic, quick change, high production tooling and fixtures generally related to one of the five basic fixture concepts (Figure 9-3).

Limitations

In spite of the many favorable characteristics of spindle finishing, there are limitations which must be recognized. Again because of the wide variety of parts and processes, limitations cannot be totally defined, however, in the general concept of which this chapter refers to, the basic spindle finishing limitations include:

1. All parts must be fixtured or "held" for processing, opposed to "batch" processing in tumbling barrels or vibratory finishing equipment.

2. Spindle finishing operations, in the majority of cases, must have full time machine operator.

3. Not practical for many types of small sized parts, particularly when the finishing operations are departmentalized.

4. Part holding fixtures are immersed in the abrasive mass as well as the parts being finished and are subject to wear, however, in most cases fixture costs are nominal.

5. Not practical, in terms of standard equipment, for parts exceeding 48 inches in diameter or weighing over 100 pounds.

6. Subject to part configuration and degree of burr, has limitations for internal burr removal.

Summary

Although no single finishing concept can satisfy the wide variety of the metalworking industries' mechanical deburring or finishing needs, improved technology in the design of basic spindle finishing machines as well as vast improvements in fixturing methods, together with recent advancements in finishing media and compounds have been contributing factors in generating increased industry awareness of the favorable potential of spindle type finishing.

The following "before and after" photos show parts that are typically and ideally suited towards spindle processing and the basic processing information at which they were run.

References

[1] J. N. Brandt, "Spindle Finishing--Capabilities and Limitations," SME Paper MR75-832, 1975.

MACHINED IDLER
REQUIREMENT: DEBURR
PROCESS: OXIDE MEDIA MIXED 12A & 4A, 2 MINUTES - SPINDLE

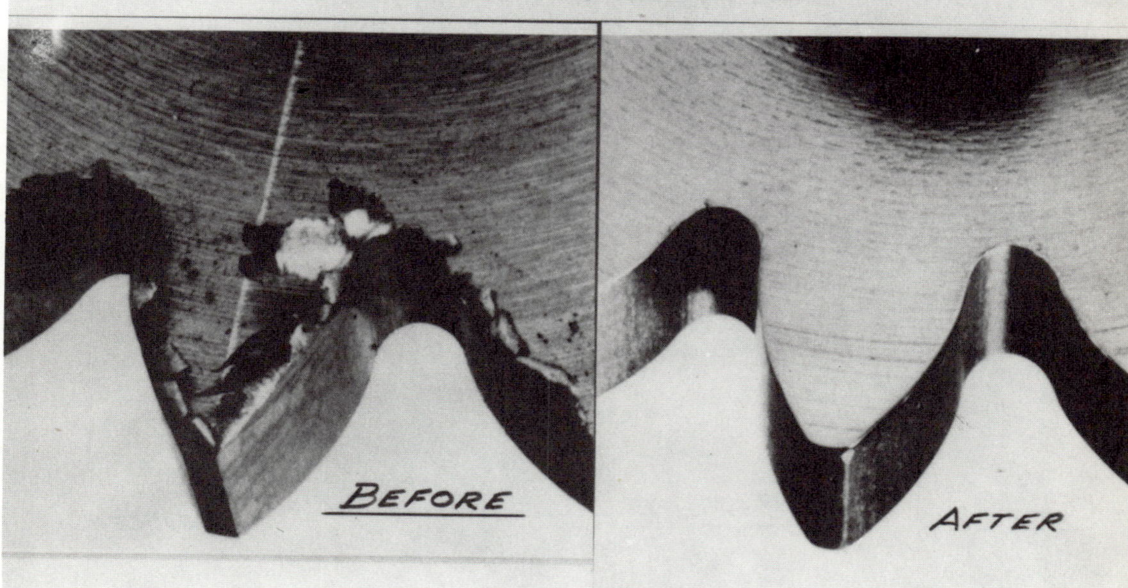

FIGURE 9-4: SPINDLE FINISHED IDLER

MACHINED GEAR:
REQUIREMENT: DEBURR AND IMPROVE SURFACE FINISH
PROCESS: SPINDLE, 12A OXIDE, 30 SECONDS.

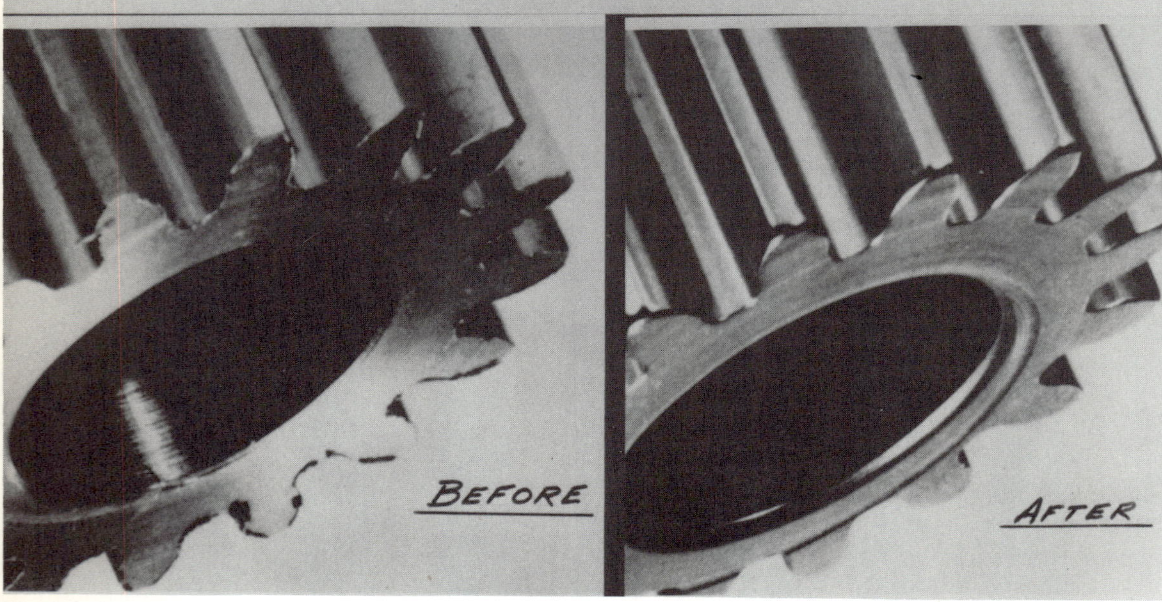

FIGURE 9-5: SPINDLE FINISHED GEAR

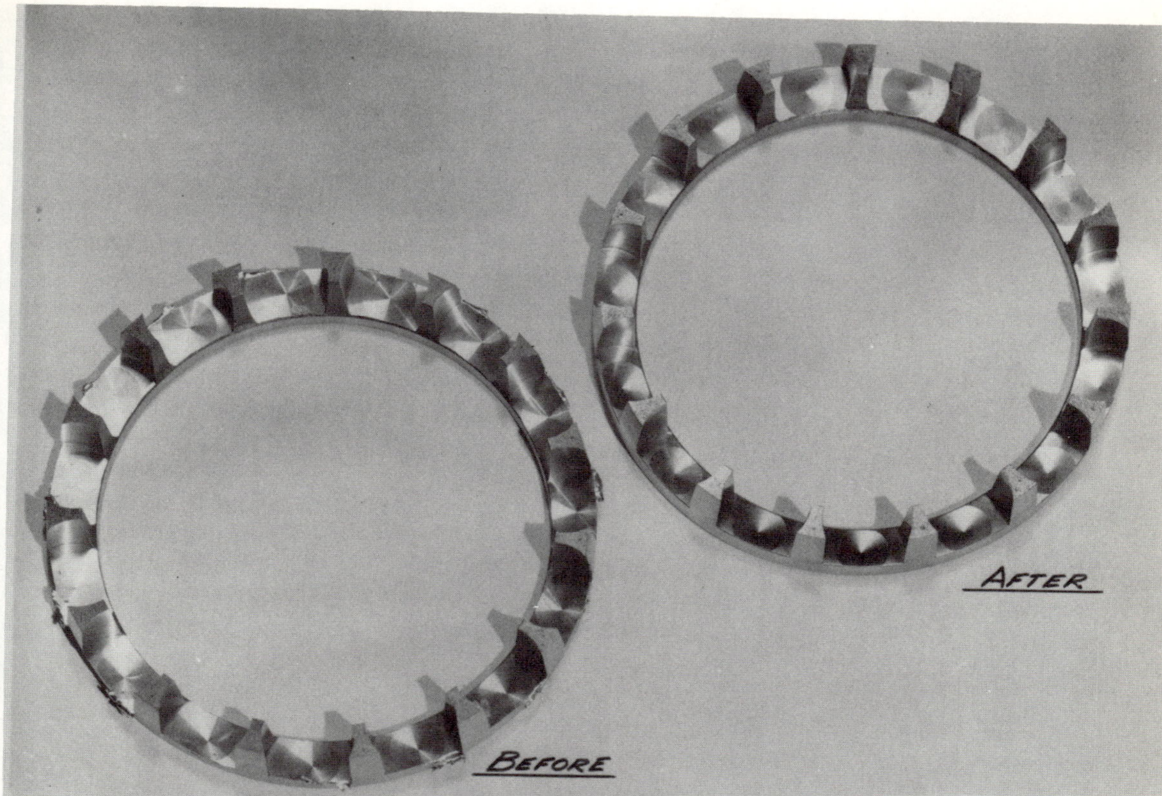

BRONZE BEARING CAGES
REQUIREMENT: DEBURR, RADIUS EDGES AND IMPROVE MICRO FINISH
PROCESS: SPINDLE, #12XM MEDIA, 2 MINUTES

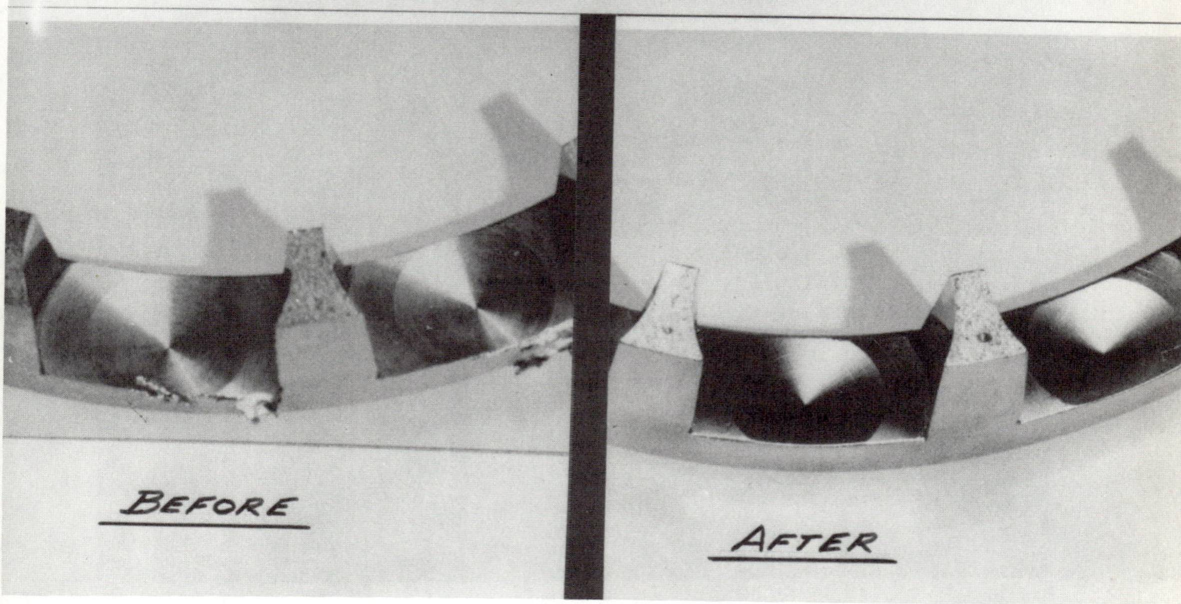

FIGURE 9-6: SPINDLE FINISHED BEARING CAGE

ALUMINUM GEAR-CAM
REQUIREMENT: DEBURR AND BLEND
PROCESS: SPINDLE, #12 XM, 40 SECONDS TIME CYCLE

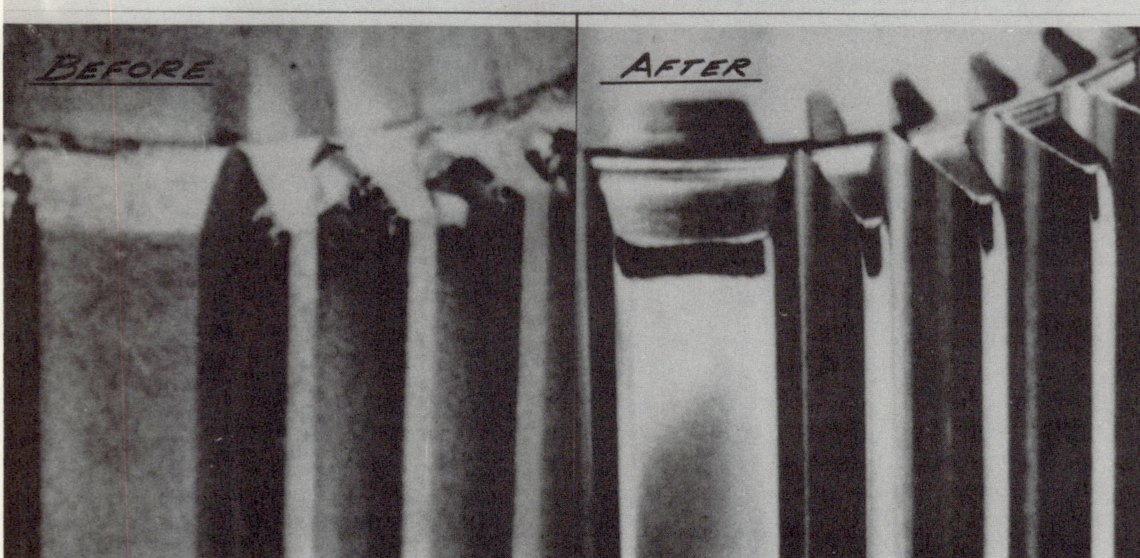

FIGURE 9-7: SPINDLE FINISHED CAM

MACHINED CAM
REQUIREMENT: DEBURR AND OBTAIN UNIFORM RADIUS
PROCESS: SPINDLE, 10 SECONDS, 12A OXIDE

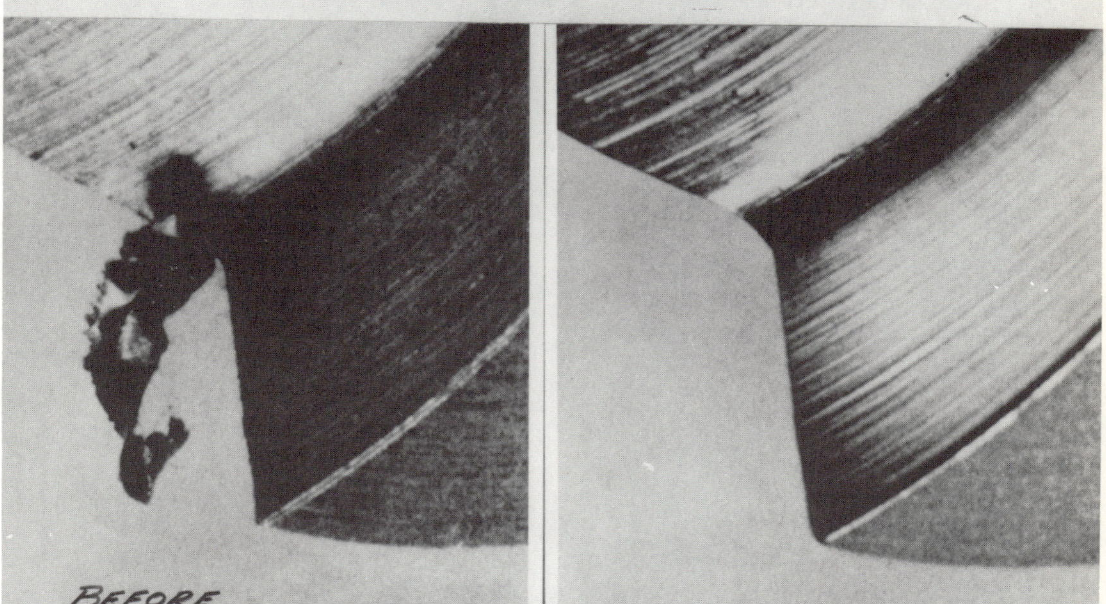

FIGURE 9-8: SPINDLE FINISHED CAM

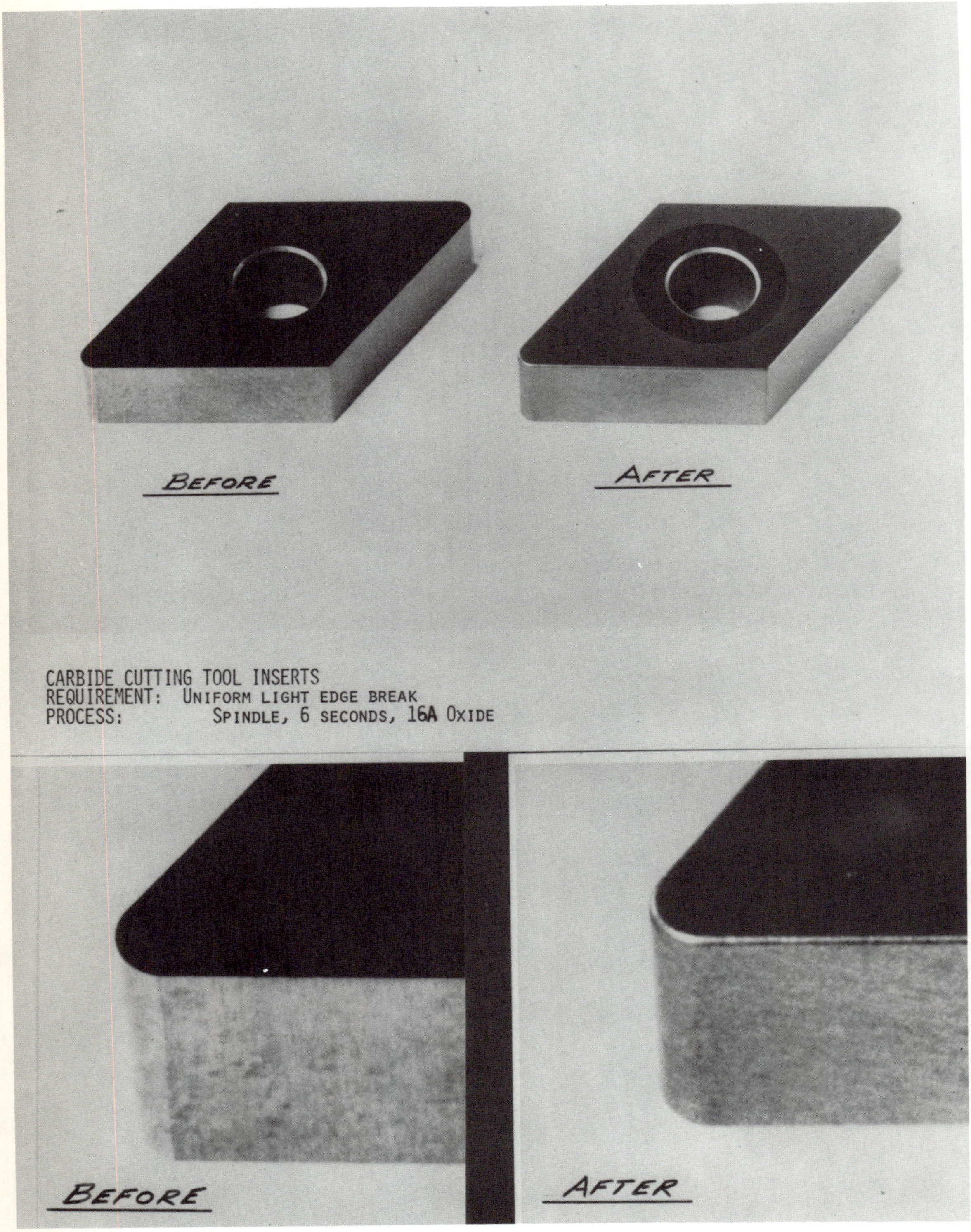

FIGURE 9-9: SPINDLE FINISHED CUTTING TOOL INSERTS

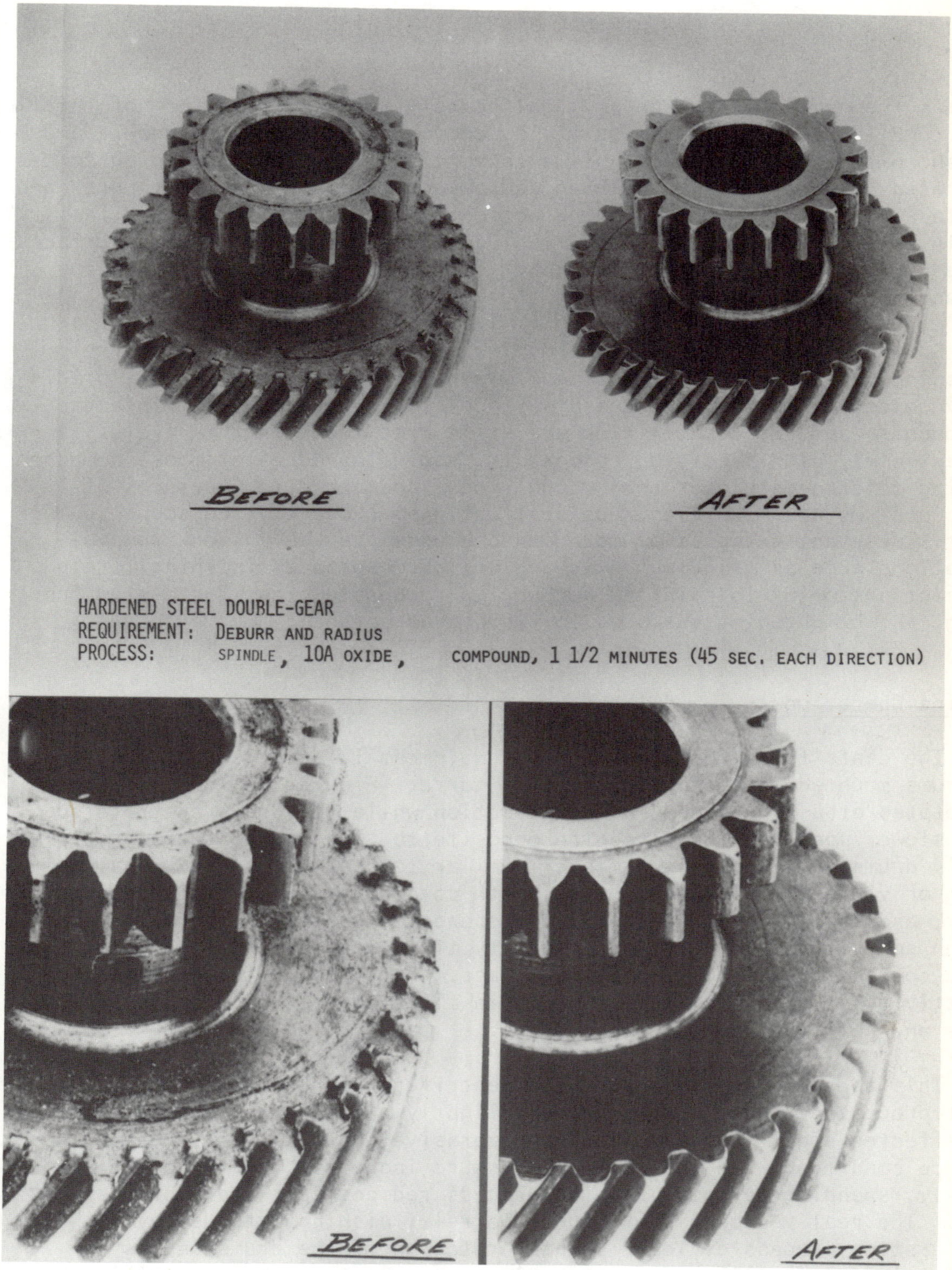

FIGURE 9-10: SPINDLE FINISHED GEAR

CHAPTER 10

CENTRIFUGAL BARREL FINISHING

As the name implies, centrifugal barrel finishing is a form of tumbling. Like tumbling, it uses abrasive media, compound, and water to deburr and finish components. However, unlike conventional tumbling, the high forces generated by centrifugal finishing equipment press media against parts and create a completely smooth sliding action. This results in a very fast, highly controllable deburring and radiusing operation--it enables surfaces to be uniformly refined and can impart high compressive stresses in the surfaces of parts.

Best known by the trade name of "Harperizing," centrifugal barrel finishing has been in use for some fifteen years. Harperizing was originally conceived for deburring and surface finishing precision miniature components, and its capabilities are still best known in this field. Deburring extremely high precision components, radiusing edges of those components, holding tolerances closer than 0.0001 inch, and achieving surfaces of 1 microinch rms are more glamorous applications. However, the scope is much wider. Relevant examples of more mundane uses are heavy stock removal operations such as grinding balls and rollers, surface finishing forgings and castings prior to painting, deburring rough stampings for easier handling, and brightening surfaces to improve appearance.

Machine Operation

The centrifugal barrel finishing equipment is comprised of a number of drums mounted on the periphery of a turret (Figure 10-1). The turret is rotated at a high speed in one direction while the drums are rotated at a slower speed in the direction opposite to that of the turret. Generally drums are loaded in a manner similar to that for conventional tumbling or vibratory finishing, i.e., with components to be finished, media, water, and some form of compound. In operation the turret rotation creates a high centrifugal force of up to thirty times gravitational weight. This force compacts the load into a tight mass. Rotation of the drums causes an activity of the load, parts and media slide against each other, removing burrs and refining component's surface.

This abrading action under high centrifugal force results in very short process cycles. A product very lightly rubbed by an abrasive is not affected significantly. The same abrasive, however, will change the surface condition when rubbed with pressure increased thirty-fold. Obviously, speeding up the finishing process reduces cost and improves work flow. The real value of centrifugal barrel finishing extends beyond just accelerating process cycles. Counter rotation of the drums causes a completely smooth slide of media against components with no possibility of one part falling or banging against another. Figures 10-2 and 10-3 are photographs of Harperizer drums while the machine is running and these show the smooth sliding action of the load. Because of this action, completely consistent and reproducible results will be achieved. High tolerances can be maintained in even the most fragile parts.

Capabilities

The main functions that Harperizing can perform in the finishing department are:

Deburring
Radiusing edges and corners
Surface finishing
Stock removal
Changing stress in the surface of the parts

For all metals, some plastics and ceramics, equipment is available to handle components weighing up to 1,000 pounds (4536 Kg) and up to five foot lengths. As with most deburring and finishing processes, components being Harperized will have their edges radiused as burrs are being removed. There will be some action on all accessible surfaces of the components. While a combination of functions occurs simultaneously, it is worth considering each of these functions separately.

Deburring

The most important rule concerning Harperizer deburring is: "If the component is presently hand deburred, consider centrifugal barrel finishing as an alternative."

The value of the process to remove accessible burrs is limited only by economic considerations and the tolerances to be maintained on components being processed. Some examples may best illustrate this:

1. Nickel squares used as anodes for electroplating are deburred for ease of handling and improved flowability. A Harperizer of the type shown in Figure 10-4 is used. Tolerances are of no importance and loss of metal is insignificant. Total process cycle is less than five minutes, giving production of 20,000 pounds (90,720 Kg) of finished parts per hour. While the improved appearance of the squares after Harperizing is of benefit, virtually any deburring process would achieve an adequate result. Choice of equipment was therefore dictated essentially by economic considerations. The extremely short cycle enables one machine to handle the total output, resulting in attractive equipment cost and minimum floor space. With automatic handling, one man runs the machine, keeping labor cost low. Deburring is achieved by a part-on-part operation (i.e., with no abrasive media), a low-priced tumbling compound being the only consumable material used.

2. Bendix, Kansas City Division, Harperizes a variety of miniature

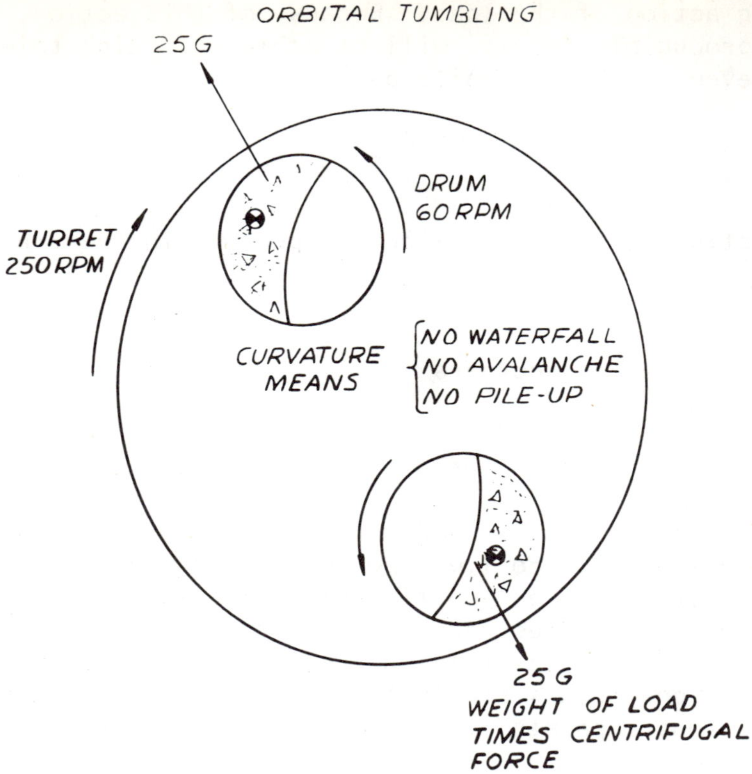

FIGURE 10-1: PRINCIPLE OF CENTRIFUGAL BARREL FINISHING

FIGURE 10-2: CENTRIFUGAL BARREL ACTION

FIGURE 10-3: CENTRIFUGAL BARREL ACTION

FIGURE 10-4: LARGE CENTRIFUGAL BARREL MACHINE

components such as those shown in Figures 10-5 through 10-8. Process costs are of secondary consideration compared with deburring capabilities. Having so many very expensive parts in small quantities means that consistent results in the first run are very important. Data has been developed for these parts relating burr removal to metal removal from the component proper for the centrifugal barrel process. The burrs on the 1/4 inch (6.35 mm) diameter aluminum spacers shown in Figure 10-5 were 0.0035 inch (88.9 μm) thick at their roots, 0.008 inch (203 μm) high. A 20 minute cycle removes these burrs, reducing the diameter of the spacers by 0.0005 inch (12.7 μm). The stainless steel parts shown in Figure 10-6 have to be machined oversize as the 15 minute Harperizer cycle removes 0.0003 inch (7.6 μm) stock from one dimension which has a tolerance of only 0.0004 inch (10.2 μm). In deburring the stainless steel gear, Figure 10-7, component thickness was reduced by up to 0.0003 inch but the tooth profile was undisturbed. In 20 minutes the stainless steel parts shown in Figure 10-8 had the journal diameter reduced from 0.025 inch (61 mm) by 0.00008 inch (2.0 μm). A dime is included in this photograph for size comparison.

3. Delavan manufactures fuel nozzles. These nozzles have to be deburred to very high standards but measurement of quality is ultimately made by measurement of fuel flow through the nozzles not by control of final dimensions of the components.

   Figure 10-9 shows nozzles typical of those made by Delavan. Figure 10-10 shows the critical groove in a nozzle before deburr. Figure 10-11 shows the deburred groove.

   Of course no loose burr can remain on the components. These would clog grooves. The dimensions of the components must be held to close tolerances. However, the rate of flow of fuel through the slots in the nozzles and the resulting spray pattern are the critical parameters which are measured and controlled. Process cycles in the Harperizer are from two to ten minutes depending on the size and type of nozzles. Here there is direct economic justification of equipment compared to old deburring methods. The real value of the process in in achieving completely uniform results and optimum fuel flow conditions.

The requirements for all three applications mentioned are to deburr. Each group of products has its burrs defined differently. Each company making the parts has quite different interpretation of the specification "deburr." The fact that Bendix can give dimensions of burrs which will be on the components when presented to the finishing department is rare indeed.

Because of the enormous variety of requirements deburring methods today typically are decided on a trial and error basis rather than by a predictable selection technique.

FIGURE 10-5: ALUMINUM SPACER

FIGURE 10-6: STAINLESS STEEL COMPONENT WITH DIME

FIGURE 10-7: GEARS BEFORE AND AFTER HARPERIZING

FIGURE 10-8: STAINLESS STEEL SHAFTS

FIGURE 10-9: FUEL NOZZLES

FIGURE 10-10: GROOVE IN NOZZLE

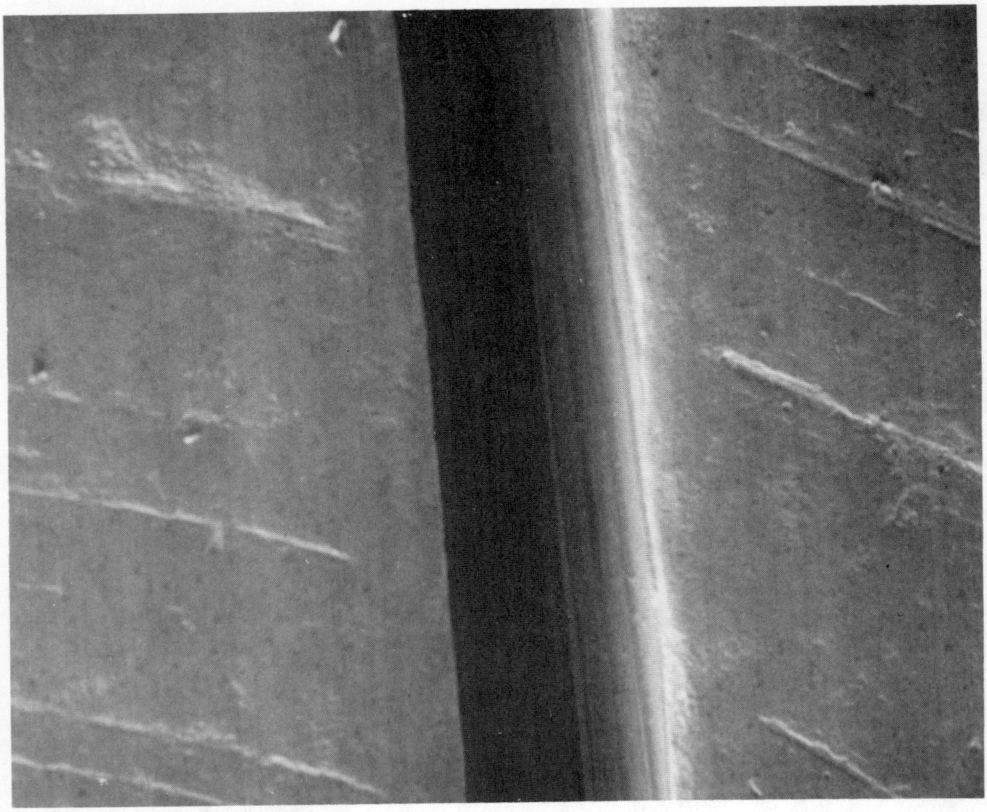

FIGURE 10-11: NOZZLE GROOVE AFTER HARPERIZING

Radiusing

It is easier to offer data relating to edge and corner radiusing than deburring. A radius is an objective measurement understood by all. However, perhaps the data which exists for different processes is of only minor assistance to manufacturing engineers in deciding process method best suited to their products.

For example, centrifugal finishing equipment can impart radii to steel pins 1 inch (25.4 mm) long, 3/16 inch (4.75 mm) diameter, 40 Rockwell C. 0.005 inch (127 µm) radius is obtained in five minutes, 0.025 inch (.64 mm) in 60 minutes, under certain process conditions. On a 1 inch square aluminum cube, an edge radii of .010 inch (0.25 mm) can be achieved in 10 minutes while width of the block at its center is reduced by 0.0001 inch (2.5 µm). Alternatively, the same radius can be achieved with a better surface finish in 30 minutes with a size reduction of 0.0007 inch (17.8 µm).

Generally a requirement to radius edges of any component will also include a need to deburr. Some examples of the use to which centrifugal barrel finishing is applied for radiusing should give an indication of its capabilities:

1. Singer Controls Company Harperize the switch followers shown in Figure 10-12 using a small Harperizer of the type shown in Figure 10-13. Forty thousand parts are processed each load with approximately a 60 minute cycle. Radii on the top edge of these parts must be 0.012 inch (30.5 μm) minimum without excessive radii on the corners. By use of small sized abrasive media this is achieved without thinning the overall part and at the same time producing the desired fine finish. The process also produces uniform results even though burrs are irregular. On occasions when tooling is worn the burrs are heavy yet the process is still consistent.

2. Aircraft turbine and compressor blades require uniform and very closely controlled radii on leading and trailing edges. The General Electric plant at Rutland, Vermont, Harperizes 3,000,000 blades per annum generating these radii during three separate Harperizing operations. While producing the radii, roots and tips of the blades are deburred and airfoil surface finish is also improved.

3. Deutsch Relays processes a wide variety of parts in centrifugal barrel equipment. For example, the discs shown in Figure 10-14 require uniform radii of .015 inch (.38 mm) on the edges of the holes while holding very close tolerances on the hole diameters. This is produced in a 20 minute process cycle while O.D. of the part is deburred and surface finish improved. Figure 10-16 shows the radius generated around the holes, whild removing the type of burr shown in Figure 10-15.

Surface Finishing

Most surface finishing operations are required for both decorative and functional purposes. In the former case, standards are essentially subjective. Even for functional finishing, measurements of surface profile and reflectivity do not always define the true requirement. Individual judgment may still be required.

Some examples of surface finishing in centrifugal barrel equipment are:

1. Outboard motor propellers are Harperized. Investment cast stainless steel propellers are Harperized after major defects and flash have been removed by hand. Process cycle is in the order of 45 minutes to produce a very smooth lustrous finish equivalent to that achieved by hand buffing. Cast aluminum propellers, such as those in Figure 10-17, use completely different media with a process cycle of between three and five minutes. This operation removes defects and flash, radiuses edges and produces a finish suitable for painting as shown in Figure 10-18.

2. Golf club heads are Harperized to produce a final lustrous finish. Here there are several happy marriages of competitive equipment where vibratory finishing techniques are followed by centrifugal

barrel finishing. This reduces hand polishing and buffing while improving product quality.

3. New Hampshire Ball Bearing uses centrifugal barrel finishing equipment for deburring and for surface finishing their precision miniature bearing components. Of course on parts such as these, tolerances are specified, generally in the order of 0.0001 inch (2.54 μm). Radii tolerances are +0.002 inch to -.000 inch (+50.8 to -0 μm) and surface finishes are invariably 3 microinches (0.076 μm) rms or better. At New Hampshire Ball, as at most other miniature bearing manufacturers, centrifugal barrel equipment is meeting exacting requirements but the technique for doing this has been developed on an ad hoc basis, not from scientific data. Each company is using somewhat different techniques. At New Hampshire Ball, the immense experience and understanding of tumbling techniques has elevated their finishing methods to an art, if not a science.

   Figure 10-19 shows results achieved by Harperizing miniature bearing races and Figure 10-20 shows a small machine suitable for this finishing operation.

Stock Removal

Centrifugal barrel finishing is being used as a means of removing metal on some components more economically than by grinding. For example, unhardened carbon steel balls 3/8 inch (9.52 mm) diameter can be Harperized to remove metal at a rate of 0.001 inch (25.4 μm) off diameter every 10 minutes. A machine of the type shown in Figure 10-21, costing less than $25,000, can handle up to 600 pounds (2669 N) of balls at a time with consumable material costing less than $.50 per hour.

During manufacture of many needle bearing rollers it is necessary to remove "knobs" from the ends of the rollers as shown in Figure 10-22. These rollers can be sizes from as small as 1/16 inch (1.57 mm) length. Typical rollers with body diameter 0.230 ± 0.002 inch (5.842 ± 0.051 mm) with knobs 0.258 0.002 inch (6.553 ± 0.051 mm) can be processed for two hours resulting in body diameter of 0.222 inch + 0.000 - 0.004 inch (5.639 ± .102 mm) and ends within 0.001 inch of body.

For grinding operations considerable heat can be generated and the equipment used will be somewhat different from standard deburring and finishing units. The shape of drum interior will affect the grinding action. Automatic equipment of the type shown in Figure 10-4 can handle up to 3000 pounds (13,344 N) of balls or rollers at a time.

Changing Stress in Surface of Parts

High compressive stress in the surface of components will increase resistance to fatigue failure. In addition if parts have edges radiused and surface finishes improved, the fatigue strength will be still further enhanced.

While investigating means to improve the stability of gage blocks, researchers have reported that some 2 inch (50.8 mm) gage blocks made from 52100 steel were hardened, stabilized, and then Harperized. The compressive stress after Harperizing was found to be -201,000 psi (-1386 MPa) parallel to the 2 inch direction and the -232,000 psi (-1599 MPa) perpendicular. This results in much improved stability of the gage blocks compared to those tumbled by conventional means.

The capability of imparting such high compressive stresses is utilized by bearing, aircraft engine, and spring manufacturers.

Flapper valves used in air conditioning and refrigerating units are generally finished by Harperizing. Being fragile parts, generally of thickness in the order of 0.015 inch (0.38 mm) and having slots cut in them makes them ideal for the process. All edges require full radii. Surface finish has to be excellent and high compressive strength is mandatory for good performance. Similarly, silent chain and roller chain which has been finished by the centrifugal barrel process will give improved chain performance as well as easier automatic assembly.

Limitations

As with other bulk mechanical finishing processes, components will have all edges and corners deburred and radiused while work is being done on the surface of those components. Relative action on edges and surfaces can be controlled by choice of media size and centrifugal force, but it is not possible to impart a radius on one edge while maintaining sharpness on another equally accessible edge.

If the finishing method is considered at the same time as the basic forming method of any component and allowance is made for this overall action, the result is most often desirable. The problem is more serious with existing products where a cost of tool modification to allow for any metal removal cannot be justified.

An unavoidable limitation occurs when deburring or finishing holes or recesses. When the depth of a blind hole substantially exceeds its diameter, action becomes limited and eventually negligible.

Finally, economics can be a limiting factor. If a part is to be finished by hand or by special purpose equipment, then centrifugal barrel equipment should certainly be investigated as there are potentially large savings to be made. If, however, entirely satisfactory results can be achieved in conventional tumbling equipment with short process cycles, say of less than one hour, and if the equipment is to be utilized for only a few different components, then centrifugal barrel units will probably not be economical.

Summary

Ultimately, choice of equipment for the finishing department has to be made by comparative tests of the processes which appear to meet basic requirements.

The deburring and finishing business is a dynamic one. All worthwhile companies marketing mechanical finishing equipment are positively searching for new problems. If they are unable to use their existing technology to solve the problem and no economic answer appears to exist elsewhere, they will at least give serious consideration to developing new technology.

Centrifugal barrel finishing offers unique benefits in speed of operation, capability to produce very fine finishes and to produce high compressive stresses while deburring and finishing. Development of the process to meet new problems and to compete with other processes are continuous. It is therefore undesirable to pigeonhole the capabilities too precisely as this could handicap future development.

References

[1] "Abrasive Process Techniques Decreases Cost, Improves Jet Blade Quality," Machine and Tool Blue Book, January, 1975.

[2] American Machinist, November 11, 1974.

[3] "Orbital Tumbling--A Means of Improving the Finishing Department," ASAM 13th Annual Conference Proceedings 1975.

[4] "Tedious Mold Hand Finishing Eliminated," Modern Machine Shop, December, 1972.

[5] M. Matsunaga and H. Kobayashi, "Some Experiments on Centrifugal Barrel Finishing," Metal Finishing (A), Volume 64, No. 5, p. 57.

[6] J. Bernard Hignett, "Capabilities and Limitations of Centrifugal Barrel Finishing," SME Paper MR75-834, 1975.

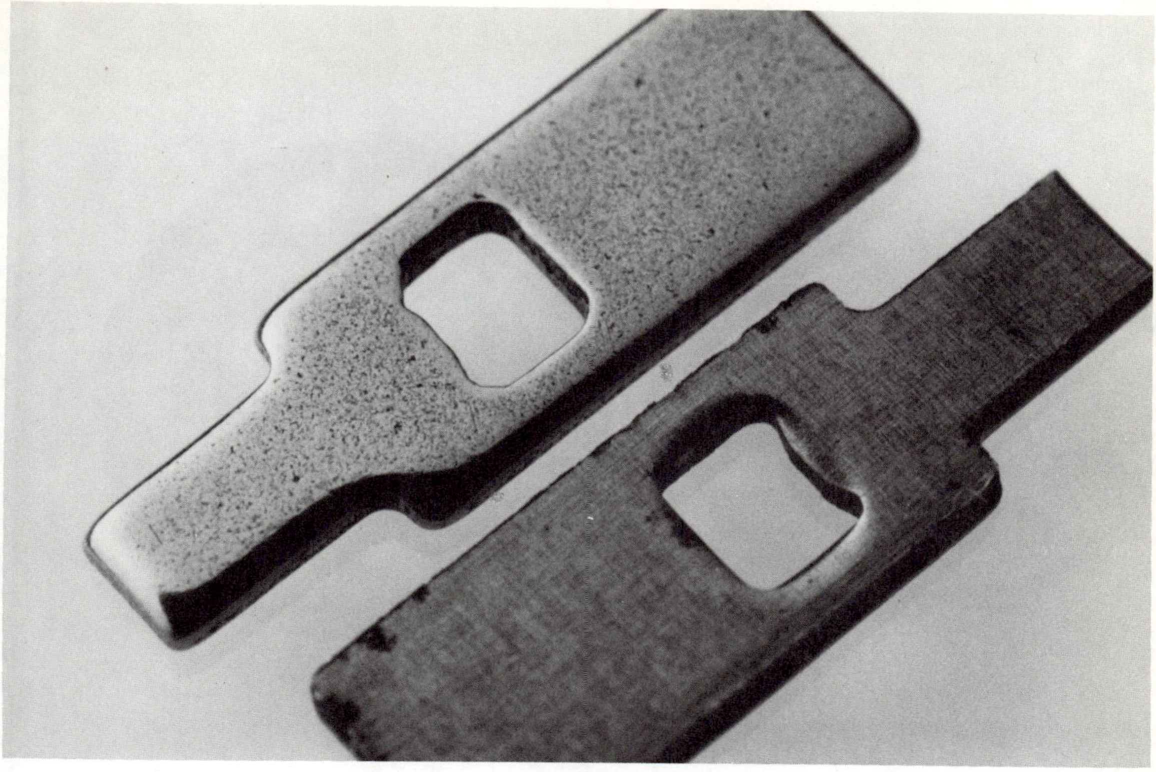

FIGURE 10-12: SWITCH FOLLOWERS BEFORE AND AFTER HARPERIZING

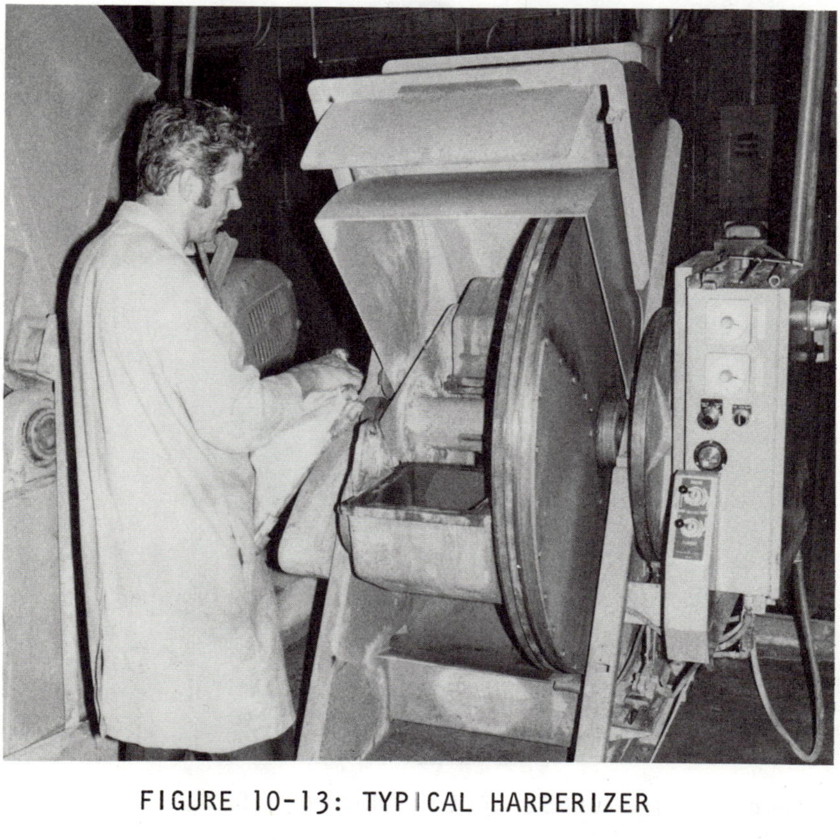

FIGURE 10-13: TYPICAL HARPERIZER

FIGURE 10-14: DISCS BEFORE AND AFTER HARPERIZING

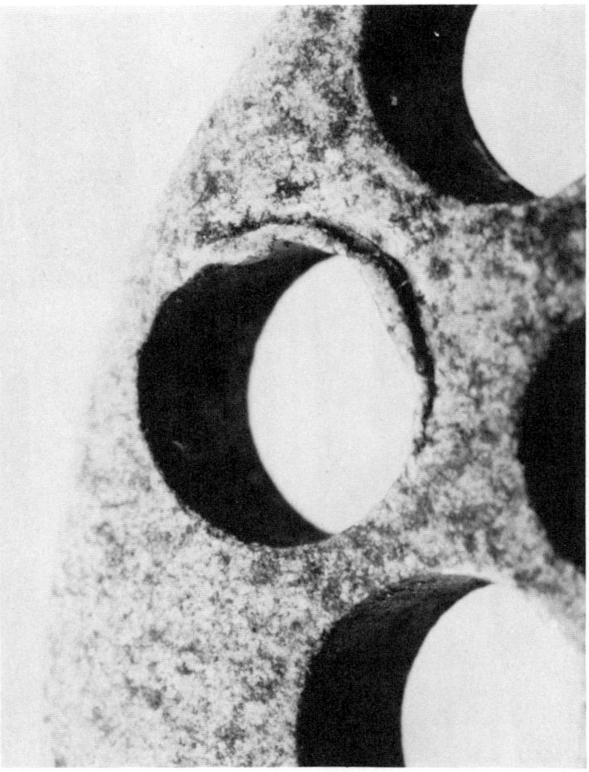

FIGURE 10-15: CLOSE UP OF BURR ON DISC

FIGURE 10-16: DISC AFTER HARPERIZING

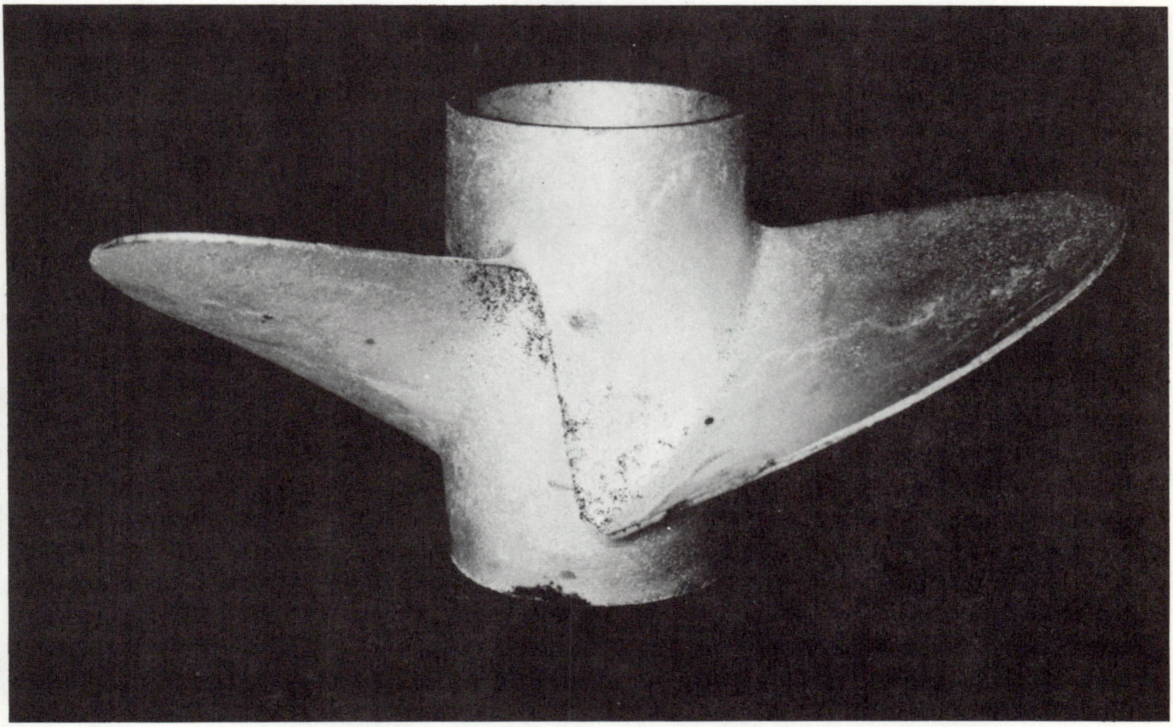

FIGURE 10-17: PROPELLER BEFORE HARPERIZING

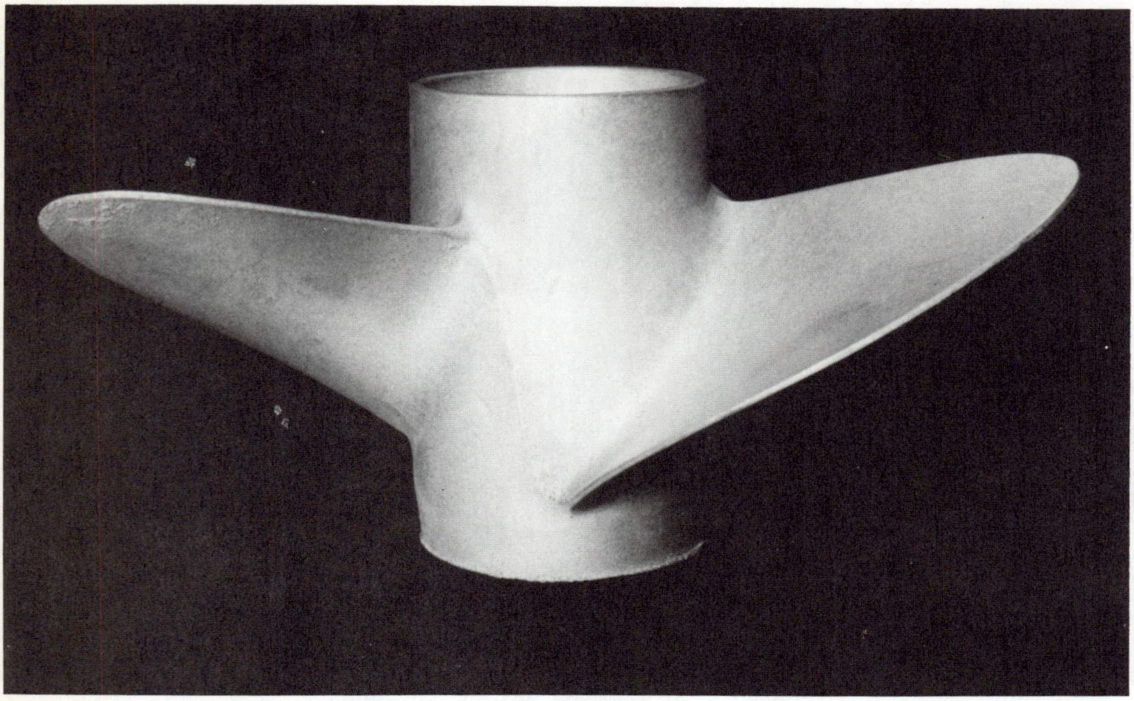

FIGURE 10-18: PROPELLER AFTER HARPERIZING

FIGURE 10-19: MINIATURE BEARING RACES

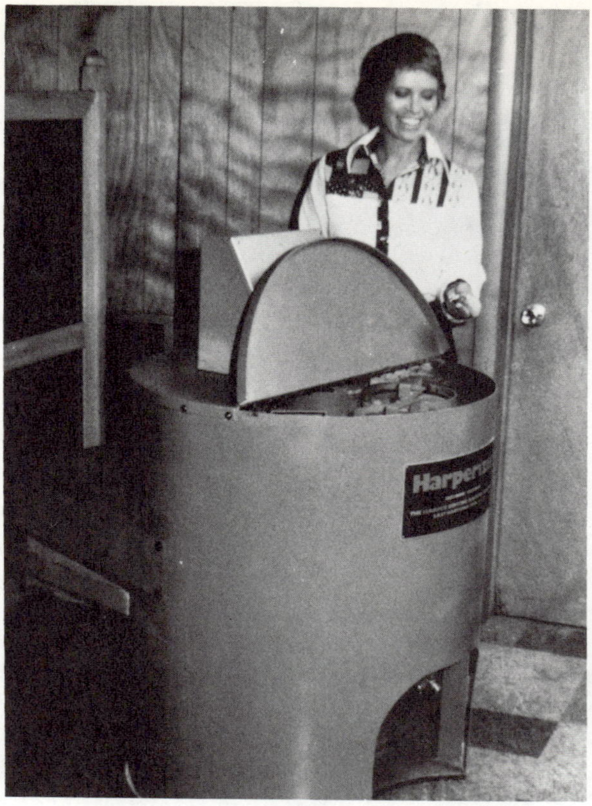

FIGURE 10-20: SMALL HARPERIZER FOR MINIATURE PARTS

FIGURE 10-21: LARGE HARPERIZER

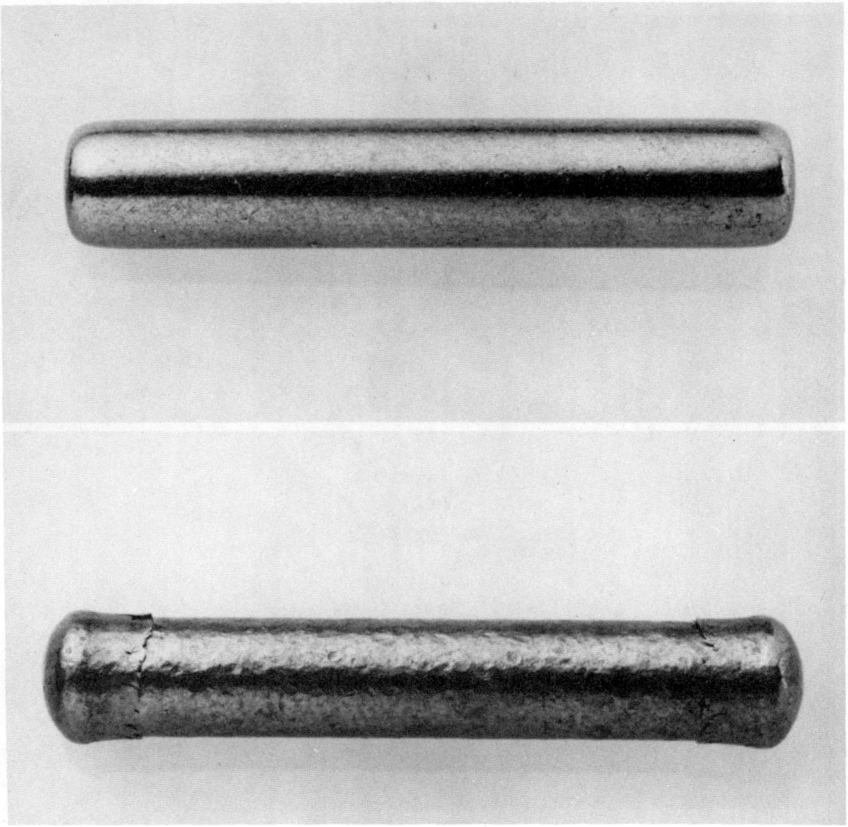

FIGURE 10-22: NEEDLE BEARING ROLLERS BEFORE AND AFTER HARPERIZING

CHAPTER 11

ABRASIVE JET DEBURRING

Abrasive blasting has been used for almost 75 years as a technique for cleaning, smoothing, and edge finishing. Despite its widespread usage on metals and plastics, little data has been published on this process.

Abrasive jet deburring can be subdivided into the following five categories:

1. Dry Blasting
   a. Conventional
      (1). Concentrated jet or stream of abrasive
      (2). Wide stream of abrasive
   b. Miniature jets of abrasive
2. Wet Blasting
   a. Concentrated jet of abrasive and water
   b. Wide stream of abrasive and water

The capabilities of the conventional dry blasting approaches and concentrated wet blasting have been covered extensively by Plaster.[1] The use of miniature jets of abrasive has been described by Kulischenko.[2] The use of multiple streams of water-carried metal abrasives, while not new, has received little public attention. This chapter describes the capabilities and limitations of the Centri-Blast® version of this process.

What is the Process?

The wet abrasive surface finishing process utilizes the impingement of an abrasive slurry on the parts to be surface finished under controlled conditions of velocity, direction, and time. Figure 11-1 is a schematic depicting the process.

1. Water containing suitable additives is pumped at elevated pressure to an injector located at the base of the abrasive hopper.

2. The abrasive which is maintained at a predetermined level is induced into the high velocity water to form the abrasive slurry.

3. The abrasive slurry is impinged upon the part at an angle of attack, pressure, and time consistent with the finishing requirements.

4. The abrasive laden water overflows the abrasive hopper and flows through a series of weirs contained in the abrasive return hopper. The concentrated abrasive is recovered at the apex of the abrasive return hopper and returned for reuse via an injector

to the abrasive hopper.

5. Abrasive free water overflows the abrasive return hopper into the clean water tank. The fine silt generated by the finishing process is permitted to settle in this tank for manual or automatic removal.

It is important to note that all usable abrasive is maintained in the abrasive hopper and that no abrasive passes through the pump. Complex parts may require multiple blast nozzles providing many angles of attack as well as differential blast velocities.[3]

Water Media Requirements

The use of water for generation of the abrasive slurries and their transport demands that specific conditions be maintained for optimum performance from the equipment.

1. Rust Inhibition: The steel construction of the Centri-Blast® machine as well as the possibility of processing ferrous parts and/or using ferrous abrasive requires the incorporation of suitable rust inhibitors.

2. Corrosion Inhibition: The processing of nonferrous metals and/or mixed metals may require the incorporation of corrosion inhibitors to avoid galvanic reactions or other undesirable reactions between the base metals and the water media.

3. Foam Control: Excessive foaming may result in loss of pressure due to pump cavitation or loss of abrasive from the abrasive hopper due to froth flotation.

4. Adequate Wetting: Preferential wetting of the abrasive as well as the processed parts will insure the formation of a hydrophillic surface necessary for the production of clean parts, machine cleanliness, and avoidance of loss of abrasive. This can be provided through the use of suitable alkaline materials capable of reducing the surface tension to approximately 35 dynes/sq cm or less. The use of oil-based emulsifiable products will invariably lead to formation of hydrophobic films; resulting in dirty equipment, dirty parts, and loss of abrasive. The use of an additive with satisfactory wetting will also promote more ready removal of the abrasive in the flushing cycle as well as rapid drainage in the drying cycle.

Abrasive Media Requirements

For optimum longevity and consistency of performance, the abrasive media should comply with specific requirements.

1. Unaffected by Water: Since water is utilized as the vehicle,

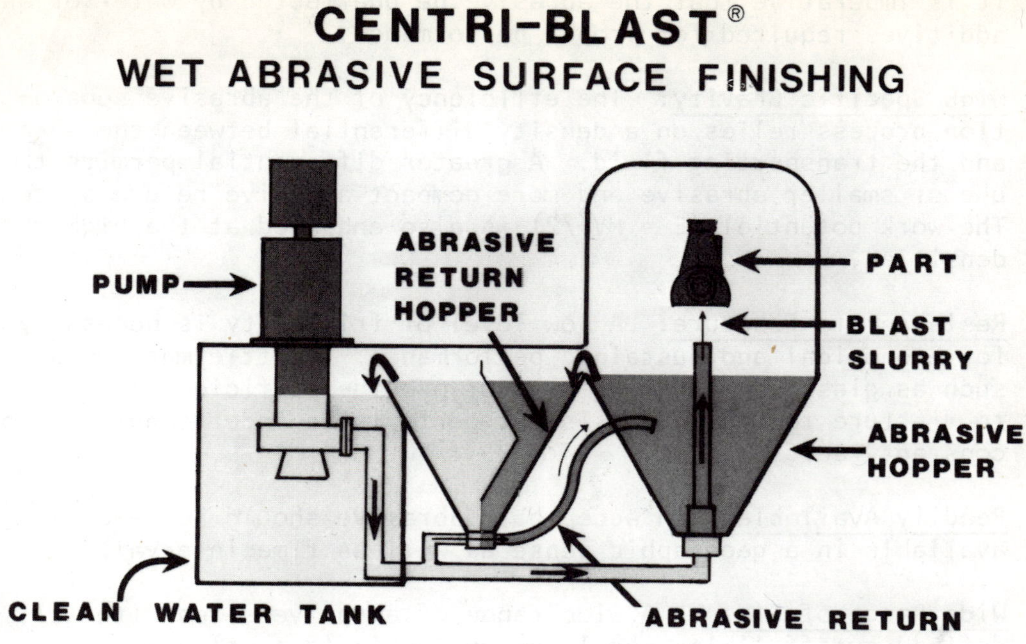

FIGURE 11-1: THE BASIC SYSTEM

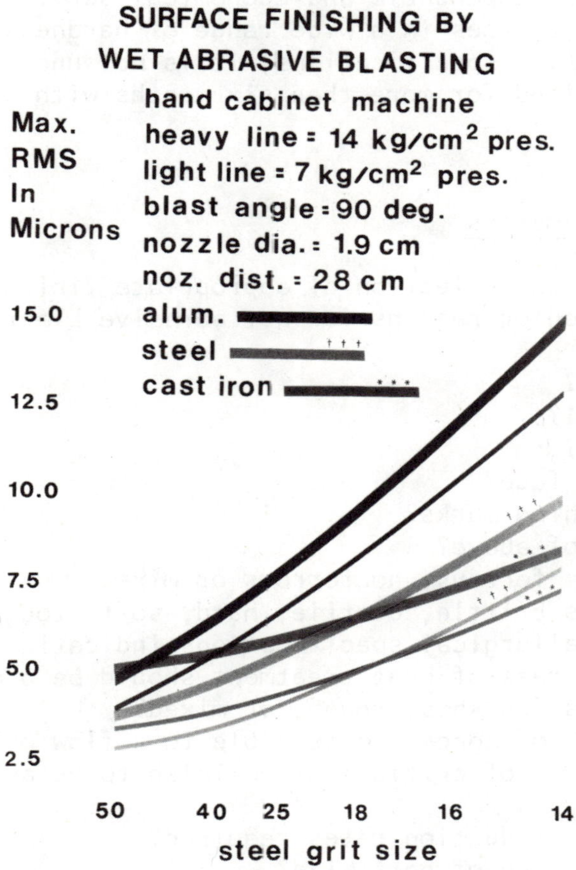

FIGURE 11-2: SURFACE FINISH PRODUCED IN VARIOUS METALS

it is imperative that the abrasive be unaffected by water or any additives required for proper performance.

2. High Specific Gravity: The efficiency of the abrasive separation process relies on a density differential between the abrasive and the transporting fluid. A greater differential permits the use of smaller abrasive and more compact abrasive return systems. The work potential ($E = MV^2/2$) is also enhanced at the higher densities.

3. Resistant to Fracture: A low level of friability is necessary for economical and sustained performance. Brittle materials such as glass and ceramics may not provide sufficient resistance to fracture to insure consistent performance under conditions of constant recycle.

4. Readily Available: An acceptable abrasive should be readily available in a geographic sense as well as time interval.

5. Wide Range of Sizes: A wide range of abrasive sizes with well defined cutoff limits should be available to preclude wedging and meet microfinish requirements of processed parts.

Over a decade of experience has proven tempered cast steel shot and grit to provide the most dependable and economical service. Readily available from quality sources in a wide range of hardness and size standards conforming to S.A.E. specifications, it is not uncommon for an abrasive charge to be utilized for more than six months with only minor make-up required.

Finalizing Process Parameters

The first step in the selection of appropriate finishing parameters is to analyze the reason or reasons for wet abrasive blasting.

Burr removal?
Flash removal?
Soil removal?
Improve seal face?
Remove machining marks?
Combination of above?
Are the parts ferrous, nonferrous or mixed?
Are the parts brittle, ductile, hard, soft, tough or mixed?
Complete metallurgical specifications indicating metal type or types and areas of heat treatment should be provided.
Are the parts finished, rough, or mixed?
Are the areas of concern accessible to a flow of abrasive slurry?
Are there areas of critical microfinish to be avoided or protected?
What are the production rates required?
What is the range of part size?
Are there areas of critical geometry that should be considered to

avoid abrasive wedging?

After careful consideration of the foregoing items, it now becomes a question of selecting the appropriate type and size of abrasive as well as velocity, angle, and duration of impact. The following generalizations[3,4,5,6] may be observed in making these selections:

1. The larger the abrasive size, the greater the work and the higher the microfinish.
2. The smaller the abrasive size, the lower the microfinish and the work.
3. The higher the velocity, the higher the work and the microfinish.
4. The lower the velocity, the lower the work and the microfinish.
5. The more acute the angle of attack, the lower the microfinish.
6. An increase in microfinish is obtained with an increase in the angle of attack.
7. The longer the blast duration, the greater the work and effect on microfinish.
8. The shorter the blast duration, the lower the work and effect on microfinish.
9. Grit will produce a higher microfinish with ductile materials then shot.
10. Shot will produce a higher microfinish with brittle materials than grit.

Through accurate control of the process variables, effective removal of burrs, flashing, rust, foundry soil, etc., can be affected without deleterious effect on part dimensions or microfinish.

Figures 11-2 and 11-3 indicate the terminal or maximum microfinishes obtainable on 3 inch x 3 inch x 1/4 inch panels of aluminum (6061-T6, BHN 96-500 Kg load); cold rolled, low carbon steel (1018, BHN 187-3000 Kg load); and grey cast iron (BHN 174-3000 Kg load) under the following conditions.

MACHINE
Centri-Blast® manual cabinet unit utilizing one 19 mm I.D. vertical blast nozzle directed downward. Nozzle-to-part distance --28 cm. Blast angle 90 degrees (perpendicular to surface of test coupon).

ABRASIVE
Tempered steel grit (Figure 11-2) R_c 42-48
Tempered steel shot (Figure 11-3) R_c 42-48

WATER SOLUTION
2 oz/gal Jem Blast® 13-S (wet abrasive deburring compound manufactured by Ajem® Laboratories Division, Centri-Spray® Corporation).

PRESSURE ON BLAST NOZZLE
14 Kg/cm^2 and 7 Kg/cm^2 as noted (1 Kg/cm^2 = 14.22 psi)

METAL COUPON MOVEMENT
Reciprocated laterally at approximately 25 mm/sec (1 inch/sec)

INITIAL MICROFINISH
All panels surface finished to an initial RMS of 2.5 microns.

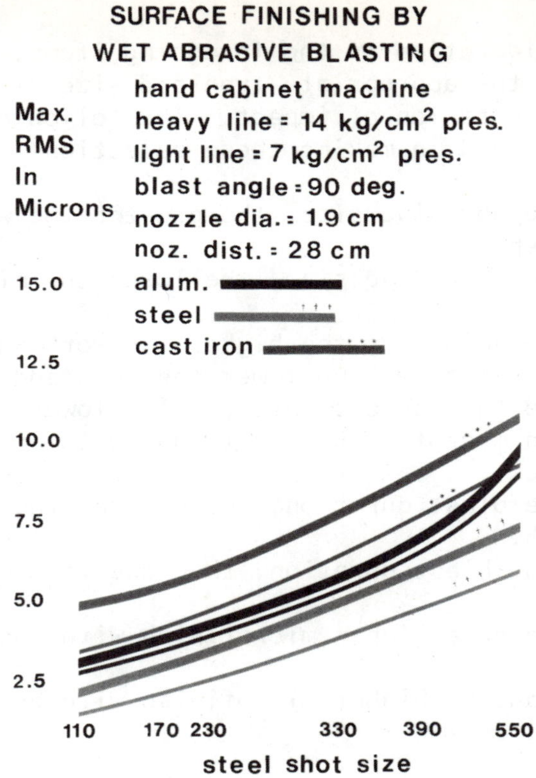

FIGURE 11-3: SURFACE FINISH PRODUCED BY STEEL SHOT

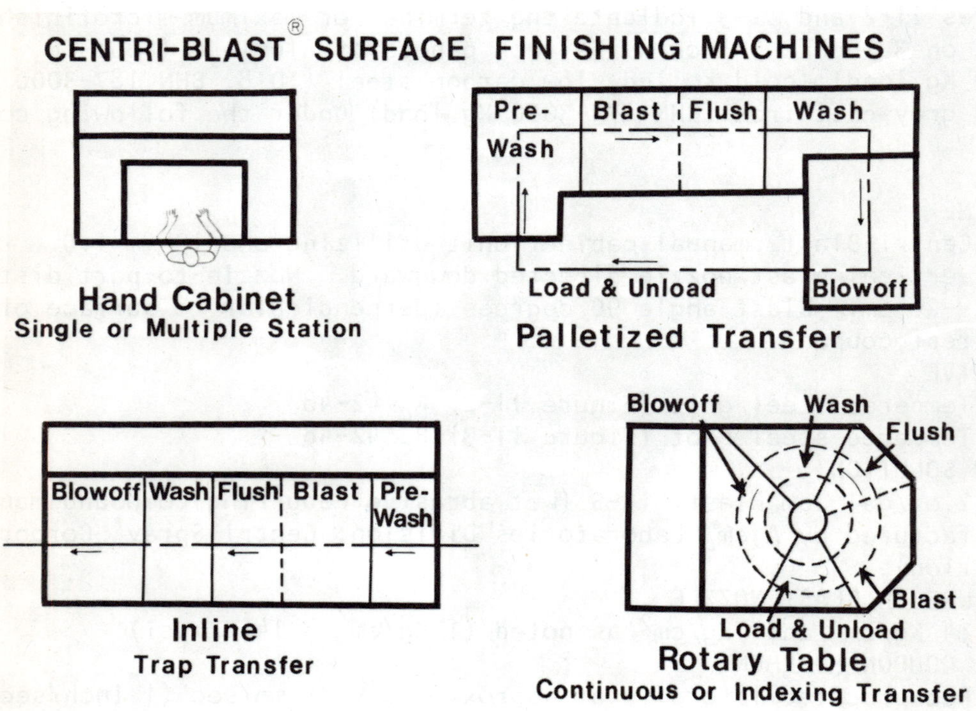

FIGURE 11-4: BASIC SYSTEM LAYOUT

All profilometry readings taken at a cutoff point of 2.5 mm (0.1 inch) and a tract speed of 7.6 mm/sec.

TIME OF BLAST

Until three consecutive profilometry checks indicate microfinish equilibrium has been attained. (Time varied with abrasive type and size, metal type, and blast pressure.)

The curves indicate the dramatic difference in response of a friable material (cast iron) and a ductile material (aluminum and C.R. steel) to abrasive type (grit versus shot). It should be stressed that the microfinishes indicated are maximums that can be obtained within the parameters of the test conditions. They could readily be altered by reducing blast exposure time, decreasing pressure, or reducing the angle of attack. In production applications, the forenamed variables exert their influence on the microfinish of the production part with the consequent marked propensity to reduce all machine tool marks to a definite matte finish. Since the contrast in performance (particularly in a sealing application) of a matte surface finish versus a finish with a unidirectional lay could be significant, the microfinish standards should be comprehensive enough to include this. It is the author's observation that manufacturing engineering specifications rarely take into consideration the direction or type of lay when a microfinish is specified on a print.

A summation of the variables affecting the wet abrasive surface finishing process would include the following:

1. Metallurgy of processed parts.
2. Work function to be achieved and production rates.
3. Accessibility and wedging characteristics.
4. Blast nozzle size, type, and distance from part.
5. Angle of attack.
6. Abrasive type and size.
7. Velocity of blast slurry.
8. Density of blast slurry.
9. Time of exposure.
10. Initial and terminal microfinish specifications.

Process Limitations

Considerations inherent to the parts to be processed may limit or preclude the use of wet abrasive blasting. These would include:

1. Materials adversely affected by exposure to water. (Reactive metals, materials subject to moisture absorption or porosity.)
2. Size limitations--extremely large parts may generate unduly large machine requirements and render the process economically unattractive. Small parts that cannot be held stationary would present untenable handling problems.
3. Extremely thin parts may be subject to distortion.
4. Abrasion resistant materials may require excessive exposure times in order to meet production requirements.
5. Materials of high malleability may be subject to excessive plastic

deformation.

6. Brittle materials (unless capable of being suitable masked) may prove to be subject to unacceptable attrition. Ceramics would be generally susceptible.
7. Inaccessibility of area to be finished to a flow of the abrasive slurry.
8. Inability to avoid wedging of abrasive due to part geometry. This can often be overcome through a change in part design.
9. Areas of extremely low microfinish generally require shielding or parallel flow of the blast stream to avoid an increase in microfinish.
10. Extremely close dimensional tolerances may be difficult to hold. Quite often this can be overcome by closely controlling the angle of attack.

Process Advantages

The wet abrasive surface finishing process is available in designs ranging from hand cabinets to completely integrated and automated units offering several or all of the following advantages:

1. Surface finishing of ferrous and nonferrous metals, singularly or mixed. Aluminum and zinc diecastings, cast iron parts (from as-cast to completely machined), steel of varying alloy and states of heat treatment, and copper alloys are typical of materials currently being processed.
2. The range of part sizes may extend from hand held to engine cylinder block size.
3. Production rates to meet requirements in hand machines or over 1200 parts per hour in automatic units.
4. Internal and external finishing capabilities through the use of standard blast nozzles, diffusers, and probes as required.
5. Pressure differential control over blast nozzles hitting different areas of the part permits removal of more difficult burrs in one area of a part while not exerting destructive action in other areas.
6. The process can be used to deburr, deflash, and derust as well as remove heat-treat scale and foundry soil.
7. Improve surface microfinish for enhancement of sealing capabilities.
8. Induce compressive stresses through peening action for improved fatigue life.
9. A standard unit can handle a variety of steel grit sizes from S.A.E. G-50 to G-10, or steel shot from S-110 to over S-930 without requirement for mechanical change. Units handling finer abrasive are available for special applications. Abrasive hardness ranges from below R_c40 to over R_c60.
10. The wet abrasive process materially extends the working life of the abrasive while continuously removing the fine particles resulting from the finishing action.
11. Elimination of pump wear due to absence of abrasive passing through pump.

12. The process will accept parts that are not clean or dry. On high production machines, handling parts containing gross amounts of machining chips or machining coolants, an integrated prewash section is normally incorporated.
13. Design flexibility permits incorporation into any production flow process, inline or off.
14. In addition to the desired surface finishing requirement, all (but the small hand cabinets) produce a clean, rust and/or corrosion inhibited, dry part ready for assembly.

Wet Abrasive Surface Finishing Systems

Schematic diagrams of several Centri-Blast® units can be seen in Figure 11-4.

1. Hand Cabinets: Permits finishing and abrasive flush off with manual load and unload. Normally does not incorporate drying capability. Pressures can be varied to accommodate application. The hands are inserted in heavy rubber gloves while manipulating the parts.

2. Palletized Transfer: This design utilizes a pallet as a transport device to carry a fixture holding the parts to be finished. This permits the retention of a common pallet while interchanging fixtures to accommodate parts of different shape or size. Thus the applicability can be extended to a variety of parts without a change in the basic machine. While the machine depicted shows integrated prewash, blast/blast-flush, wash, and blowoff sections; the system could consist of only a blast/blast-flush and blowoff. Additionally, a final rinse could be added to the system depicted prior to the blowoff. The machine requirements are determined by influent soil levels, production rates, part cleanliness specifications, part complexity, and available space.

   While the pallet transfer unit illustrated has a common load/unload station, the pallets could be inserted inline at the prewash station and immerge inline from a relocated blowoff section. This would permit automatic load and unload with an inline capability if desired. The pallets continue to index in a rectangular path, pallet-on-pallet.

3. Inline Trap Transfer: Useful for applications handling only one part design and is capable of automated load and unload. System offers complete flexibility for integration of additional sections required to meet special part requirements. Typical blast pressures may range from 5 to 20 Kg/cm^2 (71 to 286 lbs/in^2) with production rates in excess of 200 parts/hr.

4. Rotary Table: An extremely versatile design offering common point load and unload. The rotary table surface finishing machine incorporates a platform that rotates on a concentric axis. The platform may rotate continuously or index in finite

increments. It can be utilized for support of pallets with drop-in part silhouettes, perforated plate with movable pin design for random part orientation, or upright fixtures for multidirectional blasting.

A detailed view of an indexing type rotary wet blast deburr may be seen in Figure 11-5. The parts to be finished are manually loaded at the operator's position. The table automatically indexes to the right and carries the part into the abrasive blast station. After blasting for the requisite period of time, the parts are progressively indexed into the blast-flush and the blowoff sections. As each section (carrying finished parts) is returned to the operator's station, the finished parts are removed and unfinished parts are inserted. Production rates of several hundred parts per hour are attainable in this fashion.

The systems described are but a few of the many variations that have been manufactured and are possible incorporating the wet abrasive blasting principle.

Applications of Wet Abrasive Blasting

Figure 11-6 displays a small sampling of the many varieties of parts that are being processed with wet abrasive finishing machines. A list of a few applications would include:

- Aircraft engine heads, cylinders, and housings
- Bearing caps
- Camshafts
- Clutch drums
- Compressor parts
- Crankshafts
- Differential carriers
- Fuel injector bodies
- Gears
- Hydraulic pump bodies
- Hydrostatic transmission parts
- Oil pump bodies
- Pistons
- Transmission cases
- Valve bodies

While the foregoing list of applications was generated primarily through the requirement for deburring, one large manufacturer of refrigeration compressors related the additional advantages:

"A Centri-Blast® rotary table machine was purchased to effect consistent and complete deburring of compressor cylinder heads. The motivation was a cost reduction from required hand deburring, manual load and unload at a following wash station, blowout, and final inspection. In addition to meeting all of the deburring, cleaning, and drying requirements, the machine has proven to be capable of removing core

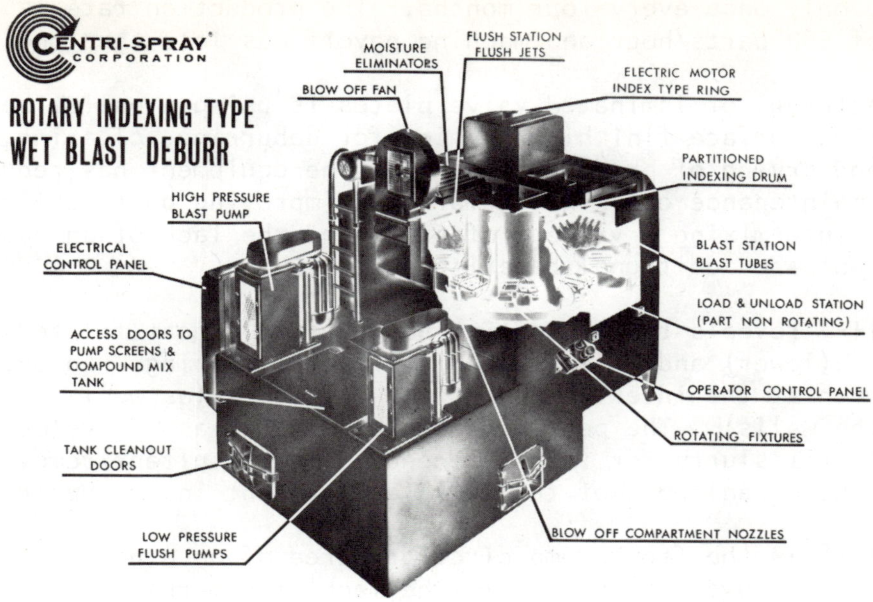

FIGURE 11-5: WET BLAST MACHINERY

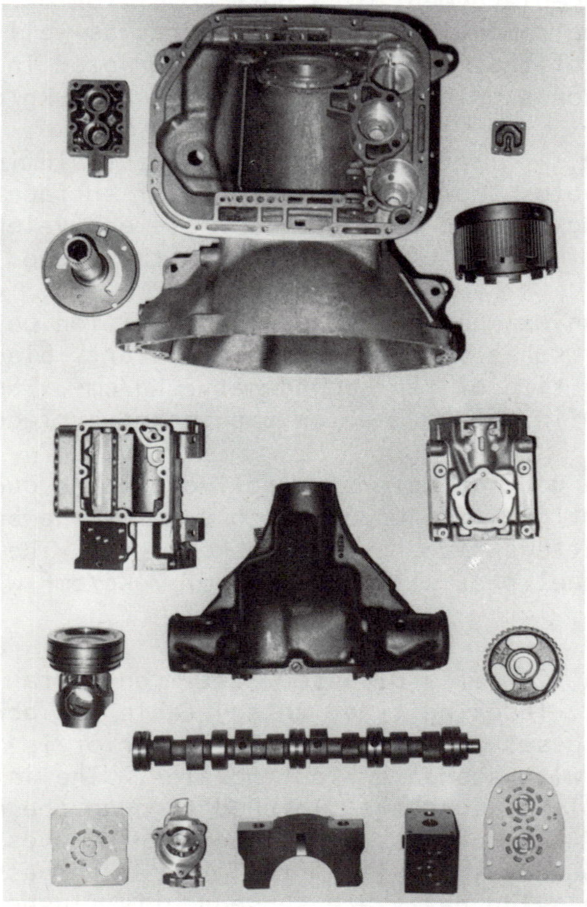

FIGURE 11-6: PARTS FINISHED BY WET BLASTING

sand, generates a superior finish in gasket seal areas, and requires cleaning only once every four months. The production rate is in excess of 500 parts/hour and machine payoff was less than one year."

A manufacturer[8] of laminated valve plates is using a palletized transfer wet abrasive surface finishing machine for deburring, cleaning, rust inhibiting, and drying of production parts. The equipment has reduced handling and maintenance costs while markedly improving part quality. The capability of intermixing a variety of parts and the lack of an additional heat requirement are additional benefits.

Figure 11-7 reveals the connecting passages of a heat treated crankshaft prior to (lower) and after (upper) surface finishing. An oxidized burr can be seen at the intersection of the oil holes (as well as scale deposits on the walls). The passage was subjected to a high velocity flow of G-16 steel grit slurry for thirty seconds. The resultant removal of the burr and the cleanliness of the walls are evident in the upper photo.

Figure 11-8 is the front pump circuitry area of an aluminum transmission case. The lower photo details the machining marks and burrs prior to deburring. The upper photo exhibits the matte finish and burr removal produced by finishing with S-230 steel shot at 7 Kg/cm^2 for ten seconds.

Figure 11-9 is an enlargement of a burr condition on machined steel compressor plates. The removal of the large burrs as well as the radius produced on the edges of the machined slot can be noted in the upper photo. Cycle time was ten seconds using S-170 steel shot at 7 Kg/cm^2.

Figures 11-10 and 11-11 are areas of a die cast aluminum compressor housing. The upper photos show the consequence of surface finishing with G-50 steel grit at 7 Kg/cm^2 and a production rate equivalent to over 500 parts per hour. Note the lack of any damage to the tapped holes.

Figure 11-12 is an example of a flashing condition on an aluminum die casting. The part was subjected to an abrasive slurry blast using S-170 steel shot for a cycle time of ten seconds at 7 Kg/cm^2. Note the flash removal and the matte finish produced on the gasket surface.

Figure 11-13 is a good example of a roll-over burr due to machining of an aluminum die casting. The upper photo details the burr removal plus the production of a matte surface on the gasket area. The finishing process utilized S-170 steel shot at a pressure of 7 Kg/cm^2 with a ten second cycle time.

Figure 11-14 is a view of a bisected cast iron hydraulic pump body. A completely automatic, indexing type, rotary table, wet abrasive surface finishing machine processes these parts. The exterior is finished while transferring position with standard blast nozzles. The internal spool bore and associated oil passages are finished through the use of automatically inserted probes. The process simultaneously removes all remnants of core sand and other foundry soils (from the oil return passage) while deburring the spool bore. The parts are subsequently flushed and dried. The process uses G-40 steel grit at 7 Kg/cm^2 and is capable of producing over

FIGURE 11-7: RESULTS IN INTERSECTING HOLES

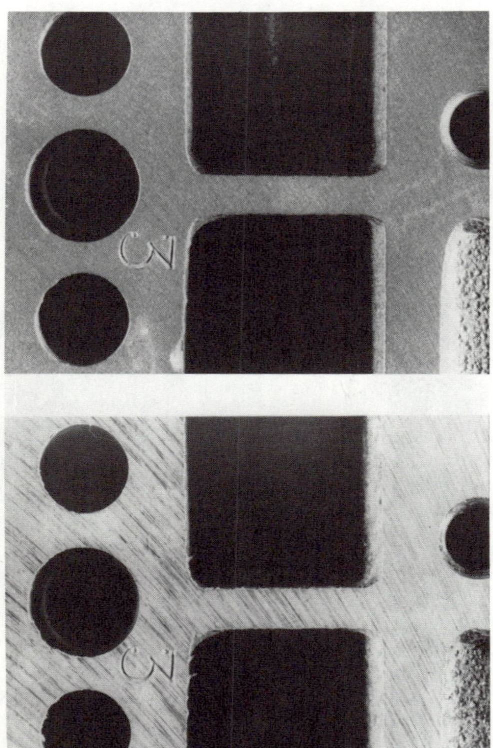

FIGURE 11-8: TRANSMISSION CASE HOLES

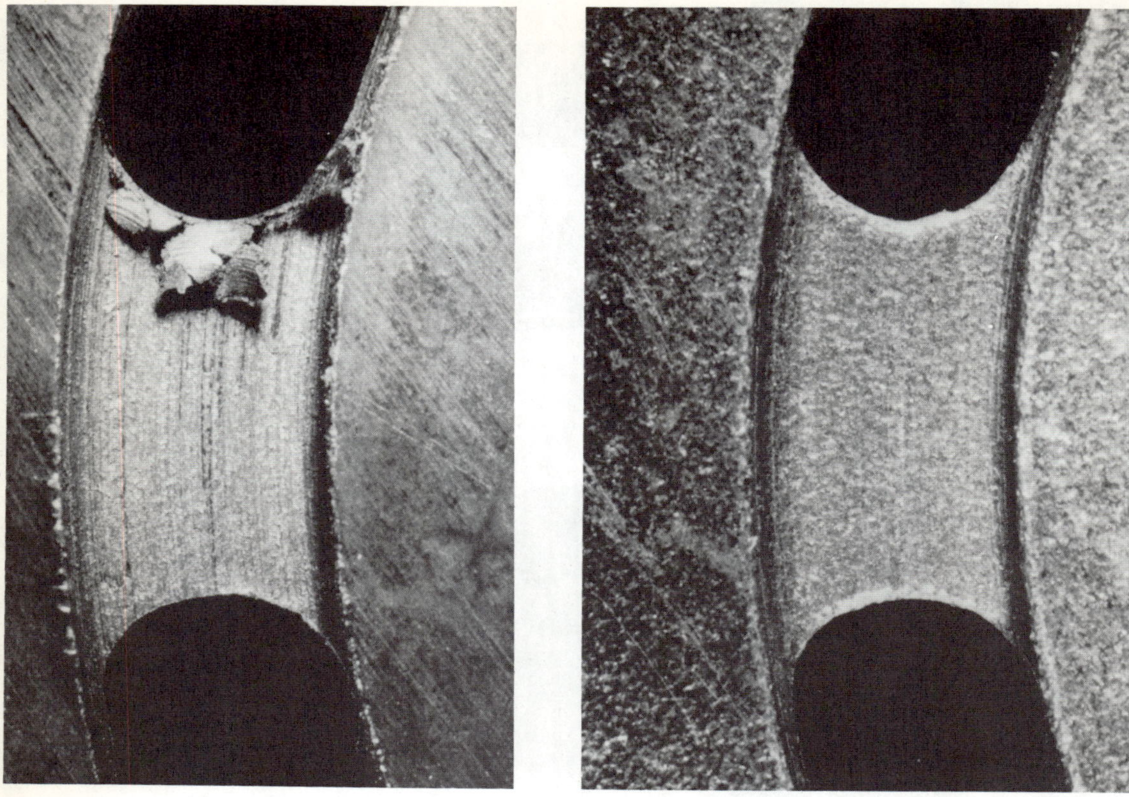

FIGURE 11-9: BURRS ON STEEL COMPRESSOR PLATES

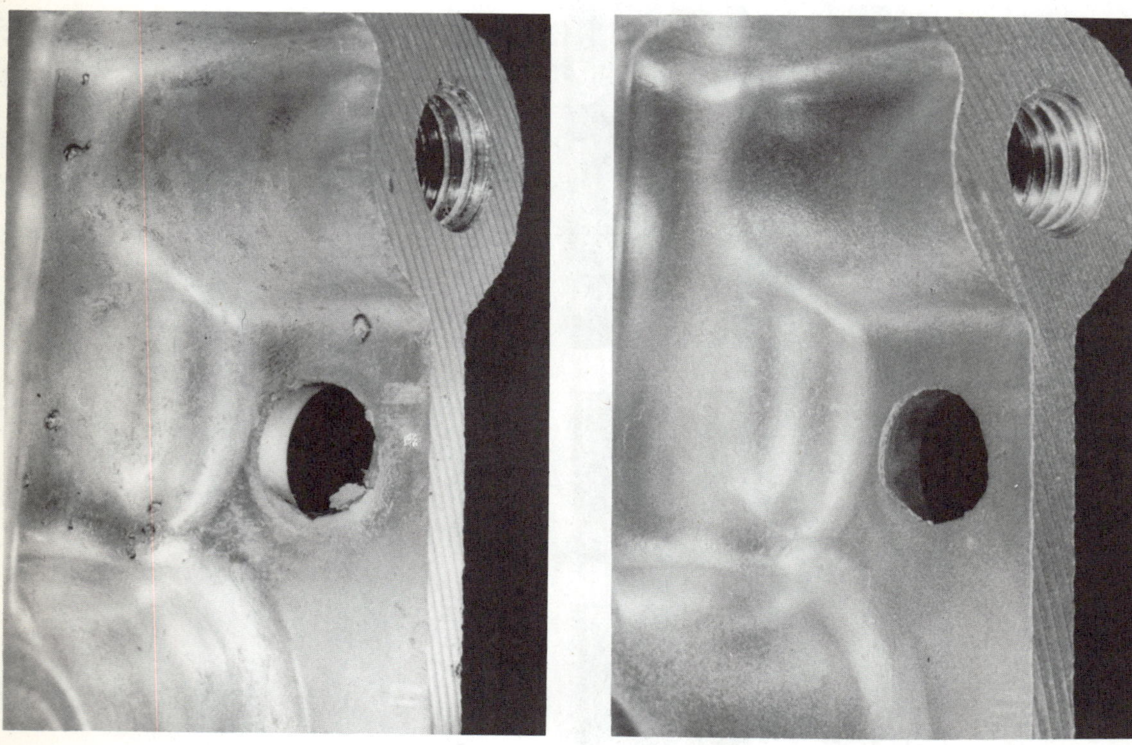

FIGURE 11-10: ALUMINUM COMPRESSOR HOUSING

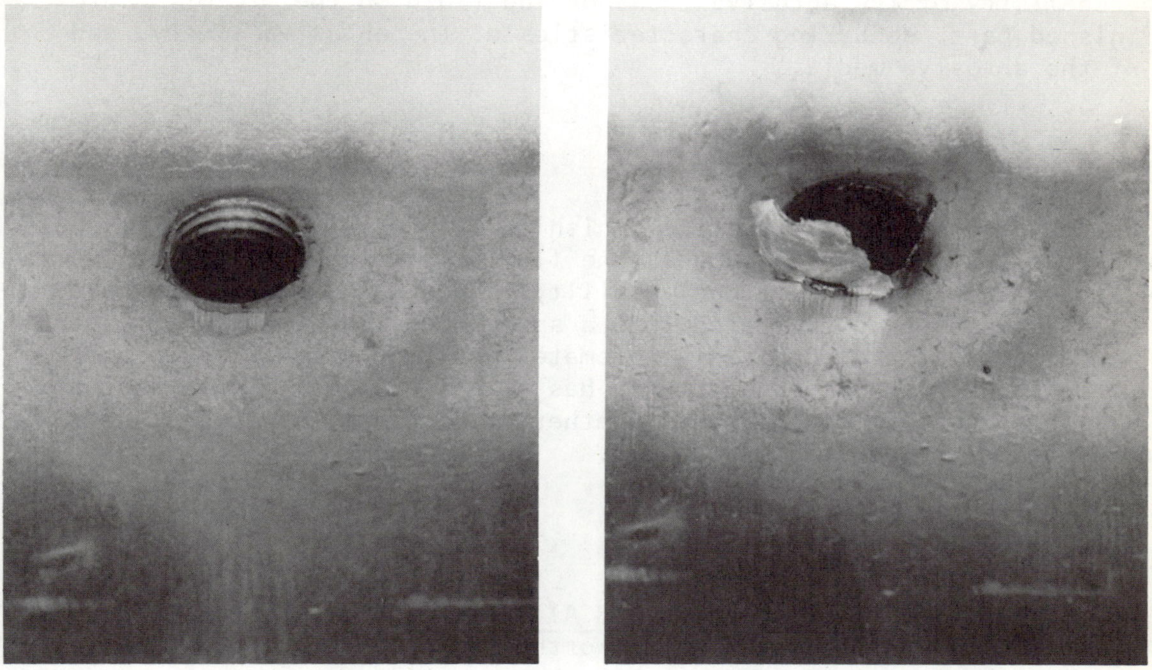

FIGURE 11-11: BURR ON ALUMINUM HOUSING

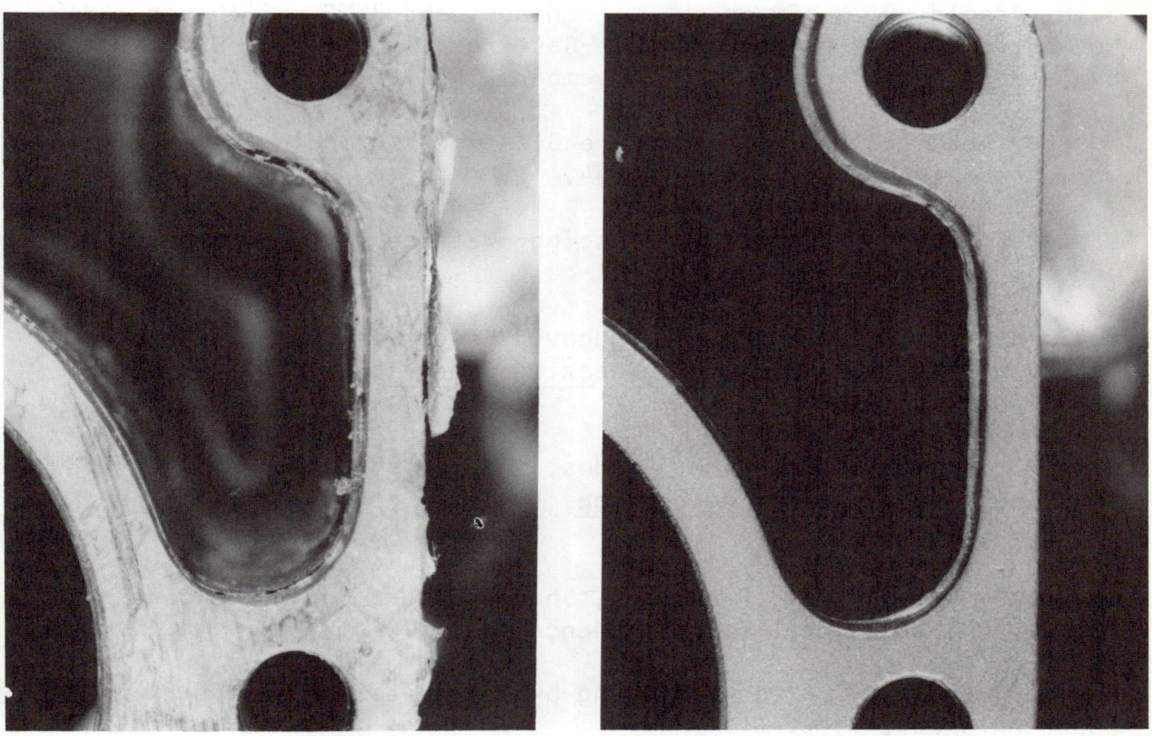

FIGURE 11-12: FLASH ON DIE CASTING

200 parts/hour.

Additional work is in progress in an endeavor to quantify the effect of the chemistry of the abrasive transporting fluid on the microfinish of the finished part, deburring characteristics of the abrasive slurry, and life of the abrasive media.

Summary

While the wet abrasive surface finishing with metal abrasives has been in commercial use for over a decade, the flexibility inherent to the process and its wide range of applicability is not well known. Wet abrasive blasting can be incorporated into a spectrum of machine types ranging from manual cabinets to completely automated, multifunctional, high production units. Thus, the manufacturer has at his disposal a process that will fit into any part-flow system, whether it be manual or automatic, inline or common point load and unload.

References

[1] H. S. Plaster, Blast Cleaning and Allied Processes, Volumes I and II, Garden City Press Limited, Letchworth, Hertfordshire (England), 1972.

[2] Walter Kulischenko, "Air Brasive Jet Machining Techniques and Related AJM Equipment," ASTME (SME) Paper MR68-524, 1968.

[3] O. M. Arnold, D. M. Evans, W. L. Johnson, and R. G. Millhiser, "New Method for Deburring Machined Die-Castings," Society of Diecasting Engineers, Detroit, Michigan, November, 1964.

[4] "Use of Abrasive Grain in Pressure Blasting," Grinding and Finishing, Vol. 9, September, 1963, pp. 41-44.

[5] Lester F. Spencer, "Abrasive Blasting--Wet and Dry," Metal Finishing, June, 1967, pp. 97-100, 104.

[6] D. G. Beletskii and V. N. Zhemchugov, "Basic Parameters of the Vapour Blasting Process," Russian Engineering Journal, Vol. 47, No. 5, May, 1967, pp. 53-56.

[7] John R. Jones and Michael N. Gardos, "An Investigation of Abrasive Blasting," Lubrication Engineering, Vol. 27, November, 1971, pp. 393-400.

[8] Sharon J. Hudson, Jr., President, Sharon Manufacturing Company, Toledo, Ohio, Personal Correspondence, July 16, 1975.

[9] Raymond Harbin, "Surface Finishing by Wet Abrasive Blasting," SME Paper MR75-828, 1975.

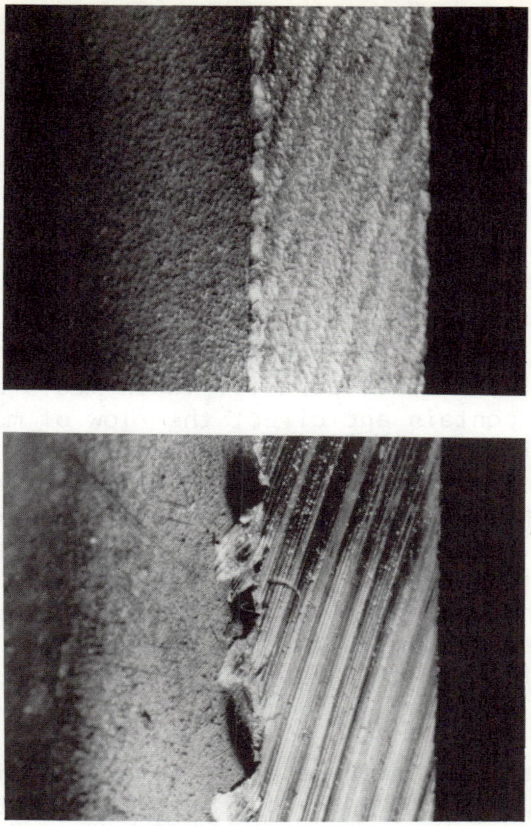

FIGURE 11-13: ROLL-OVER BURR PRODUCED BY MILLING

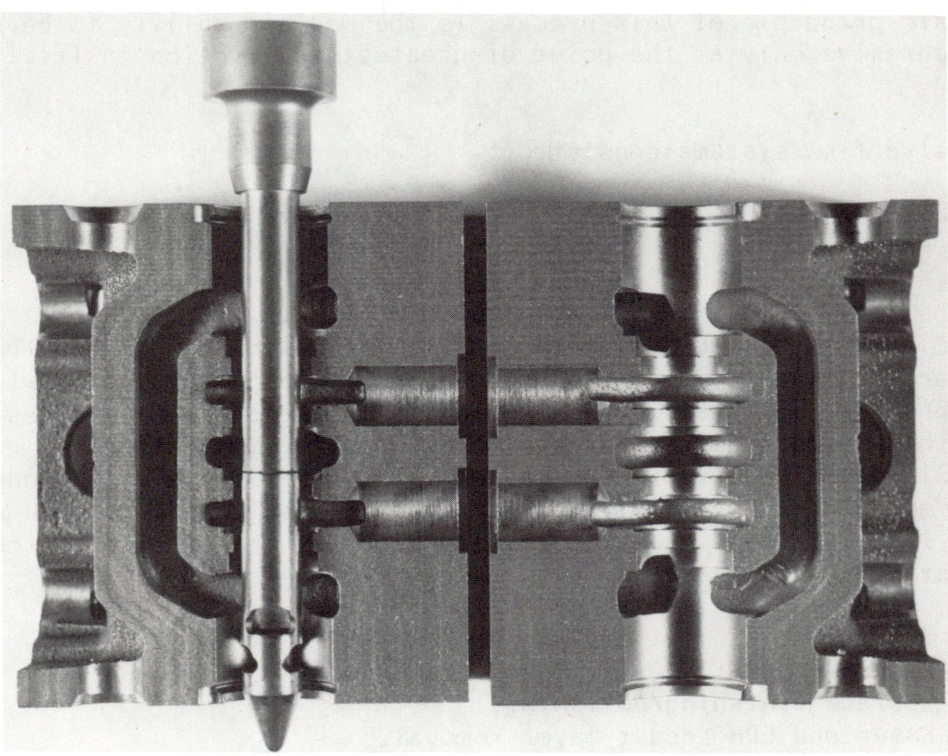

FIGURE 11-14: HYDRAULIC PUMP BODY

CHAPTER 12

ABRASIVE FLOW DEBURRING

Basic Fundamentals

The abrasive flow deburring process is a method of abrading surfaces by extruding a viscous, semisolid abrasive laden media across areas or surfaces. In order to contain and direct the flow of media, the workpiece is confined within a fixture between the two vertically opposed media chambers (or cylinders) of the machine. Under hydraulic pressure, the media is extruded from one media chamber through the fixtured workpiece and into the opposite media chamber. In most applications, the media is then extruded back through the part into the original media chamber. The amount of abrasion that is accomplished with a given media formulation and process pressure is directly related to the quantity of media flow. Depending upon the application, adequate flow quantities range from a few cubic inches to many complete cycles.

The process can be thought of as a flowable file with capabilities ranging from a light buff to coarse stock removal. By using various compound formulations, machine adjustments, and fixture designs, selective processing can be accomplished on many types of components.

A basic principle of this process is the media's ability to be significantly abrasive only at the point of greatest restriction in its flow path.

Abrasive flow systems consist of:

Machine,
Media,
Fixtures.

BASIC MACHINE CONFIGURATION (Figure 12-1 through 12-3): Abrasive flow machines are designed to accomplish and control media flow over a wide range of process parameters. The machines are offered in many sizes and styles with many optional features to meet specific manufacturers' requirements. All units are direct displacement systems designed to extrude the semisolid media through a fixtured workpiece at a selected pressure and velocity. Automatic and manual operation machines are available with accurate, adjustable controls for:

Deburring,
Radius generation,
Surface finishing/polishing,
Laser and EDM recast layer removal,
Post processing of conventionally machined features to eliminate residual surface stresses.

Manufacturer	Model	Distance Between Pillars	Maximum Daylight	Maximum Pumping Capacity	Main Motor Drive	Operation Mode	PSI	Pressure Kg/cm²
Dynetics	G-18-18	18" (45.7cm)	12" (30.5cm)	18 gpm (68 l/min**)	10 HP	Auto	150-475	10.5-33.4
Dynetics	G-36-30	36" (91.4cm)	22" (55.9cm)	30 gpm (114 l/min)	15 HP	Auto	150-475	10.5-33.4
Dynetics	G-36-60	36" (91.4cm)	22" (55.9cm)	60 gpm (228 l/min)	30 HP	Auto	150-475	10.5-33.4
Dynetics	HL-15	14¼" (36.2cm)	8½" (21.6cm)	15 gpm (57 l/min)	7.5 HP	Semi-Auto	150-475	10.5-33.4
Dynetics	HL-60	37" (94.0cm)	26¼" (66.7cm)	60 gpm (228 l/min)	30 HP	Semi-Auto	150-475	10.5-33.4
Extrude Hone	7100	22" (55.9cm)	13" (33.0cm)	5 gpm (19 l/min)	5 HP	Manual	100-1600	7.0-112.5
Extrude Hone	7400	22" (55.9cm)	21" (53.3cm)	10 gpm (38 l/min)	10 HP	Manual	100-1600	7.0-112.5
Extrude Hone	Mark VIIA	29" (73.7cm)	22" (55.9cm)	15 gpm (57 l/min)	15 HP	Auto	100-1600	7.0-112.5
Extrude Hone	1107	35" (88.9cm)	24" (61.0cm)	15/50 gpm* (57/190 l/min)	15 HP	Auto	100-1600*	7.0-112.5
Extrude Hone	Mark VIII	48" (121.9cm)	30" (76.2cm)	30 gpm (114 l/min)	40 HP	Auto	100-1600	7.0-112.5

*Machine can be switched to high flow, low pressure, or standard flow and pressure.
**liters/minute

TABLE 12-1: CAPACITY OF AVAILABLE ABRASIVE FLOW MACHINES

Table 12-1 provides some technical details of available standard machine sizes and capabilities.

1. Media Flow Pressure: The machines are capable of flowing the specially formulated media at any pressure within the specified range. Some systems are limited to low media pressures while others can accommodate either low (100-500 psi) or high (500-1600 psi) pressure ranges so the manufacturer can successfully process a wider range of parts. The manufacturer can select the low pressure setting and flowable media for one application and then merely change media and pressure setting to take advantage of the more aggressive, semisolid media flowing at higher pressures.

2. Media Flow Rate: The rate of flow of media (velocity) through a given workpiece is controlled by adjusting media formulation, media pressure, and the flow control valves provided on most machine models.

3. Cycle Timer and Counter: Both a timer and counter are provided on some automatic machine models to permit the operator to select a cycle length and quantity of cycles to process a given workpiece.

   Once the timer or counter is preset, the machine will automatically repeat its process cycle when the operator places the part holding fixture into position and depresses the clamp close buttons.

These generally firm-bodied abrasive laden chemical substances are the key element for successful application of abrasive flow deburring.

The pressure range used to flow these compounds vary from 100 to 1600 psi (7 to 112.5 Kg/sq cm). This relatively low range enables delicate and complex components to be processed as well as large sturdy parts. Fixturing to contain these hydraulic pressures need not be of massive construction, nylon and aluminum being common fixture body materials.

Part cleaning after abrasive flow deburring consists of compound retrieval for reuse by an air gun. Most of the compound will be removed from the part in this manner even from blind and threaded features. Final cleaning can be performed with commercial solvent or detergent cleaning systems. Ultrasonic agitation of small parts is also a convenient and very effective method.

BASIC MEDIA CONSIDERATIONS: The rheological characteristics of the flowable abrasive laden media are important in the successful application of the process. All media formulations are developed in the laboratory to obtain the following characteristics:

1. Wide range of flowability,
2. Uniform suspension of abrasive grain,
3. Stable--long life without breakdown or separation,
4. Nontoxic for operator safety,
5. Noncorrosive to parts,

6. Maximize aggressiveness to minimize process cycle times,
7. Cohesiveness--to improve handling and cleanliness of operation.

The basic media consists of abrasive grain uniformly suspended in a semisolid/semifluid extrudable matrix. The matrix is carefully formulated to obtain the desired rheological characteristics. Proprietary additives are utilized to obtain the desired consistency, stability, and cohesion. Various sizes, types, and ratios of abrasive grain are utilized in media formulations depending upon the given application. Aluminum oxide, silicon carbide, boron carbide, and diamond grain are utilized in the process. Silicon carbide is the preferred grain for deburring applications. Size ranges from #8 to #600 grit with larger sizes selected for faster stock removal and larger radii and smaller sizes selected for finer surface finishes.

BASIC FIXTURE CONSIDERATIONS: Fixtures are utilized in abrasive flow deburring to:

1. Hold the workpiece in position between the media cylinders of the machine.
2. Direct media flow to selected holes and/or part surfaces requiring finishing.
3. Meter media flow.
4. Protect critical edges and surfaces.
5. Assist in loading, unloading, and cleaning operations.

Fixturing can be either very simple and inexpensive or complex and comparatively expensive depending upon the specific part requiring deburring, the manufacturer's deburring standards and requirements, and production rate required.

Urethane and/or nylon is often used on fixture surfaces in contact with the workpieces. This is done to prevent impinging the workpieces with loose abrasive grain during clamping. Steel and aluminum are normally selected for major fixture details due to their low cost and good machinability. Wear inserts of abrasive resistant tool steel are utilized in wear prone areas of the fixtures.

Fixtures have been designed, built, and used in production for workpieces as small as a precision gear with a pitch diameter of 0.100 inch (2.54 mm) and as large as 47 inch (119 cm) diameter turbine hub. Production rates on small parts of up to 1000 parts/hour have been obtained through proper fixturing techniques.

Media in Action

In abrasive flow deburring, abrasive particles are held somewhat less than rigidly in the flowable, semisolid media. As the media is extruded through the part (which is clamped between the machine's two opposed media extrusion cylinders) it enters the restricting passage within the part to be abraded. If we think of the media passing through this restricting passage as a "solid slug" then the maximum total force which is applied to the media "slug" is simply the media extrusion pressure times the cross sectional area of the passage. If the media were indeed truly solid, all of that force would be transferred to the abrasive grains at the outer surface of the media "slug" in contact with the surfaces or edges of the restricting passage being abraded. But the media is less than solid and as a result less than the total force applied to the "slug" is transferred to the cutting grains on its outer surface. The less solid the abrasive media is, the less force will be transferred to the cutting grains, and the less deep the cut made by each grain will be. The remainder of the force applied but not transferred to the cutting grains is expended moving the media within itself.* As the viscosity of the semisolid matrix decreases, its ability to carry the abrasive grains along in the extruding slug decreases. Abrasive grains contacting the surface slow down when they encounter resistance, and if they are to do any cutting, they must overcome the resistance of the metal to being cut. A pure Newtonian fluid does not have a sufficient bond between incremental units of itself to even overcome the friction of the fluid moving along the walls of the passage through which it flows.

Therefore, the magnitude of the force on the cutting grain moving along the surface (or across the edge) of the workpiece being abraded is dependent upon the viscosity of the media matrix as well as the extrusion pressure. A pure solid applies the most force; a pure fluid applies the least.

Of course, a pure solid will not flow at all, so we must deal with semisolids. The lower the viscosity of the semisolid, the faster it will flow. Assuming that it continues to flow more or less like an extrusion--exerting sufficient force on the cutting grains at the surface to overcome the resistance of the metal to being cut and pulling these grains along with its flow--lowering the viscosity of the semisolid increases its flow rate through a given size passage at a given extrusion pressure and therefore increases the number of individual cutting grains forced across the surface (or edge), and therefore the number of cuts. However, in order for an increase in media flow rate to cause a portional increase in stock removal rate:

1. The depth of cut made by each grain must remain the same.
2. The flow rate of the media along the cutting surface must increase at the same rate as the total flow through the passage.

As the semisolid media becomes more like a fluid, it will more readily flow within itself transferring less force to the surface cutting grains which decreases the depth of cut and does not proportionally increase the speed of those grains moving along the workpiece surface.

Therefore a reduction in viscosity of the media matrix resulting in an increase in flow velocity through a restricting passage will not produce a <u>proportional</u> increase in cutting efficiency. It may nevertheless produce an increase--perhaps a substantial one. Additional reduction in viscosity

---

*This unproductive use of energy creates heat unnecessarily and very likely causes the abrasive grains to wear prematurely due to the unnecessary rubbing of abrasive particles against one another.

FIGURE 12-1: EXTRUDE HONE MARK VII-A MACHINE

FIGURE 12-2: AUTOMATIC ABRASIVE FLOW MACHINE

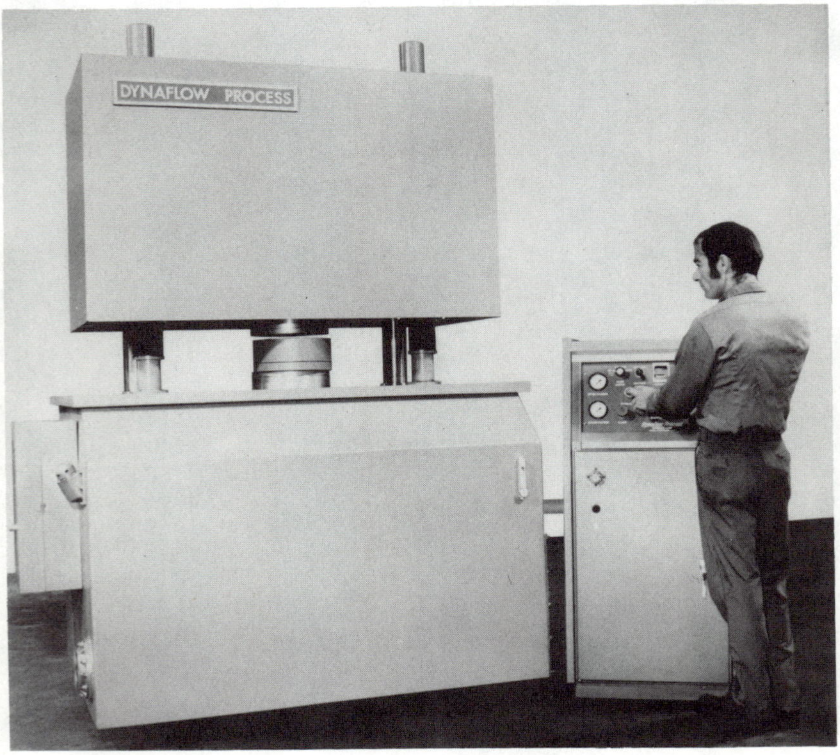

FIGURE 12-3: SEMIAUTOMATIC ABRASIVE FLOW MACHINE

and increases in velocity will provide ever diminishing increases in cutting efficiency until at some point additional reductions in viscosity and increases in flow velocity can no longer overcome the consequent reduction in force applied to the surface cutting grains and cutting efficiency decreases.

What this optimum viscosity is depends greatly on the size of the restricting passage and the maximum extrusion pressures available. "Extrusion" flow can be achieved with a much lower viscosity media in a .020 inch diameter hole than in a .200 inch hole or a 2.000 diameter hole. Consequently media suppliers offer a wide range of media viscosities for all types of applications.

As in traditional machining, there is a play off between "speeds" (the number of cuts per time unit) and "feeds" (the depth of each cut).

1. "Speed"--the number of cuts per time unit--is a function of volumetric flow rate and uniformity of flow rate through the cross section of the matrix "slug."
    a. Volumetric flow rate is increased by increases in extrusion pressure and decreases in viscosity.
    b. Uniformity of flow rate through the cross section of the matrix slug increases with an increase in media viscosity.
2. "Feed"--depth of each cut--increases with increases in extrusion pressure and viscosity of media.

Some equipment offers media extrusion pressures--ranging from 200 to 1600 psi--and media delivery capacities up to 100 GPM (in low pressure configurations). Some media can be used at high or low pressure and cover a spectrum from very high to very low viscosities. This generous selection of abrasive flow process variables permits the user to optimize the process for his particular application and has contributed substantially to the success achieved by over 200 machines currently in the field on hundreds of diverse applications.

Typical Process Cycle Details

Parts requiring deburring are manually loaded into a fixture and the loaded assembly then placed into process position on the machine. The operator then initiates the preset automatic process cycle by pressing the clamp close buttons.

During the automatic process cycle, the operator is available to unload processed parts, blow off fixtures and parts, and reload new parts into the fixture.

Upon completion of the automatic cycle, the machine will open and the operator can remove the processed fixture and position the second fixture in the machine. The operator then repeats the sequence listed.

Utilization of two or more fixtures reduces production "dead" time to a few seconds and thus maximizes production rate.

Numerous options are available such as rotary tables, refeed units, and unloaders to increase production rates.

One manufacturer of abrasive flow equipment uses polyurethane diaphragms as a piston to drive media back and forth. These diaphragms are mounted one above and another below the fixture, inside steel domes that act as chambers for the compound load. An advantage to using this design is that there are no mechanical moving parts to wear or jam with abrasive particles.

In this equipment, parts are placed in suitable fixtures and clamped in the fixture holder. When the automatic cycle is actuated, compound is fed from the reserve tank containing fifteen gallons of compound into the processing area via the hose to the top machine dome (Figure 12-2). After the processing area is fully loaded with compound then the compound is pumped by the diaphragms back and forth through the parts. While the machine is cycling a second fixture is loaded with parts to be processed. When the first fixture load is processed, the fixture holder returns to the load station, the fixtures are exchanged and a repeat automatic cycle initiated. Sliding machine components are polyurethane coated to resist wear and jamming caused by the abrasive particles in the compound. A full flow inline check valve is incorporated in the reserve tank hose to prevent compound flowing back to the reserve tank during processing. Specifically designed antisurge valves eliminate hydraulic shocks which could damage fragile parts and worm gear screw jacks clamp fixtures by torqueing out through a friction clutch.

A low reserve tank level stops machine functions and triggers an audible signal. Whenever the process is stopped, restart continues precisely from the point the action was suspended. These features ensure uniform treatments of parts according to the predetermined program.

Figures 12-4 and 12-5 show a Model G-36-60 machine fitted with a large fixture load system and hoist. The load station for this unit is at the rear of the machine. When not in use normal fixture processing can take place at the front of the machine. When large fixtures are to be processed the standard fixture load system is disconnected from the auxiliary hydraulic tank and the hoses from the large fixture load system are then connected at the same quick-disconnect fittings. Fixtures up to 36 inch diameter can be accommodated using this unit and it is ideal for parts such as turbine disks. The lower half of the fixture is secured to the unit and the part is loaded into it. The upper half of the fixture is then located in position and clamped. The assembly can then be moved to the process station on the antifriction bearing blocks by winding a handle, a chain, and sprocket arrangement providing the required mechanical advantage. Once in the process station a handle is twisted to actuate the automatic cycle. Before compound filling and cycling commences three small hydraulic cylinders located under the fixture plate allow the fixture to rest on the lower machine dome and the moving platen clamps the fixture. At the end of the preset automatic cycle, the moving platen raises and the three small hydraulic cylinders raise the fixture allowing the unit to be moved back to the load station.

FIGURE 12-4: LARGE FIXTURE LOAD SYSTEM

FIGURE 12-5: LARGE FIXTURE IN MACHINE

Cooling units for cooling media and hydraulic systems are used in some high production systems. A compound reclamation cabinet is used to minimize loss of compound. Parts are normally purged of compound using an air gun. All compound reclaimed is reusable. A fixture separation device can also be used for initial separation of fixture halves. This is mounted to the work table of the machine for convenience.

Practical Fixture Considerations

Each part normally requires its own fixture but in some cases similar parts can be accommodated in "shuttle" fixtures using various replaceable inserts.

As indicated earlier, fixtures are necessary to provide correct orientation of parts in relationship to machine compound chambers with the necessary porting in the fixture to allow compound through flow. Figure 12-6 shows a tubular component with holes drilled through sidewalls which have to be deburred. The left view shows arrows denoting compound flow from lower chamber to upper at which stage the primary action is deburring internal intersections. Right view shows flow reversal and primary action is deburring outside intersections.

In many parts, the holes or features to be deburred or otherwise processed form a natural restriction for the compound flow. It should be noted that no abrading action occurs where compound is flowing at a relatively

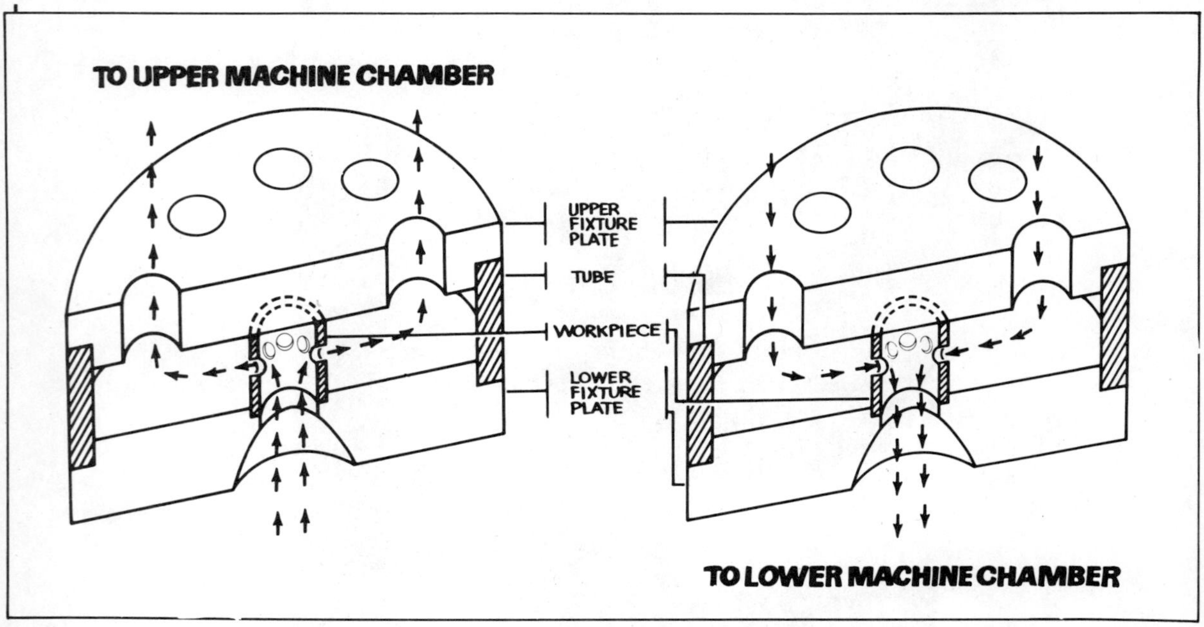

FIGURE 12-6: FIXTURING FOR SIMPLE PART WITH INTERSECTING HOLES

low velocity. This relativity is based against the flow rate at the greatest restriction in the part/fixture arrangement.

Greatest Action at Greatest Restriction

Least Action at Least Restriction

From this simple rule it is apparent that all configurations of components that have to be deburred or otherwise processed do not provide natural restrictions. In these cases, a restrictor must be an integral part of the fixture design. In production tooling these are made from polyurethane because of its abrasion resistant properties.

For maximum machine utilization, dual fixtures are used on the Automatic Series machines, one being unloaded/cleaned/reloaded with workpieces while the other is being processed.

Two-way processing of parts (as shown in Figure 12-6) is most common as both entrance and exit of holes are processed. In situations where action is required on only one side of a hole then a simple flap or piston check valve can be incorporated in the fixture. This enables flow through the part in one direction and bypassing the part by allowing flow through the check valve in the opposite direction.

The feasibility for processing is normally determined using temporary or "soft" tooling. This is usually fabricated in a single cavity configuration. After successful tests, compound formulation, process times, pressure and flow volumes are recorded. These process parameters are then extrapolated into realistic production rates, production fixturing costs, and ideal type of machine. These optimized elements enable cost evaluation of the complete deburring/finishing facility.

Terminology

For the purpose of simplifying the presentation of the mechanics involved in abrasive flow deburring and due to the lack of standard definitions, a few descriptive terms must be considered.

The generalized typical machining burr (Figure 12-7) is composed of:

1. Flag--the section of the burr which extends visibly beyond the surface of the workpiece. It is characterized by large surface area and small cross sectional area.
2. Trunk--the section of the burr just beyond the workpiece. This area of the burr is composed of extruded and laterally displaced metal. It exhibits a larger cross sectional area than the flag.
3. Hinge area--the area of transition from the flag to the trunk areas.

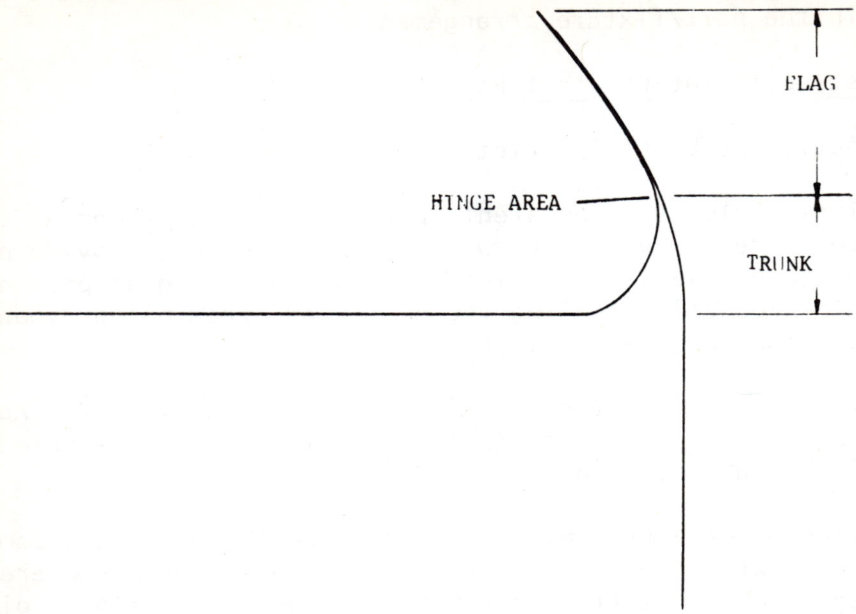

FIGURE 12-7: GENERALIZED TYPICAL BURR

Burr Removal Theory

The abrasive flow process removes burrs by directing the flow of abrasive laden media against the burr (Figure 12-8). The media flows against the burr and the force is applied against the relatively large surface area of the flag. Once the force exceeds the rigidity of the hinge, the flag bends in the direction of media flow. When the direction of media flow is reversed, the same thing occurs and the flag bends back. The series of flow reversals causes the hinge area to fatigue and the flag to break off leaving the trunk section. The pressure, flow time, and number of cycle reversals required depends upon the workpiece material, burr geometry, and burr location. Removal of the flag normally is done in a matter of seconds. In many applications, this meets manufacturers' requirements for deburring.

In most applications, manufacturers require removal of the trunk section of the burr. The trunk exhibits a larger cross sectional area than the flag therefore it normally will not bend under the force of media flow. The trunk is removed by the grinding action of the abrasive laden media as it flows across the trunk. The rate of this grinding action can be carefully controlled and varied from extremely aggressive to very gentle depending on workpiece conditions and manufacturing requirements. Generally speaking the aggressiveness of a media is dependent on several factors which include:

- Hardness and size of abrasive grain,
- Sharpness of abrasive grain,
- Unit pressures between abrasive particles and workpiece surfaces,
- Speed and direction of movement of the abrasive grain in relation to the workpiece,
- Consistency of the media matrix,

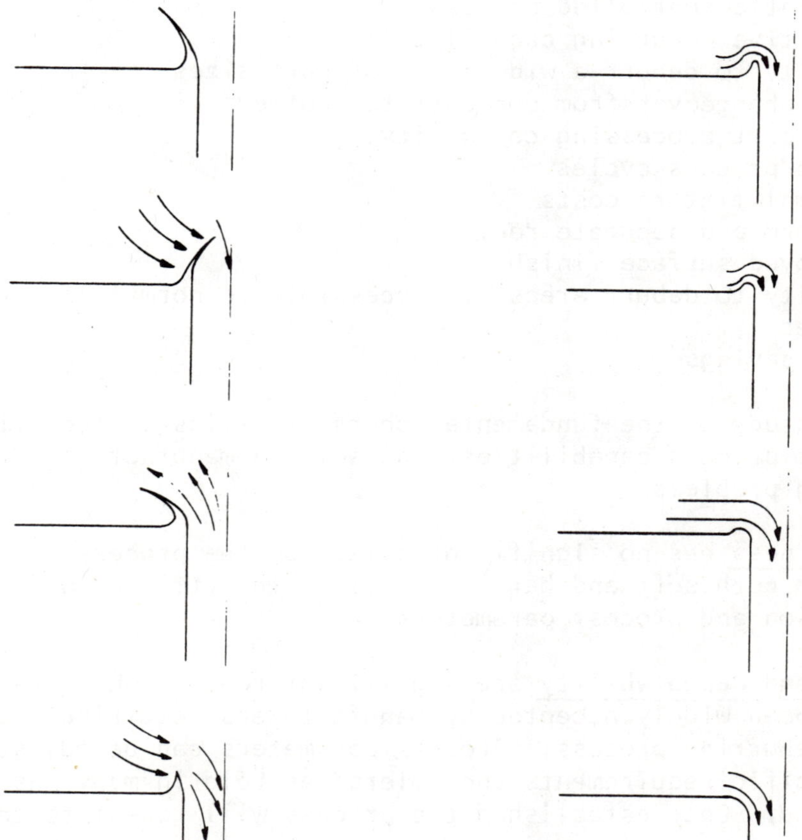

FIGURE 12-8: THE REMOVING OF A BURR BY ABRASIVE FLOW DEBURRING

. Process pressure.

As discussed previously, the abrasive flow deburring process is extremely versatile and flexible in all the above areas and therefore it can be adjusted to meet almost any deburring requirements.

Deburring Characteristics, Advantages

As a relatively new deburring process, abrasive flow deburring has gained a wide acceptance by manufacturers because it offers several advantages for deburring applications. These include:

1. Ability to deburr all metals
2. High degree of accuracy and repeatability
3. System flexibility
4. Automatic controlled process
5. Selective deburring capability
6. Ability to deburr a wide range of part sizes
7. Easy changeover from one part to another
8. Multipart processing capability
9. Short process cycles
10. Minimal fixture costs
11. Uniform and accurate radii
12. Improved surface finish
13. Ability to deburr areas not accessible to normal deburring techniques
14. Cost savings

A closer study of the fundamental characteristics listed will reveal the significant process capabilities available to manufacturers to meet their deburring problems.

Metal Hardness has no significant effect on the process. This process can be used for both soft and hard materials with little or no change in media formulation and process parameters.

Accuracy and Repeatability are significant reasons why abrasive flow deburring has been widely accepted by manufacturers requiring a uniform high quality deburring process. Process parameters can be adjusted to manufacturers' specific requirements and tolerances held in many cases to 0.005 inch as required. Once established the process will repeat to the preset parameters.

System Flexibility is unmatched by any other commercial deburring process. All process parameters can be easily adjusted within a wide range to provide the desired process cycle. Media formulations can be adjusted to meet almost limitless part requirements and fixtures can be designed to accomplish desired finishing action.

Automatic Controlled Process that eliminates the possibility of operator error and reduces scrap. Once the process parameters have been set on the machine all the operator is required to do is load parts and initiate the automatic cycle. Unskilled labor can operate the machine with minimum training and with the same accuracy and repeatability expected of skilled people.

Selective Deburring of specified areas of a workpiece can be easily and economically accomplished by using fixturing to direct media flow only to the desired area. All other areas of the part will be unaffected.

Ability to Deburr a wide range of part sizes is another positive feature of the process. Precision miniature components to large 47 inch diameter turbine hubs are currently being abrasive flow deburred with speed and accuracy.

Easy Changeover from One Part to Another is especially important to the manufacturer with low volume runs on several different parts. The complete changeover of media and fixtures and the resetting of process parameters can normally be accomplished in minutes.

Multipart Processing Capability--depending upon the size of part and machine, many parts can be loaded into a fixture and deburred simultaneously. Thus high production rates can be achieved and per part cost minimized.

Short Process Cycles are a characteristic of abrasive flow deburring. Actual process cycle times are dependent upon workpiece condition and manufacturers' deburring requirements. Cycles of less than one minute meet most deburring requirements.

Minimal Fixture Costs are often realized due to the simple, straightforward designs normally utilized. Manufacturers can easily design and build their own tooling after a brief period of training provided by one of the equipment manufacturers. Complete tool design and tool building services are also available.

Uniform and Accurate Radii are a side benefit obtained from the process. The flow of media across the edge generates uniform radii unmatched by any other conventional deburring process.

Improved Surface Finish and high surface integrity are other significant side benefits of the process. The process produces a "stress free" surface with a very fine finish. Surface finishes down to 2 microinch have been produced on steel workpieces.

The process avoids the possible problem areas of intergranular attack with electrochemical deburring, oxide coatings with thermal deburring, and heat affected zones of EDM.

Cost Savings are possible in many deburring applications. The aforementioned items highlight many cost savings and advantages of abrasive flow deburring. However, often traditional accounting techniques overlook the intangible savings, many of which could be directly attributed to this process. For example, what value does a company place on improved part quality and service life? What is the value of extreme reliability and consequently greatly reduced inspection and rework?

Limitations

All deburring processes have limitations and this process is no exception. The process is automatic and will produce uniform repeatable results only if the burr conditions presented to the process are relatively consistent. The process is flexible and can accommodate changes in burr conditions but the operator must then make decisions about process parameters.

As in all deburring processes, there is a limitation as to the size of burr that can be removed without affecting part dimensions significantly. Process parameters can be adjusted to minimize this factor and on close tolerance parts, allowance can be made for stock loss which accompanies deburring. The process can remove any size burr as long as part tolerances and process time available permit.

Blind holes and cavities are difficult and often impossible applications for the process since it is based on the controlled flow of media, and blind holes prevent conventional flow.

The machine, fixtures, and media normally are more expensive than vibratory or tumbling equipment and, therefore, if satisfactory deburring can be obtained utilizing these processes, it is more economical to do so.

Applications

As seen in the following illustrations, abrasive flow deburring has been applied to a wide variety of materials, shapes, and workpiece sizes.

Figure 12-9 illustrates some configurations ideally suited to abrasive flow deburring.

Figure 12-10 shows a stamping which requires burr removal, radiusing, and polishing of the thread contact areas. A production rate of 500 parts/hour can be achieved processing 16 parts/fixture load. Quality, uniformity, and reliability are dramatically improved with about a two-thirds time savings over the previous hand method.

Figures 12-11 and 12-12 provide before and after views of a drilled retainer deburred by abrasive flow deburring. Absolute burr removal was accomplished using Extrude Hone media and equipment in 30 seconds flow time. Up to 20 pieces can be processed simultaneously.

Removal of all burrs, generous radiusing, and fine polishing is required on the tiny part shown in Figure 12-13. Anything less than an excellent deburring and finishing job makes the bracket extremely irritating when worn on the tooth. Three rings, each holding 20 brackets, are abrasive flow deburred in 90 seconds--about 1.5 seconds per bracket.

Removal of all burrs and generation of .005 inch to .009 inch radius on all edges is required on the retainer shown in Figures 12-14 and 12-15. Up to four parts (depending on size) are abrasive flow deburred in under one minute.

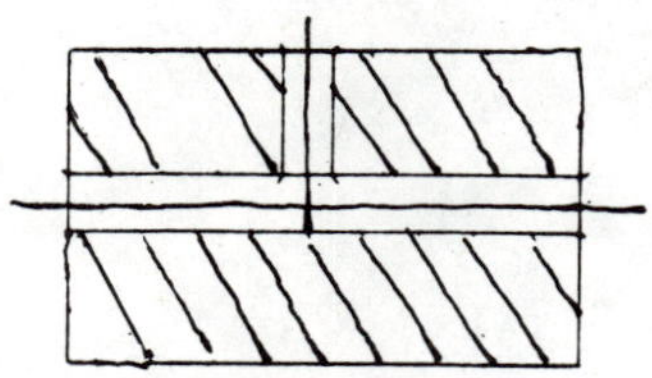

Internal intersecting holes

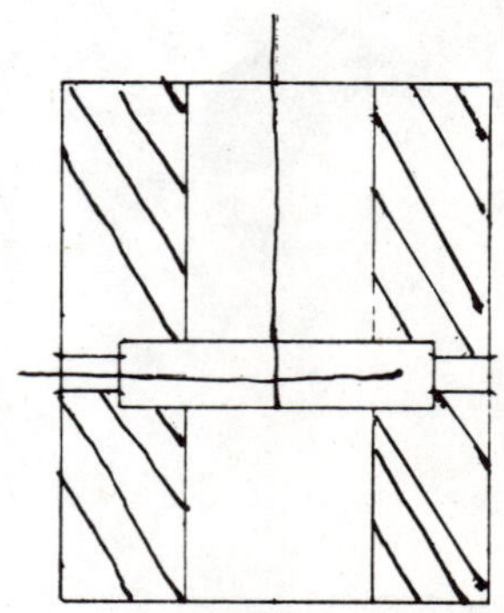

Hard to reach areas

Series of holes

FIGURE 12-9: GOOD EXAMPLES OF ABRASIVE FLOW APPLICATIONS

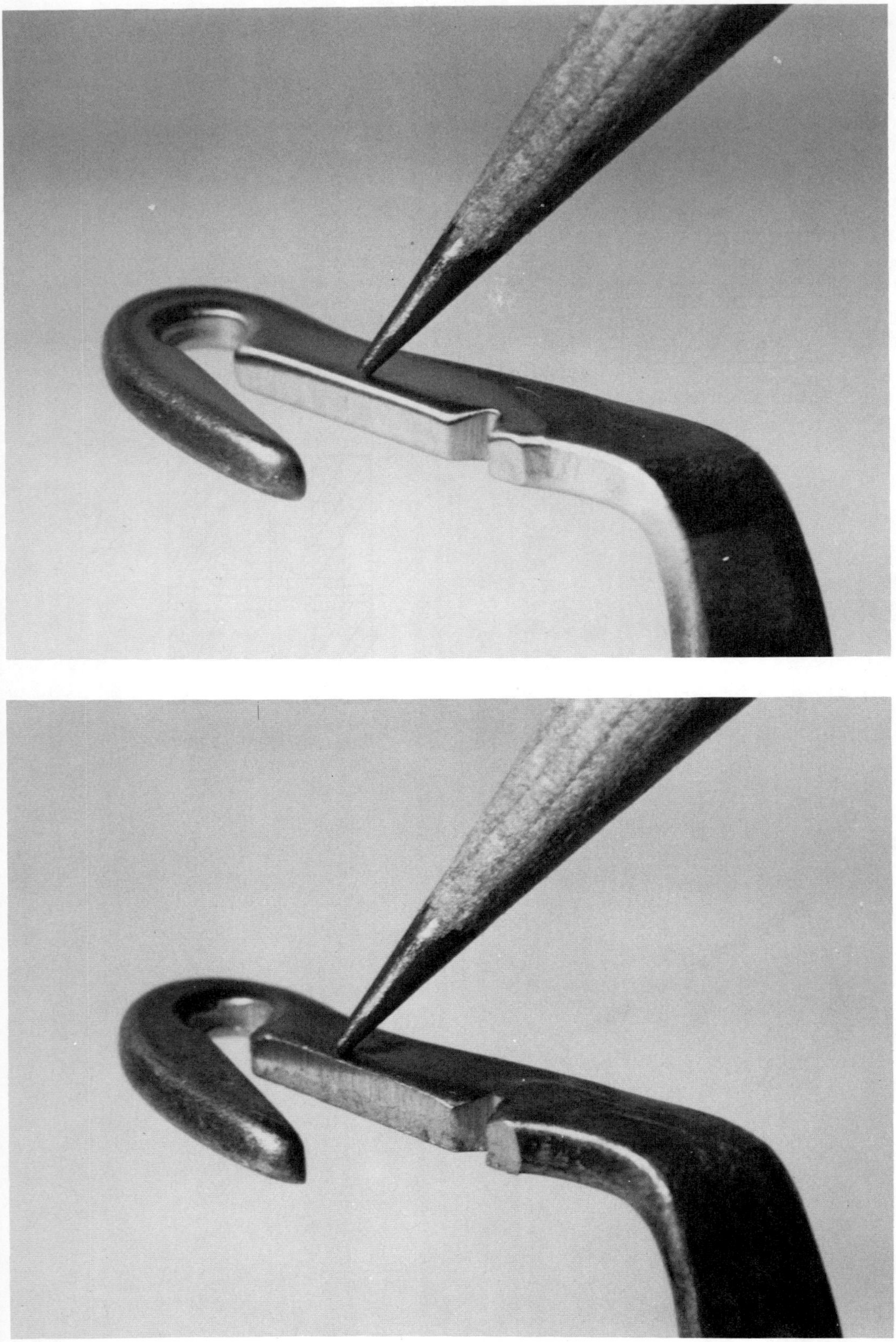

FIGURE 12-10: SEWING MACHINE TAKE-UP LEVER (STEEL)

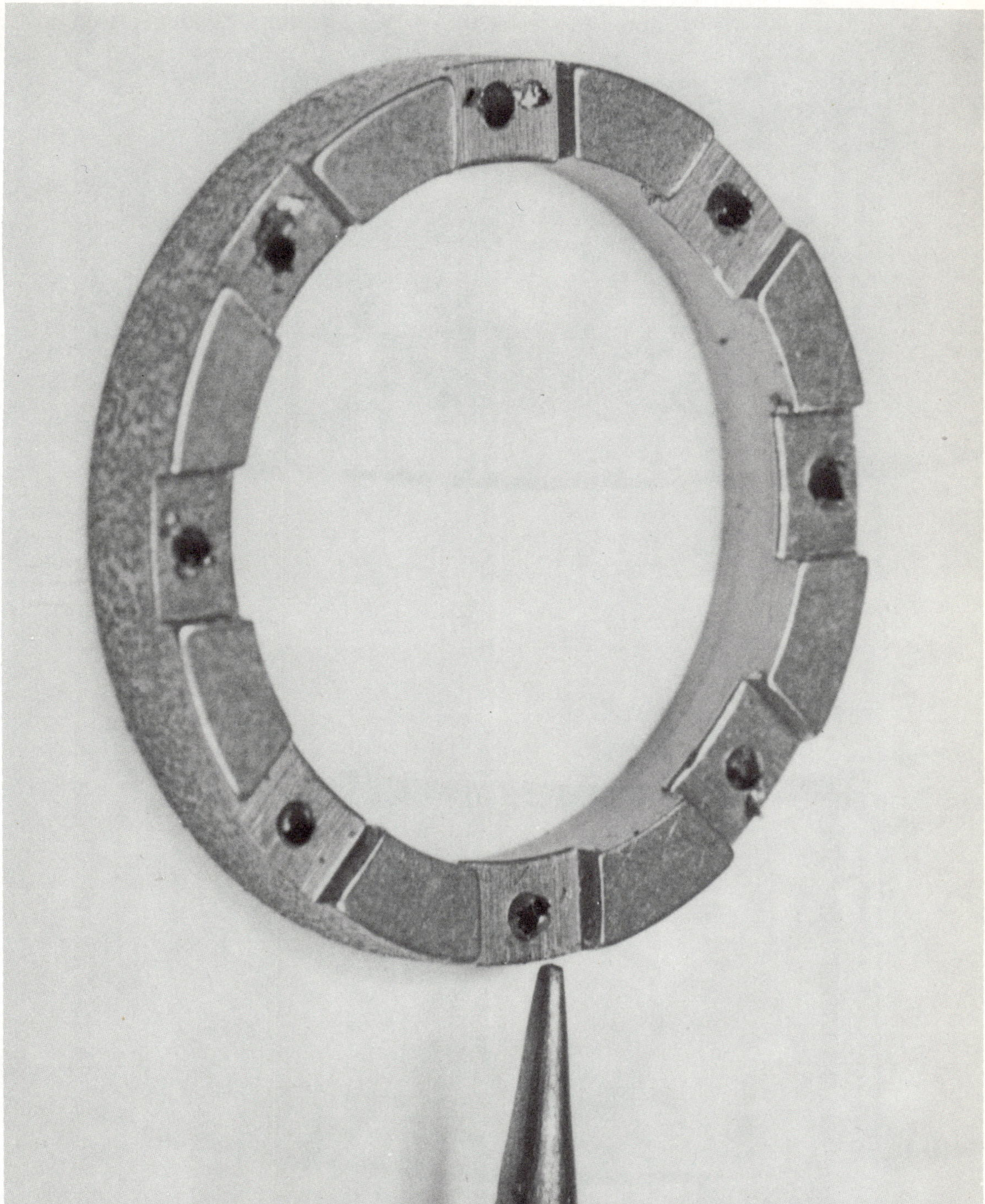

FIGURE 12-11: RETAINER RING AND PENCIL POINT

FIGURE 12-12: RETAINER (4340 STEEL) AT HIGH MAGNIFICATION

FIGURE 12-13: DENTAL BRACE BRACKET (STAINLESS STEEL)

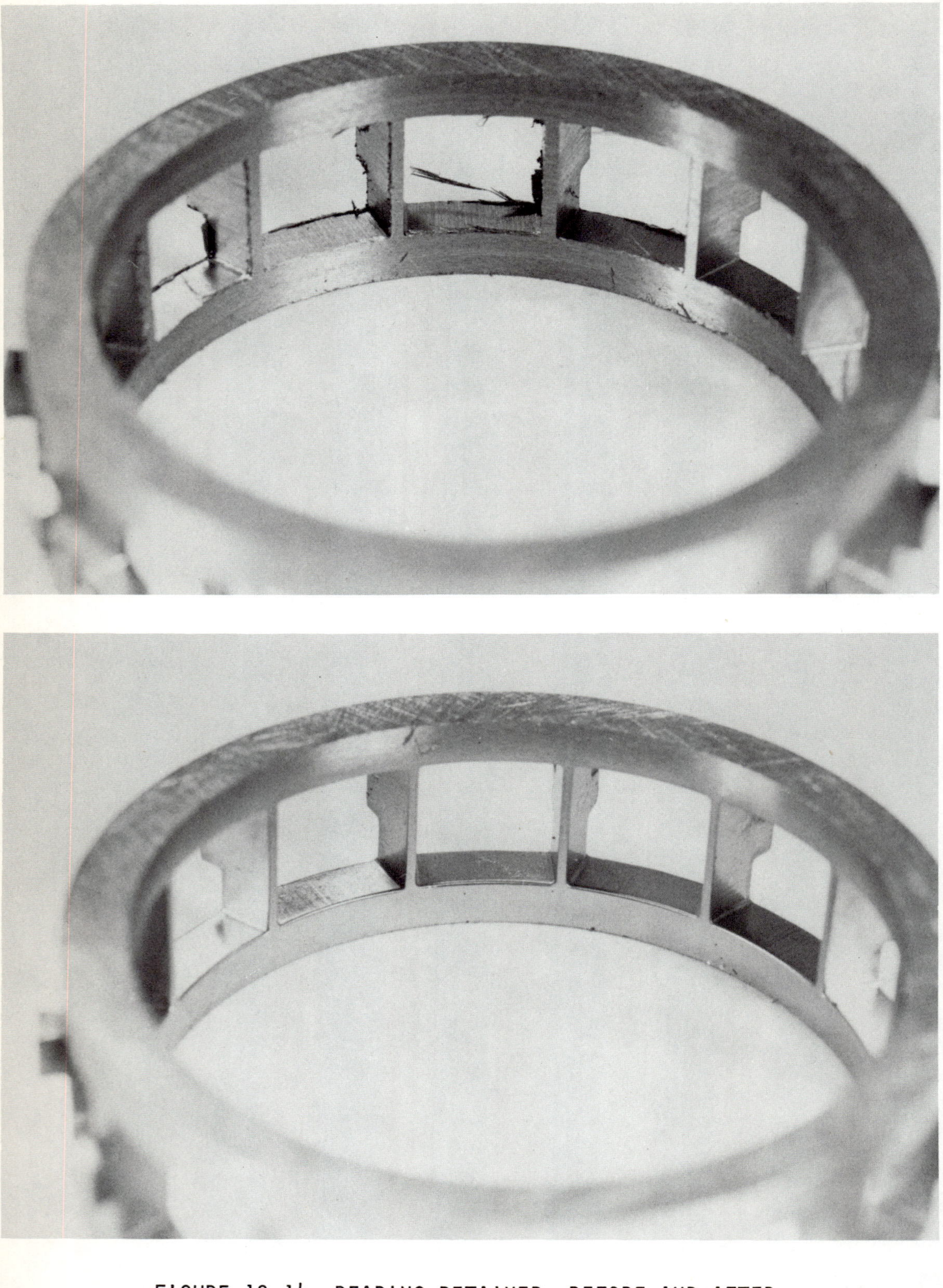

FIGURE 12-14: BEARING RETAINER--BEFORE AND AFTER

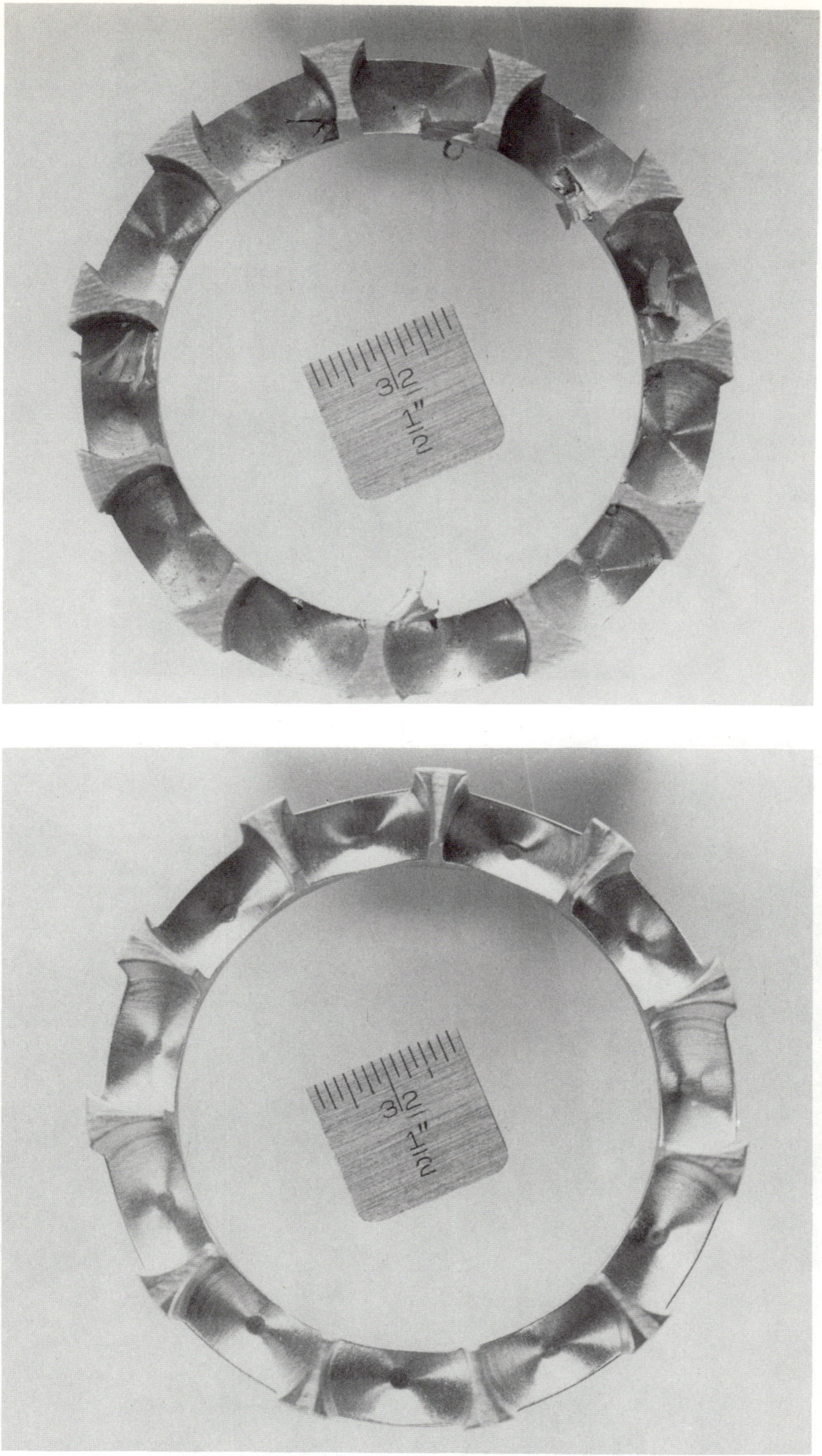

FIGURE 12-15: CROSS SECTION OF BEARING RETAINER (BRASS)

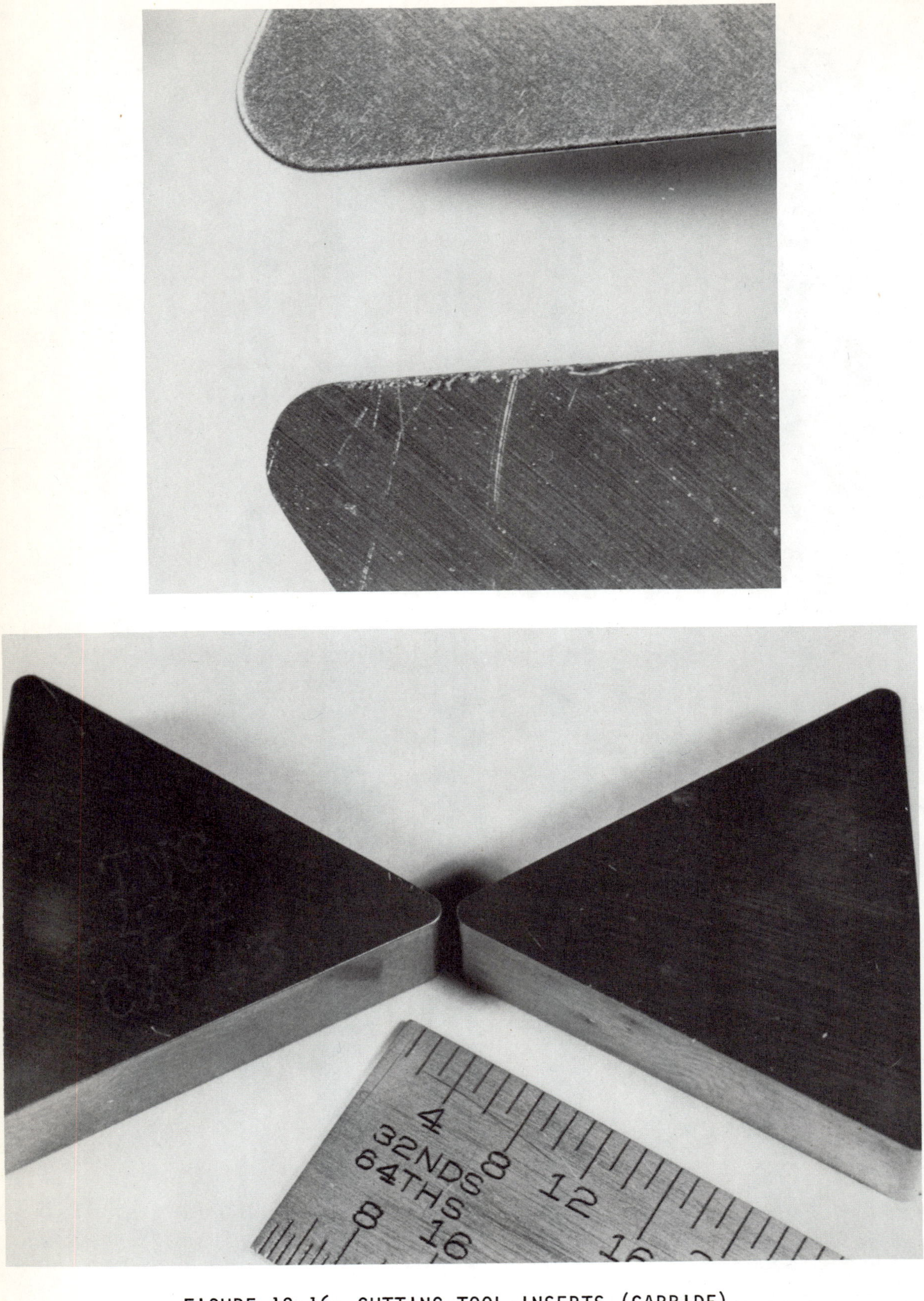

FIGURE 12-16: CUTTING TOOL INSERTS (CARBIDE)

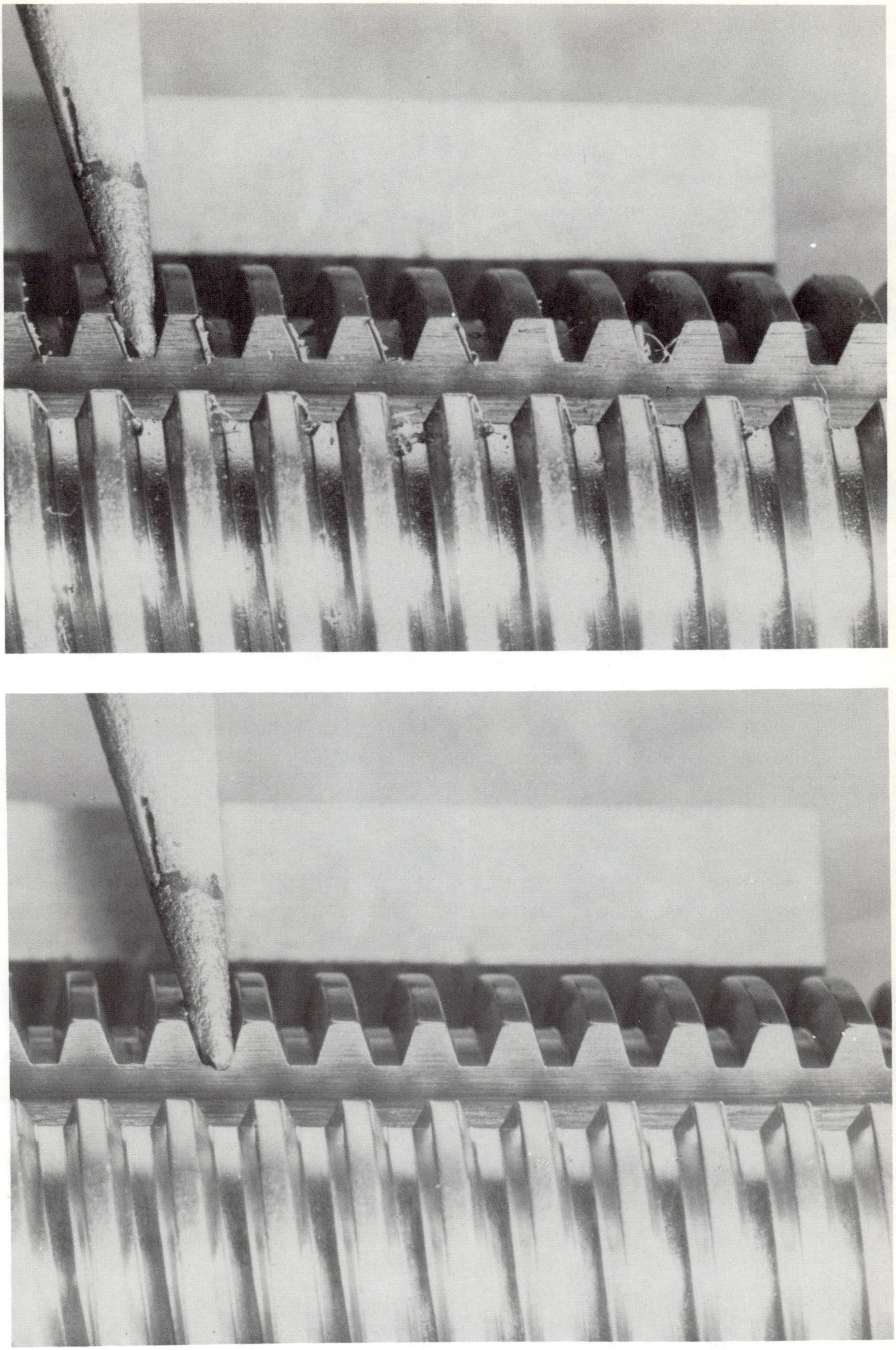

FIGURE 12-17: LEAD SCREW (COMPUTER APPLICATION) (STEEL)

FIGURE 12-18: CLOSE UP OF SCREW

Removal of sharp edges and production of an extremely uniform radius of up to .004 inch is required on the inserts in Figure 12-16. Twelve parts are processed in under three minutes and the flank of the insert is polished in the direction of tool movement at the same time.

The screw in Figures 12-17 and 12-18 is 18 inches long and approximately 1/2 inch in diameter. Milling of the slot along the length of the screw produces a burr at each thread intersection. The ground thread has a very close tolerance. Burrs are removed (to pass 20X inspection), controlled radii are produced on the edges of the thread teeth, and dimensional and surface finish requirements (16 microinch) are maintained. Two lead screws are processed simultaneously in three minutes.

In Figure 12-19 removal of burrs thrown into the .024 inch acetylene preheat holes and the .036 inch diameter oxygen hole in the center is required. Tolerance on the oxygen hole is .0005 inch. Holes are simultaneously polished to improve gas and oxygen flow resulting in a very high quality cutting flame with the "stinger" 30 percent longer than on an unpolished tip.

Figure 12-20 illustrates an aluminum metering sleeve which must be deburred while producing a radius between 0.001 and 0.003 inch (25.4 to 76.2 μm). This part has 16 holes 0.033 inch (0.84 mm) in diameter which intersect the center 0.315 inch (8 mm) honed bore. The holes are machined to low limits to allow for 0.002 inch (0.05 mm) stock removal during processing and a pin is passed through each hole prior to processing to weaken burrs and ensure uniform radius generation. Seven parts are processed in a fixture, compound being allowed to enter and exit from both ends of main bore for uniform radiusing, and pressures of 300 psi (21 Kg/sq cm) are used. Process time is 12 seconds per fixture loading using an easily flowable No. 220 silicon carbide compound.

FIGURE 12-19: OXYGEN/ACETYLENE TORCH TIPS (BRASS)

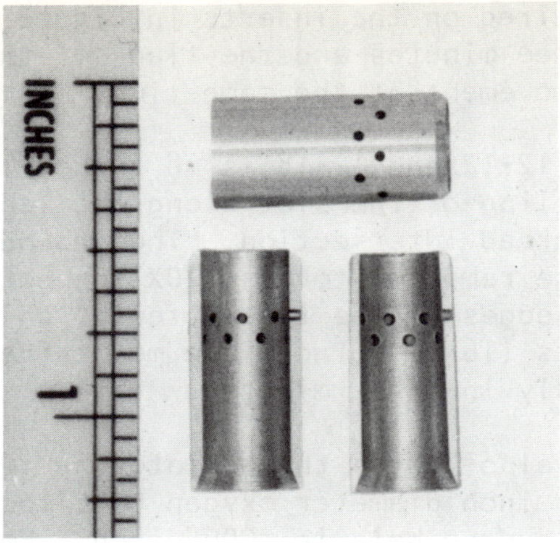

FIGURE 12-20: ALUMINUM SLEEVE

FIGURE 12-21: STEEL MICROWAVE COMPONENT

FIGURE 12-22: MICROWAVE COMPONENTS IN FIXTURE

A steel microwave component is shown in Figure 12-21. The requirement on this part is to deburr ends of internal vanes after facing operations, remove all traces of metal "smear" and improve surface finish through the vanes. These results are obtained by using the four cavity nylon fixture shown in Figure 12-22. A number 36 grit silicon carbide is used at the media pressure of 475 psi (33.4 Kg/sq cm). The total cycle time of six minutes saved twenty-nine minutes/part over former hand methods.

Figure 12-23 shows a jet engine carburetor part from which core flash had to be removed and air flow passages polished. This investment casting was salvaged after flowing compound around the airflow passage as shown. A restrictor forms an integral part of the fixture to concentrate flow at core flash area. The part measures 2.6 inch (6.7 cm) diameter and the airflow passage is inaccessible for hand finishing or other types of deburring processes.

An aircraft hydraulic system swivel connector is shown in Figure 12-24. The twelve drilled hole intersections have to be deburred on this part. The small holes are 0.152 inch (3.9 mm) in diameter. All feather edges have to be removed.

This stainless steel component is processed conventionally, i.e., back and forth through the drilled holes. Total process time is 90 seconds using a compound containing No. 20 silicon carbide at a pressure of 475 psi (33.4 Kg/sq cm). Consistent burr removal and radiusing of all sharp edges result along with general surface finish improvement in the machined passages.

Figure 12-25 illustrates a similar component. The requirement is for burr free edges with a controlled edge break around the holes through the wall of the component particularly in the bore. For this application the following three methods were used in the following sequence:

Hand finishing
Electrochemical deburring (ECD)
Abrasive flow machining (AFM)

Hand finishing resulted in high rejection rate and lack of uniformity, inspection costs were high, and some infield problems were traced to burrs that remained on the components becoming loose in service.

Electrochemical deburring was the second method used and showed some drawbacks with this part. They are made in small batches along with a family of similar parts and changeover time along with fixture cost made it less economical. The large hanging burrs had to be removed before processing and the outside intersections had to be deburred and chamfered by conventional means. Discoloration of the part also took place.

Abrasive flow processing was then evaluated. Using very simple fixturing, sample parts were processed. As with the ECD method, the large hanging burrs had to be removed to allow for controlled radiusing but other advantages became apparent. They were:

1. The outside intersections were also deburred and radiused.

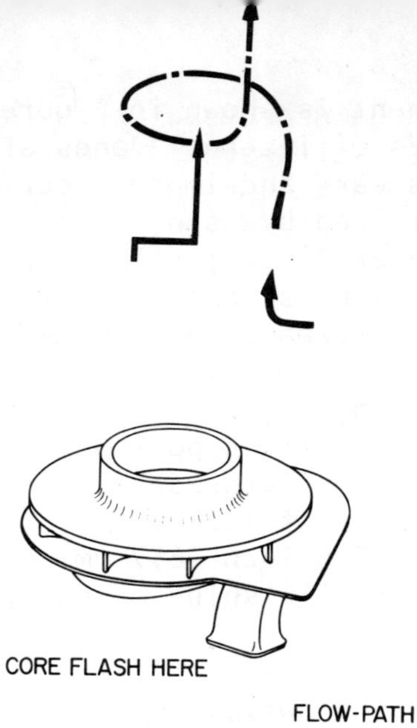

FIGURE 12-23: CARBURETOR COMPONENT

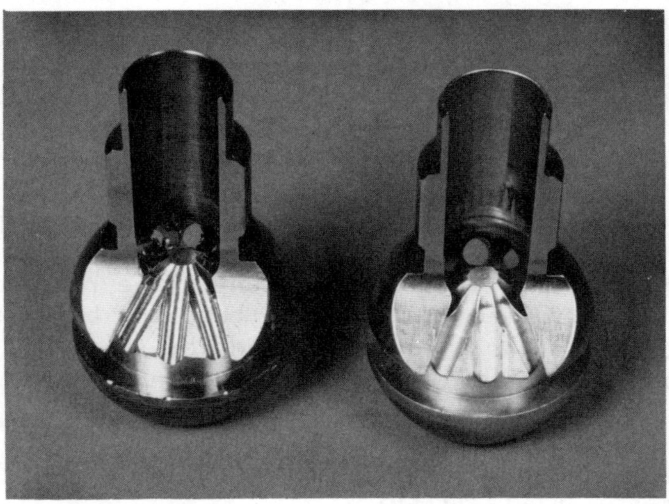

FIGURE 12-24: HYDRAULIC SYSTEM SWIVEL CONNECTOR

FIGURE 12-25: HYDRAULIC SERVO COMPONENT

2. Similar parts could be processed in the same fixturing simply by replacing inserts to suit each part number.
3. Changeover from one part number to another was fast.
4. Uniform, tangentially bended radii were produced.
5. Absolute burr removal was assured.

Figure 12-26 illustrates the four cavity production fixturing which was used. Note the removable polyurethane coated inserts.

Figure 12-27 illustrates slide tracks from an armament assembly. These investment cast components have a surface finish requirement around a "U-shaped" slot and this forms the track of a feeding mechanism for a high speed armament assembly. The following methods were considered:

Conventional machining
Tumbling
Hand finishing

Conventional machining (milling, grinding) was eliminated as the actual form of the track varied considerably. Burrs would also be produced by these methods which would mean another operation.

Tumbling media large enough to be effective would not enter the channel to be finished thus eliminating this method.

Hand finishing seemed feasible where the track is open on one side but where it is encased on both sides it was a time consuming and relatively nonuniform method.

The manufacturer of these components then became aware of low pressure AFM. Trials were conducted on experimental fixturing using process parameters from similar investment casting applications as a baseline. Because of the high L/D (Length over Diameter) ratio a compound of a high velocity was required. The predominant cutting grit was No. 36 silicon carbide. Process pressure and number of cycles were empirically derived and fixture design optimized to prevent leakage and erosion across the component face.

The finish produced on the experimental fixturing was approved and production fixturing was designed. This is shown in Figure 12-28, separated.

The two halves shown in Figure 12-28 are mated together for processing. The components are nested in polyurethane which is cast to conform to the profile of the part.

Feed orifices are machined through each half of the fixture to allow compound flow from one machine compound chamber, around the track of component, and into the opposing machine compound chamber. In the fixture shown, two parts (one right-hand, one left-hand) are shown. On another the right and left-hand parts can be mated together and four components are processed simultaneously.

The components are finished with a resultant fine directional finish in a short cycle time.

FIGURE 12-26: TOOLING USED FOR HYDRAULIC SERVO COMPONENT

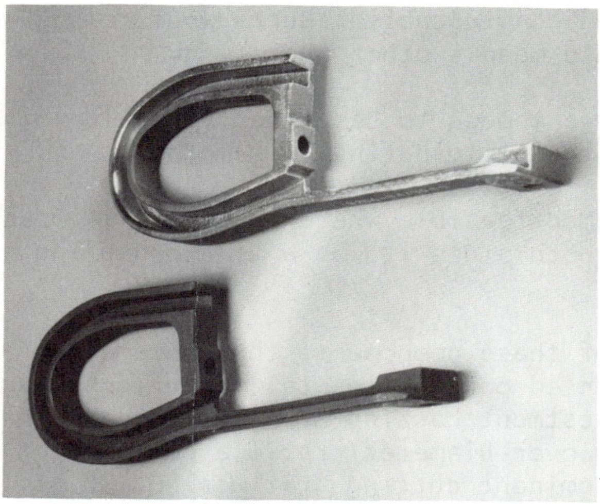

FIGURE 12-27: ARMAMENT SLIDE TRACKS

FIGURE 12-28: FIXTURING FOR ARMAMENT SLIDE TRACKS

A major manufacturer of torch tips is currently using abrasive flow process to deburr and polish the 0.026 inch (.066 cm) diameter acetylene holes and the 0.032 inch (.0813 cm) diameter oxygen hole in the torch tips seen in Figures 12-29 through 12-31. The deburring and polishing of the holes improves quality of flame and length of "stinger."

Fixturing (Figure 12-32) has been designed and built to process 20 parts/load. The production rate for the system is in excess of 1000 parts/hour.

The abrasive flow finishing of the tips produces a more uniform and higher quality part with a 30 percent longer "stinger." Semiautomatic inspection is now possible and rework has been reduced dramatically.

Hand deburring the fifty-four 0.250 inch (.635 cm) diameter holes in the stainless steel valve plate in Figures 12-33 and 12-34 was difficult due to the holes breaking out at the bottom of the grooves. The hand operation required more than 20 minutes and produced nonuniform marginal quality results.

A simple fixture was designed to hold the valve plate between the media cylinders of the machine and abrasive flow directed through the holes. Uniform 0.010-0.012 inch (.0254-.0305 cm) radii were generated on the holes in three minutes. Improved part quality with significant reduction in rework combine with a more than 80 percent reduction in production time to make this a very cost-effective application.

The aluminum pendulum (Figure 12-35) is a central component of an ultra-critical inertial guidance system for aircraft and missiles. The thimble sized part has a wall thickness of only 0.006 inch (.0152 cm) with 180 EDM'd slots 0.006 inch (.0152 cm) wide by 0.006 inch (.0152 cm) long as well as drilled holes on the tapered section and stem.

FIGURE 12-29: OXYGEN/ACETYLENE TORCH TIPS (BRASS)

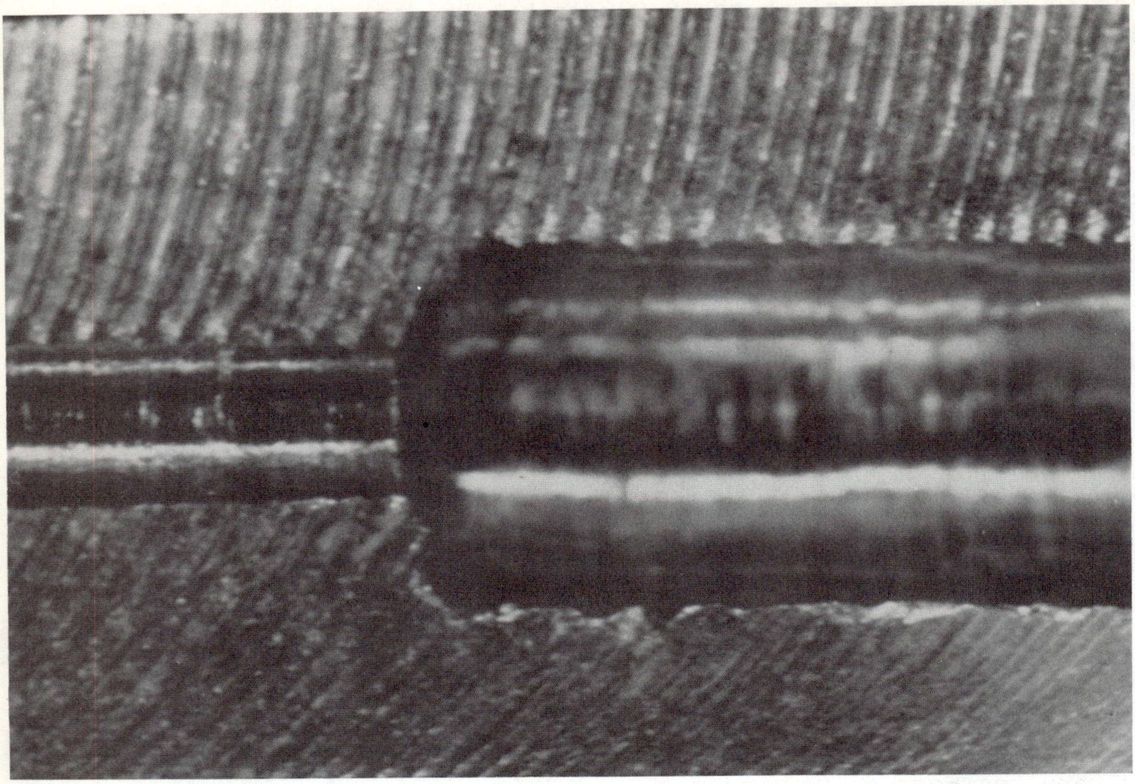

FIGURE 12-30: OXYGEN HOLE BEFORE ABRASIVE FLOW DEBURRING

FIGURE 12-31: OXYGEN HOLE AFTER ABRASIVE FLOW DEBURRING

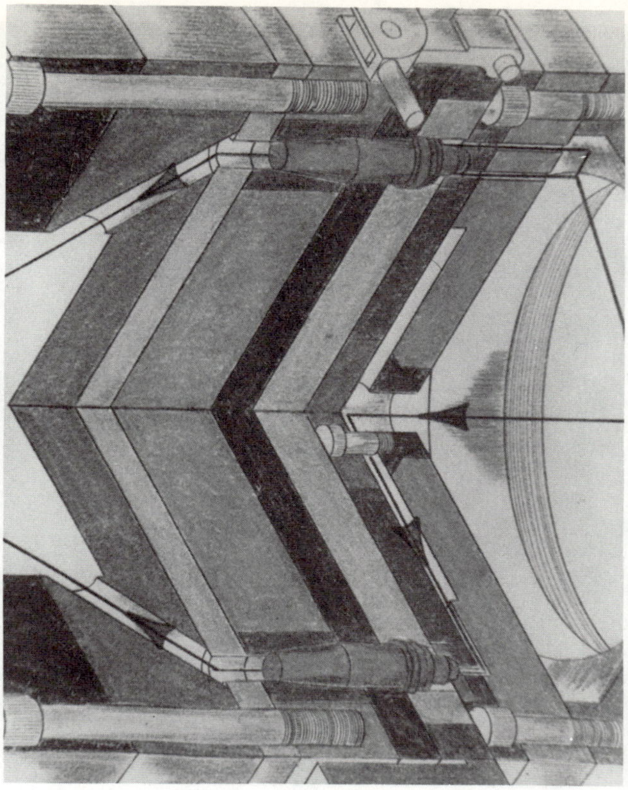

FIGURE 12-32: MEDIA FLOW PATH FOR ABRASIVE FLOW DEBURRING OXYGEN AND ACETYLENE HOLES

FIGURE 12-33: COMPRESSOR VALVE PLATE BEFORE ABRASIVE FLOW DEBURRING

FIGURE 12-34: COMPRESSOR VALVE PLATE AFTER ABRASIVE FLOW DEBURRING

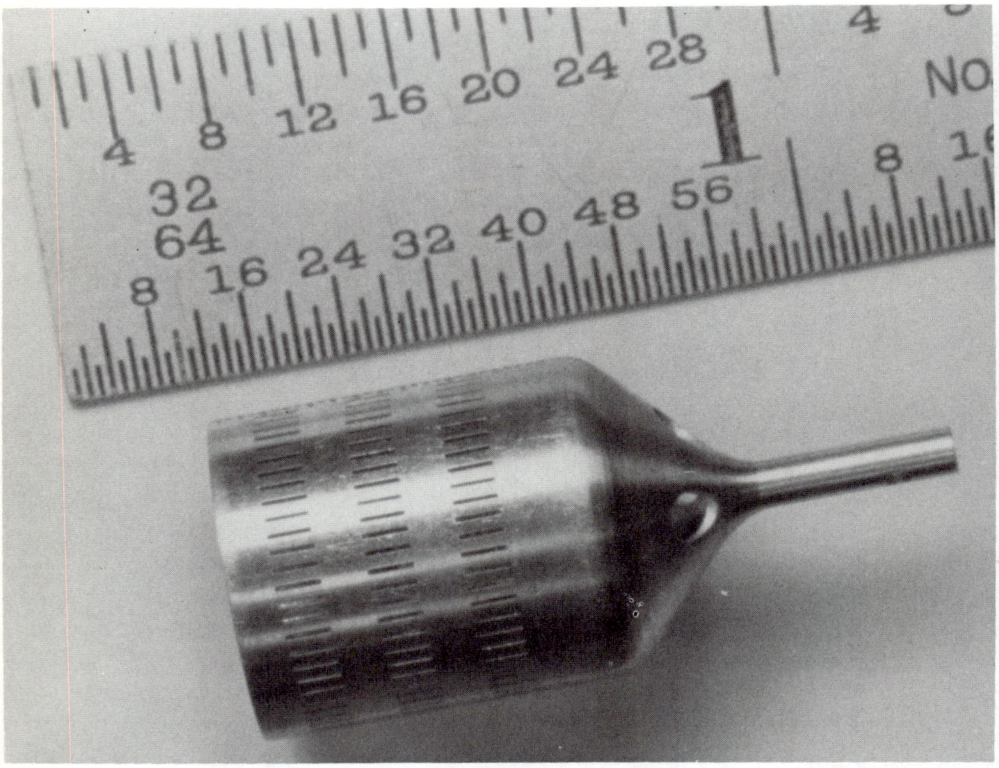

FIGURE 12-35: INERTIAL GUIDANCE PENDULUM

FIGURE 12-36 AEROSPACE COMPONENT OVERALL

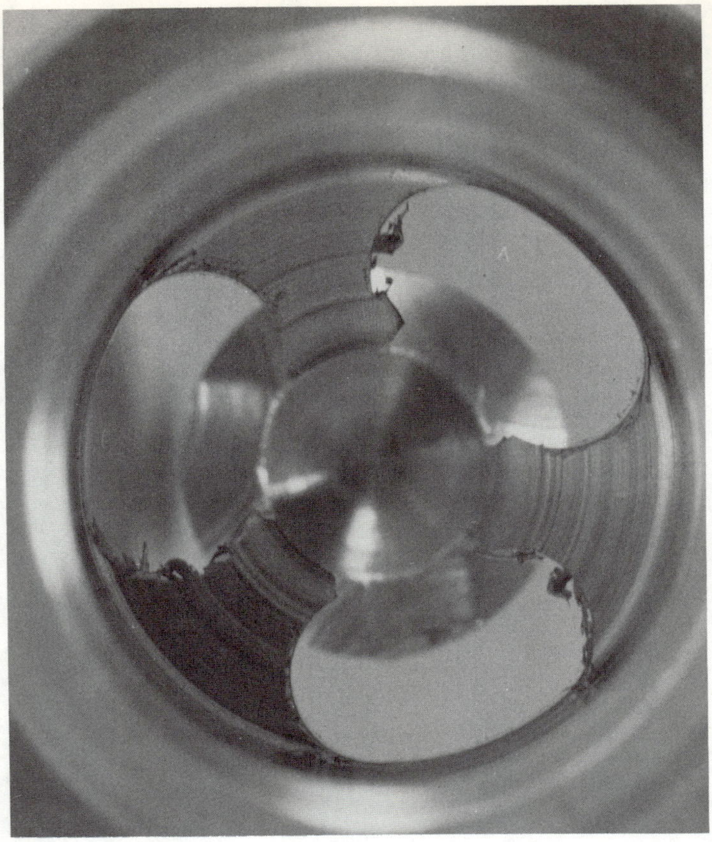

FIGURE 12-37: AEROSPACE COMPONENT CLOSE UP BEFORE ABRASIVE FLOW DEBURRING

FIGURE 12-38: AEROSPACE COMPONENT CLOSE UP AFTER ABRASIVE FLOW DEBURRING

The delicate part was fixtured in small, inexpensive nylon fixtures and a flowable media was extruded through the slots and holes at low pressure (250 psi). The flow of media was carefully controlled to remove the EDM surface from the slots, radius the I.D. and O.D. edges of the slots and deburr the drilled holes. Total process time for the pendulum was three minutes.

Figures 12-36 through 12-39 illustrate another hard to deburr part which has been successfully abrasive flow deburred.

Research on Media

Since much of the deburring action in abrasive flow deburring is the result of media properties, it is essential to understand some basic characteristics of the media. The following paragraphs present results of tests on Dynaflow® media. While the test results are only directly applicable to the media tested, the basic trends appear to be valid for most media available today.

The Dynaflow® media employs a polymeric material which is brittle and friable at room temperature. This polymer is blended (at high temperatures and shear) with a family of special gels developed for this purpose.

The blended polymer, which is referred to as the media base material or more simply as a base, is a viscous fluid. The viscosity of a given

TABLE 12-2: DENSITIES OF MATERIALS

Material	Density	
	g/cc	lb/gal
Base 050	0.88	7.3
Base 100	0.87	7.3
Base 200	0.84	7.0
Base 300	0.80	6.7
Aluminum Oxide[3]	3.9	32.5
Boron Carbide[3]	2.5	20.9
Silicon Carbide[3]	3.2	26.7

base is largely determined as expected by the quantity of gel blended into the polymer. In the next section, the rheological plots will be presented for four different bases which are called Base 050, 100, 200, and 300 respectively. Although the nomenclature is somewhat arbitrary, it is systematic. Lower numbers correspond to more viscous bases, and higher numbers correspond to less viscous bases. One may prepare bases more viscous than 050 and less viscous than 300, but nearly all applications suitable for abrasive flow machining may be accomplished using bases within this range.

The colors of the bases can be compared to lightened coffee, ranging from a fairly light brown for Base 050 to a deeper brown for Base 300. The densities of the bases are given in Table 12-2.

The densities of the most commonly used abrasive materials are also given for reference. The density of a Dynaflow® media may be found by simply summing the weighted densities of the appropriate bases and abrasives. Although other intermediate bases can be manufactured from basic chemicals, it is simpler and equally effective to blend two of the bases listed in the Table to obtain an intermediate base.

Experimental Methods

Rheology has been defined as the science of deformation and flow. In this section, we shall be specifically concerned with the relationship between shear stress rate for the different bases. Shear stress is simply the pressure applied to a fluid while shear rate is the velocity which the fluid achieves as a result of the shear stress.

To determine these rheological relationships, the bases were tested on a Dynaflow® Machine, Model HL15. This consists basically of two opposed cylinders with hydraulically driven, coupled pistons. A nylon fixture containing a cold rolled steel blank having a single hole of known length and diameter was clamped between the cylinders. The HL15 is illustrated schematically in Figure 12-40.

The base being tested was forced through the hole in the blank for a preset number of cycles at a preset pressure. The average cycle time for each test is then used to compute the velocity using the relationship

$$v = \frac{2 \quad V}{A(t_{av}-1.6)} \qquad (1)$$

where "v" is the velocity of a base at a given pressure, "V" is the volume of compound forced through the test hole each stroke (there are two strokes per cycle), "A" is the area of the hole, "t_{av}" is the average time per cycle, and "1.6" (seconds) represents the time of one cycle for an HL15 which contains no media. This time is a function of the maximum hydraulic capacity of the pump on this model machine and remains constant over the pressure range of our tests.

The vital statistics of the HL15 as set up for these tests are as follow: cylinder diameter, 5-1/6 inch; cylinder length with piston fully retracted, 11-1/4 inch; effective piston stroke during tests, 7.75 inch; amount of base forced through the test hole each stroke, 154 cu in (2/3 gallon). In order to define the pressure dependent characteristics of the media, tests were made using pressures of 200-500 psi. Although a maximum working pressure of 475 psi is recommended for these media, most tests in this paper were extended to 500 psi for the sake of round numbers and to insure that the most commonly used pressure range, 450-475 psi, did not lie near a transition point in the rheological curves of some media. Because the velocity of the bases depends slightly on their temperature, each base was adjusted to 30° C (86° F) before each test.

Rheological Results

Figure 12-41 has two plots of velocity as a function of pressure for Base 050, the most viscous base. It is readily apparent that for Base 050 the plot of velocity versus pressure is a curve rather than a straight line. In fact these curves are nearly textbook examples of pseudoplastic materials. A pseudoplastic is defined as being shear thinning[6], that is, its viscosity decreases as the pressure on it increases. Also note that different curves are obtained for the two different hole diameters. (Please note also that the length of the two holes is the same.) Since Equation 1, the velocity equation, takes the size of the hole into account, the existence of two separate curves indicates the turbulent flow of the base through the hole.

Figure 12-42 has four plots of velocity as a function of pressure for Base 100. It is immediately noticeable that Base 100 achieves much higher velocities than did Base 050. Although Base 100 is less viscous than Base 050, the large increase in velocity that was observed is more than would be expected from a simple visual comparison of the two bases. For the 0.375 inch diameter hole, a test which was included only for a direct comparison with Base 050, the curve shape was merely inferred from the other tests since by 300 psi the velocity has already approached the upper limit set by the test machine's hydraulic capacity. A cursory examination shows however that the velocity through a 0.375 inch diameter hole at 300 psi increases by a factor of over 700 from Base 050 to Base 100. The curve from the tests using the 0.250 inch diameter by 0.250 inch long hole shows smooth pseudoplastic behavior from 250-500 psi, and a 'hump' between 200 and 250 psi. This feature suggests a transition between the two types of flow. Perhaps below 250 psi the flow is less turbulent, but given the velocity of the base, flow must be turbulent above 250 psi. Because the increase in velocity is large at the transition point, we shall refer to this as a pseudoyield value. One can infer from this curve that a media made from Base 100 should not be operated at pressures at or below the pseudoyield value pressure if components are to be efficiently, accurately, and repeatably machined. The two curves obtained using 0.125 inch diameter x 0.250 inch diameter x 1.125 inch long holes show the effect of hole area and hole length on velocity. The 0.250 inch diameter hole has four times the area of the 0.125 inch diameter hole and should thus create less of a restruction on the base. However, the 0.250 inch diameter hole is also 4-1/2 times as long as the 0.125 inch diameter hole. The near coincidence of the two curves indicates that hole

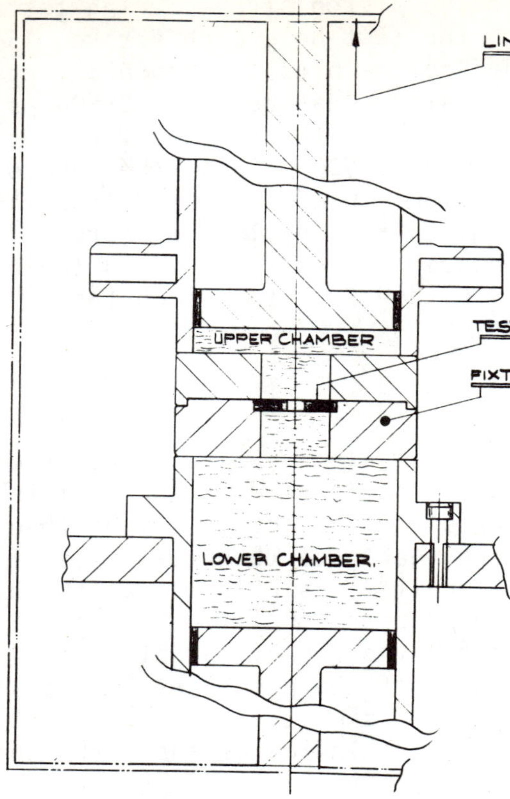

FIGURE 12-40: SCHEMATIC OF HL15 MACHINE

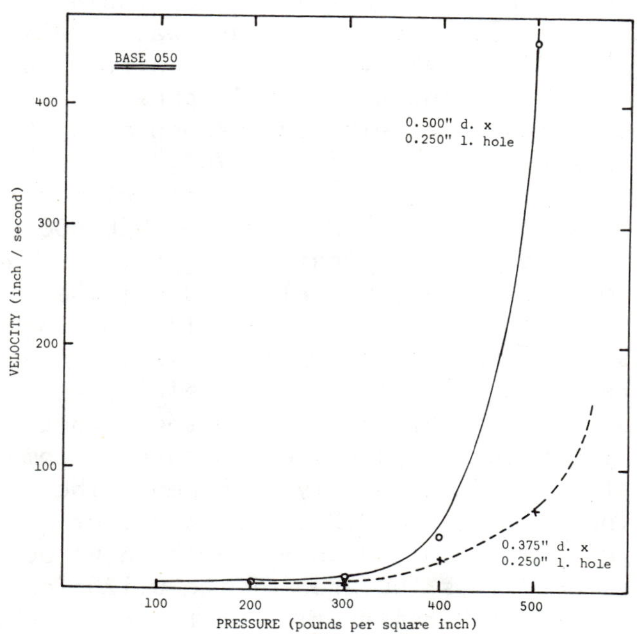

FIGURE 12-41: BASE 050 FLOW CHARACTERISTICS

length and hole area have nearly equal effects upon base velocity. Note that there is no discernable yield value in the curves from the 0.125 inch diameter x 0.250 inch long hole and the 0.250 inch diameter x 1.125 inch long hole. This apparent lack of pseudoyield values for these curves is probably due to instrumental limitations. The velocity of Base 100 at 300 psi through the more restricting holes is so near the bottom of our graph that the existence of a pseudoyield value at pressures below 300 psi would not be easily detected.

Figure 12-43 contains three plots of velocity against pressure for Base 200. If the curve from the 0.125 inch diameter x 0.250 inch long hole is compared with the 0.125 inch diameter x 0.250 inch long curve for Base 100, several differences are noted. The slope of the Base 200 is less than that of the Base 100 curve. Thus, at 300 psi Base 200 has a velocity nearly seven times as great as Base 100 while at 500 psi, Base 200 is only ten percent faster than Base 100. In addition, the Base 200 curve is much flatter than that of the Base 100 curve. Thus, as the polymer is more and more diluted with gel, as in going from Base 040 through Base 200, the material becomes less viscous and faster (especially at lower pressures), and the pressure against velocity plots approach straight lines. As before the near coincidence of the curves for the 0.063 inch diameter x 0.250 inch long and the 0.125 inch diameter x 1.125 inch long holes again shows the nearly equal effect on velocity of hole area and length. For Base 200, pseudoyield values were found or were indicated for all the curves. For the more restricting holes, this transition was found between 250 and 300 psi. The existence of a pseudoyield value below 200 psi in the curve obtained using the 0.125 inch diameter x 0.250 inch long hole is strongly implied by the following reasoning. If that curve is extrapolated to the vertical axis, it intersects that axis at approximately 900 in/sec. However, at zero pressure, the base must also have zero velocity. Thus, there must be a "drop" in the curve below 200 psi.

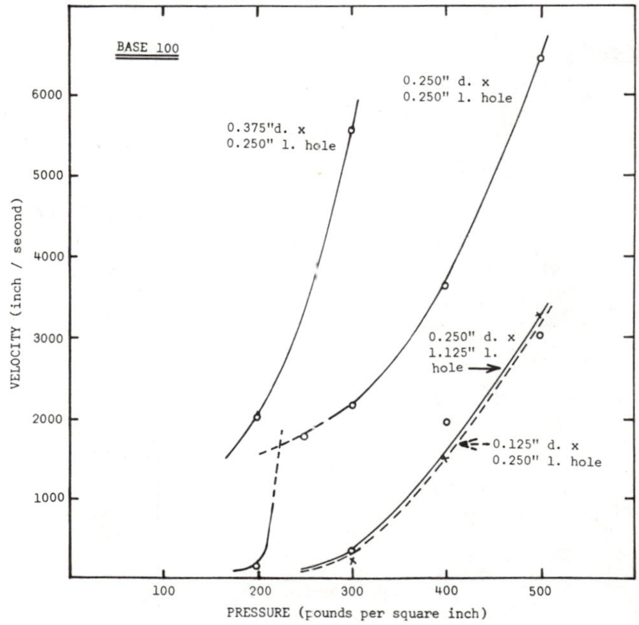

FIGURE 12-42: BASE 100 FLOW CHARACTERISTICS

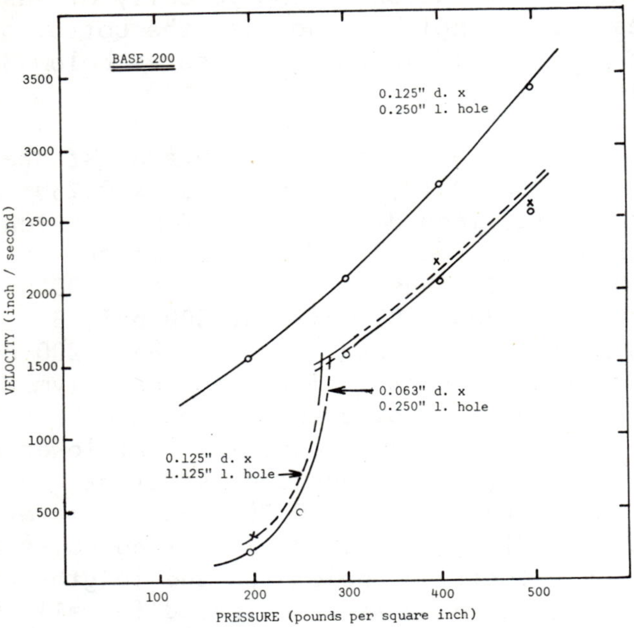

FIGURE 12-43: FLOW CHARACTERISTICS OF BASE 200

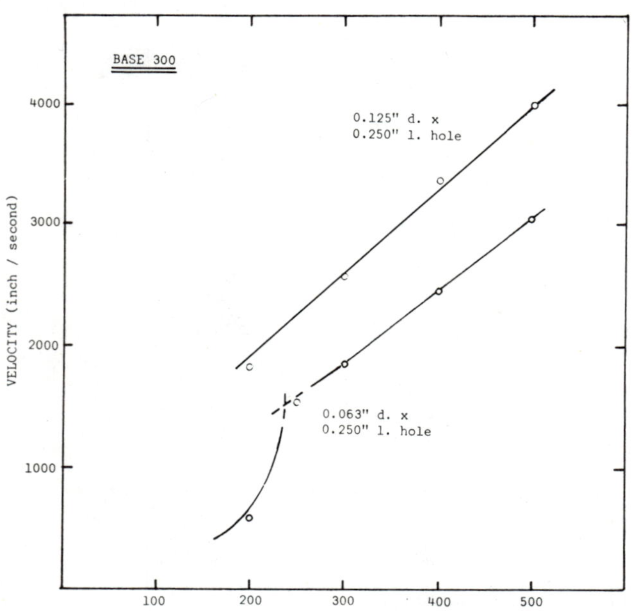

FIGURE 12-44: FLOW CHARACTERISTICS OF BASE 300

Figure 12-44 shows the results of tests made on Base 300. The velocities of Base 300 are somewhat higher than the corresponding velocities for Base 200. The curve from the 0.125 inch diameter x 0.250 inch long hole is nearly linear in the range of pressures tested. The curve from the 0.063 inch diameter x 0.250 inch long hole is nearly linear above 300 psi, but has a pseudoyield value between 200 and 250 psi. Like Base 200, the curve from the 0.125 inch diameter x 0.250 inch long hole almost certainly has a pseudoyield value below 200 psi.

The lowering of the pressure at which the pseudoyield value occurs provides much of the purpose for less viscous bases like Bases 200 and 300 because for small or for long holes the pseudoyield point of a more viscous base can be in the desired range of working pressures. Since relatively high velocities are important for the efficient use of low pressure abrasive flow machining, it is important to choose a media for a given job which will be working in an efficient region of the rheological curve.

Results Using Different Media

Although a knowledge of the rheology of the bases is necessary in order to properly understand the characteristics of low pressure abrasive flow media, the interest in and applications of this process pertain to finishing operations. These require the addition of abrasive grain to the bases. In the following paragraphs, the effects produced by different abrasive materials, different abrasive compositions, and different pressures on the cutting and radiusing abilities of low pressure abrasive flow media are examined.

Three abrasive flow media were prepared with the following composition: 56.5 percent Base 100, 37.8 percent abrasive grain size 36, 5.7 percent abrasive grain size 700. These can be thought of as "moderately fast" media. In the first media, the size 36 abrasive grain was white aluminum oxide; in the second, black silicon carbide; and in the third, boron carbide. Black silicon carbide was used for the size 700 abrasive grain for all three media. The inclusion of a small amount of fine grain (size 320 or finer) into a media containing coarse grit helps to form a matrix or lattice which improves media handling characteristics. However, as one expects, the presence of small quantity of size 700 grit has little effect on the cutting aggressiveness of a media containing size 36 grit. Each of the three media was tested at three working pressures: 300, 400, and 500 psi. Cold rolled steel blanks having holes with the nominal dimensions of 0.125 inch diameter x 0.250 inch long were selected as test pieces. This hole size was selected so that each media would have a velocity below the recommended range at 300 psi, but within the recommended range of velocities for Base 100 at 400 and 500 psi. Each test consisted of ten cycles on the HL15 using the same settings as were used for testing the bases. Also initial media temperature was again 30°C. A summary of the results of these nine tests is given in Table 12-3.

Several conclusions are apparent after even a casual inspection of the data in Table 12-3. First, for all three media the amount of cutting increased dramatically with increasing pressure. When the decrease in total time required for ten cycles is also considered, the increase in media efficiency at 400 psi through 500 psi compared to that around 300 psi is impressive. Referring back to the rheological curves in Figure 12-41, one sees that 300 psi lies in the flat region of the curve (for the appropriate 0.125 inch diameter x 0.250 inch long hole). Apparently in these flat regions of the curves where turbulent flow is minimized, low pressure abrasive flow media does not work efficiently. Because media efficiency increases so dramatically as a function of pressure, a natural question is, why stop at pressures below 500 psi? Several factors weigh against using higher pressure. First, the cutting action of a media may be increased or decreased by varying abrasive size, material, and/or base. Second, high operating pressures complicate the designing and manufacturing of fixturing and often cause sealing problems. Third, high operating pressures damage and distort many types of components. Fourth, it is easier to maintain close tolerances at low pressures. Fifth, and last, media are insulators. High pressures which result in extremely high velocities cause heat to be generated at a rate faster than can be conducted away by conventional cooling systems. Thus for all of these reasons plus machine design and safety, operating pressures below 500 psi yield the greatest overall efficiency.

The second obvious conclusion is the superior efficiency of boron carbide abrasive grain compared to aluminum oxide or silicon carbide. This clearcut superiority of boron carbide would not be expected simply from the various hardness scales which are given in Table 12-4. Indeed from this data one would expect boron carbide and silicon carbide to be comparable with aluminum oxide being slightly inferior. In fact, the silicon carbide is only slightly more efficient than the white aluminum oxide (and that advantage disappears near 500 psi). However near 500 psi, boron carbide is nearly twice as effective as the other two abrasives as measured by the weight of metal removed. It should be stressed at this point, however, that all three media were "fresh." As working time accumulates on media, the experience at Dynetics suggests that silicon carbide will deteriorate more slowly than aluminum oxide. Boron carbide should last at least as long as silicon carbide, and it is up to the user to decide whether the superior performance of boron carbide justifies its higher cost.

The final set of data presented in this chapter compares similar media, specifically media having identical abrasive grain composition but being made from different bases. Again the tests were all taken on an HL15 but these tests consisted of 30 machine cycles. For all of these media, the abrasive grains used were black silicon carbide. Three media were made with size 60 abrasive grain and the following composition: 64.5 percent of base, 21.3 percent of size 60 grit, and 14.2 percent of size 700 grit. The bases used for these three media were Bases 300, 200, and 100 respectively. The rather low amount of abrasive grain in this composition, only 35.5 percent of the total weight, was intended to produce high velocity media that would cut a large radius for a given amount of metal removed from the bore of a hole. Two media were also made having the same composition as the media in the previous section; namely 56.5 percent base, 37.8 percent size 36 grit, and 5.7 percent size 700 grit. These are considered to be heavier, more aggressive media than the previous composition and were made using Bases 100 and 050 respectively. The test pieces for these tests were again cold rolled steel discs which were 0.250 inch thick. For these tests, how-

TABLE 12-3 RESULTS FROM THE 10 CYCLE TESTS

Pressure[a]	Total Time[b]	Initial Dia.[c]	Final Dia.[c]	Increase[c]	Radius[c]	Stock Removal[d]
ALUMINUM OXIDE MEDIA						
300	18.0	.127	.141	0.014	.020	118
400	3.8	.127	.145	0.018	.020	176
500	1.8	.127	.153	0.026	.040	293
SILICON CARBIDE MEDIA						
300	22.7	.126	.144	0.018	.020	168
400	4.2	.127	.149	0.022	.025	239
500	2.0	.127	.153	0.026	.030	293
BORON CARBIDE MEDIA						
300	19.3	.126	.160	0.034	.020	328
400	3.8	.125	.164	0.039	.020	344
500	1.6	.126	.173	0.047	.040	540

a: in. psi; b: in minutes; c: in inches; d: in milligrams of steel

TABLE 12-4: HARDNESS VALUES FOR ABRASIVE MATERIAL

Material	Hardness Value	
	Mohs Scale[5]	Knoops Scale[5]
Aluminum Oxide	9	2100
Silicon Carbide	9	2480
Boron Carbide	9	2750

ever, the HL15 was set to 450 psi pressure, a very common pressure in Dynaflow® production applications. For each test, initial temperature was 25°C (77°F). The data from these 30 cycle tests are summarized in Table 12-5.

A comparison of the data for the first three tests (of the size 60 grit media) suggests several conclusions. As expected, Base 300 media required the least time for 30 cycles, while the Base 100 media required the most time. Also note that the Base 300 media opened the hole bore the least amount while giving the greatest edge break of the three size 60 grit media. Thus as suggested by this data, it is possible to formulate a media to give some desired ratio of edge break to bore enlargement for a given size hole. This feature of low pressure abrasive flow media gives design and production engineers considerable flexibility in deburring and finishing even complex, costly components.

The superior cutting efficiency of size 36 grit compared with size 60 grit is evident upon comparing the two tests on 0.125 inch diameter holes using the Base 100 media. In comparing the total times for the two tests, 4.1 and 7.7 minutes respectively, it is important to realize that a 0.189 inch diameter hole has one and a half times the area of a 0.153 inch diameter hole. Although not evident in Table 12-5, the surface finish from a size 36 grit media is very nearly as good as that from a size 60 media. Thus one may exploit the superior cutting efficiency of the size 36 abrasive grain without sacrificing surface finish. The data from the final two tests are consistent with one's expectations from the previous tests. Note that although the Base 050 media had lower velocity, it did remove a considerable amount of stock per cycle if not per minute. Actually, a Base 050 media would be more appropriate for a 0.500 inch diameter x 0.250 inch long hole.

Summary

Abrasive flow deburring is not a panacea for the world's deburring problems. It does, however, solve many of the problems where burrs are difficult to reach. As seen in this chapter, it can deburr miniature parts as well as large parts. In almost every case, however, one will note that accessibility of the burr, whether because of size or configuration, was the key to the success of this process and the failure of others. As a liquid-like material it can reach into most of the problem crevices.

Media properties can be systematized. This enables a user to select a media for a given job intelligently without resorting to "black magic" or guessery. By understanding the characteristics of the base material and of abrasives, by understanding how to select operating pressures and abrasive grain size for a given component, the maximum capabilities of abrasive flow processing may be fully exploited.

References

[1] Lawrence J. Rhoades and David E. Siwert, "Extrude Hone Deburring: Some Quantitative Observations," SME Paper MR75-488, 1975.

[2] John Stackhouse, "Deburring by Dynaflow," SME Paper MR75-484, 1975.

[3] John Stackhouse, "Application of Low Pressure Abrasive Flow Machining," SME Paper MR75-840, 1975.

[4] Lawrence J. Rhoades and David E. Siwert, "Extrude Hone Deburring: Theory and Application of Abrasive Finishing," SME Paper MR75-842, 1975.

[5] Winfield B. Perry, "Properties and Capabilities of Low Pressure Abrasive Flow Mediae," SME Paper MR75-831, 1975.

[6] O. C. C. Lin, Chem. Tech., 1975, p. 51.

[7] R. C. Weast, ed., Handbook of Chemistry and Physics, 52, The Chemical Rubber Company, Cleveland, 1971, p. F-18.

TABLE 12-5: RESULTS FROM THE 30 CYCLE TESTS

Media	Total Time[a]	Initial Dia.[b]	Final Dia.[b]	Radius[b]
Size 60 grit, Base 300	4.0	.125	.136	.045
Size 60 grit, Base 200	4.4	.125	.138	.040
Size 60 grit, Base 100	7.7	.126	.153	.020
Size 36 grit, Base 100	4.1	.125	.189	.040
Size 36 grit, Base 100	1.6	.249	.267	.055
Size 36 grit, Base 050	22.0	.373	.408	.040

a: in minutes; b: in inches

CHAPTER 13

WATER JET DEBURRING

Water jet deburring is being used extensively in the automatic transmission production industry. Its strong point is high reliability removal of burrs that could be missed by manual removal methods previously used. Dislodging of such burrs during operation of the transmission (by circulating hydraulic fluid) would result in erratic valve movement in the transmission control mechanism if the burr becomes caught between the valve and its mating bore. Erratic valve movement results in improper shifting and transmission failure. Utmost care is exercised in the manufacture of automatic transmissions to preclude the possibility of valve malfunction due to contamination with resultant transmission failure, cutomer dissatisfaction and complaints, and warranty claims and cost. Water jet deburring is one of the technologies helping to reach that objective.

Introduction

Water jet deburring is a high production process used extensively in the automotive industry to remove machining burrs and thin die casting flash from nonferrous automatic transmisison control mechanism components. The process was first used in production in 1971 by Ford Motor Company and has since been adopted by most major automatic transmission manufacturers. The process consists of ultra high velocity, high pressure (5000 psi--34.5 MPa) fan-shaped blades or conical streams of water impinging on the surface of the workpiece and cutting, eroding, breaking or otherwise disintegrating the burrs from the parent metal and removing the metal residue with the used water.

Water jet deburring has replaced primarily expensive hand deburring and abrasive blasting with walnut shells. The use of walnut shells had inherent disadvantages that were overcome with the use of water jet:

1. Corner deformation of close tolerance bores. Heavier than normal burrs (caused by dull cutting tools) would tend to be deformed by the walnut shells back into the original bore causing a reduction in diameter near the end of the bore. Water jet deburring leaves square corners.

2. Abrasive residue. Any walnut shells left in the workpiece could result in the same type of operational problems as the presence of the burrs they were intended to remove and would cause assembly problems if not removed from blind threaded holes. No foreign material of any sort remains on the surfaces with water jet.

Description of the Process

A water jet deburring machine consists of a conveyor belt that conveys workpieces in front of a series of jets through each of which is flowing a high velocity stream or fan-shaped blade of water powered by a pressure head of approximately 5000 psi (34.5 MPa). As the workpiece moves in a controlled path past these jets, as shown in Figures 13-1 and 13-2, the water impinging on the surface of the material disintegrates the most fragile and brittle portions of the workpiece, the burr. The longer the exposure to the jet, the greater the erosion or disintegration, and consequently the greater the size of the burr that can be removed. Typical automotive application exposure times are determined by test of actual workpieces that have typical sharp cutting tool burrs. The machines are not designed to remove burrs caused by dull cutting tools, laminations in die castings, or heavy die casting flash as these types of burrs are not apt to come loose in part service life anyhow. Complete removal of all burrs could only be accomplished with very long exposure times that would negate the economic justification for the machine and would result in erosion and damage of surfaces other than desired surfaces.

Economic Factors of Water Jet Deburring

- Deburring of die cast aluminum with typical machining burrs formed by sharp cutting tools is accomplished upon exposure to the water jet. Cycle time must be determined by the configuration of the part and the ability to expose all potential burr sites to within 1 inch to 4 inch (25.4 to 101.6 mm) of a jet. Typical small automatic transmission components are deburred at 350 pieces/hour.

- Performance deterioration is noted when the jets wear enough to reduce the pressure head to 4400 psi (30.3 MPa) (from 5000 psi operating pressure) on nonanodized aluminum die castings. Anodized aluminum burrs are somewhat harder and easier to disintegrate and thus can be removed at pressure down to 4000 psi (27.6 MPa).

- Manual deburring and/or abrasive media removal can be omitted, although most components will still be visually inspected for gross types of defects. Visual inspection requires much less labor content than manual deburr and/or media removal.

- Warranty cost savings may be substantial especially in areas with high average cost per unit repaired such as automatic transmission.

- Jet replacements in a properly maintained machine should not be needed until after approximately 5000 operating hours (1,750,000 parts at 350 parts/hour).

- Floor space required approximates 1000 sq ft (93 sq m) per machine. Figure 13-3 shows a recent vintage machine on the supplier's floor.

- Electrical power requirements are approximately 300 kw depending on the size of the machine. Annual electrical costs in the Detroit industrial region for this power consumption exceed $30,000.

- Water can be recirculated, filtered and chilled for minimal water consumption.

- Use of one to three percent water soluble oil is usually employed to prevent rust in the system and improve wear life of the hydraulic high pressure pumps.

Capabilities of Water Jet Deburring

- Removing burrs and flash from aluminum, cast iron, steel and phenolic moldings. Figures 13-4 through 13-9 illustrate examples of parts being deburred and typical burrs.

- Removing burrs through narrow openings (1/16 inch--1.6 mm).

- Removing internal burrs in direct line with an external opening.

- Removing burrs in blind holes through passage of water jet not necessary.

- Does not affect accuracy or surface finish of adjacent machined surfaces when removing thin burrs on aluminum parts.

- High production rate, low cycle time when removing thin burrs on aluminum parts.

- Can accommodate parts approximate size of a 12 inch (305 mm) cube (such as automatic transmission cases).

- Can accommodate a large variety of anticipated burr locations in various workpieces, including round workpieces that are rotated while exposed to the jet.

- Can be expected to reliably remove designed-for burrs until pressure falloff is noted on the machine pressure gage.

Limitations of Water Jet Deburring

- Removal of thicker than designed-for aluminum burrs, cast iron burrs, and steel burrs must be accomplished with either higher pressure (machines with pressures up to 100,000 psi (689 MPa) are available) or extended exposure time.

- Cannot deburr where there is no direct access to an external opening perpendicular to the burr.

- Requires accurate positioning of the workpiece relative to the jets and the jets relative to the anticipated burr locations on the workpiece.

- Access by the water jets to the workpiece is limited to three sides in one machine.

- Machines must be individually designed for each workpiece and cannot be conveniently changed over.

Future Work

Inasmuch as a definition of size for a burr has yet to be formulated, precise definition of the burr removal capabilities of water jet deburring do not exist in terms of the parameters of this process that affect burr removal, i.e., flow rate (GPM), pressure drop (psi), distance from jet to burr, time exposure of burr to jet, angle of jet relative to burr and burr material.

Determination of whether or not a burr can be removed by water jet deburring must be done experimentally by manufacturers of the water jet machines (most have test facilities). This is a shortcoming in that the individual manufacturing engineer cannot specify the use of a water jet machine with the same degree of confidence that he has for specifying a metal cutting machine. Future work should be coordinated with work on specifying burr dimensions to result in a numerical assessment of the water jet deburring process' capability of removing a specified burr.

References

[1] William D. Campbell, "Capabilities and Limitations of Water Jet Deburring," SME Paper MR75-827, 1975.

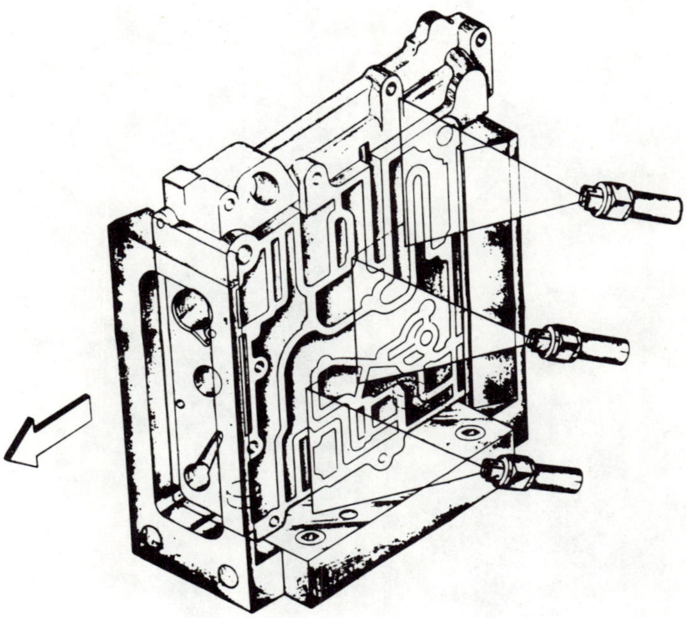

FIGURE 13-1: FAN SHAPED BLADES OF WATER USED FOR DEBURRING

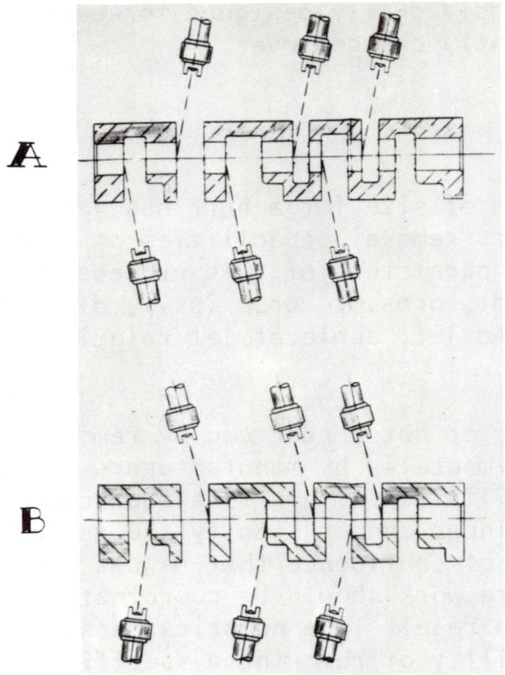

FIGURE 13-2: REMOVAL OF BURRS FROM BOTH SIDES OF WORKPIECE

FIGURE 13-3: TYPICAL WATER JET DEBURRING MACHINE

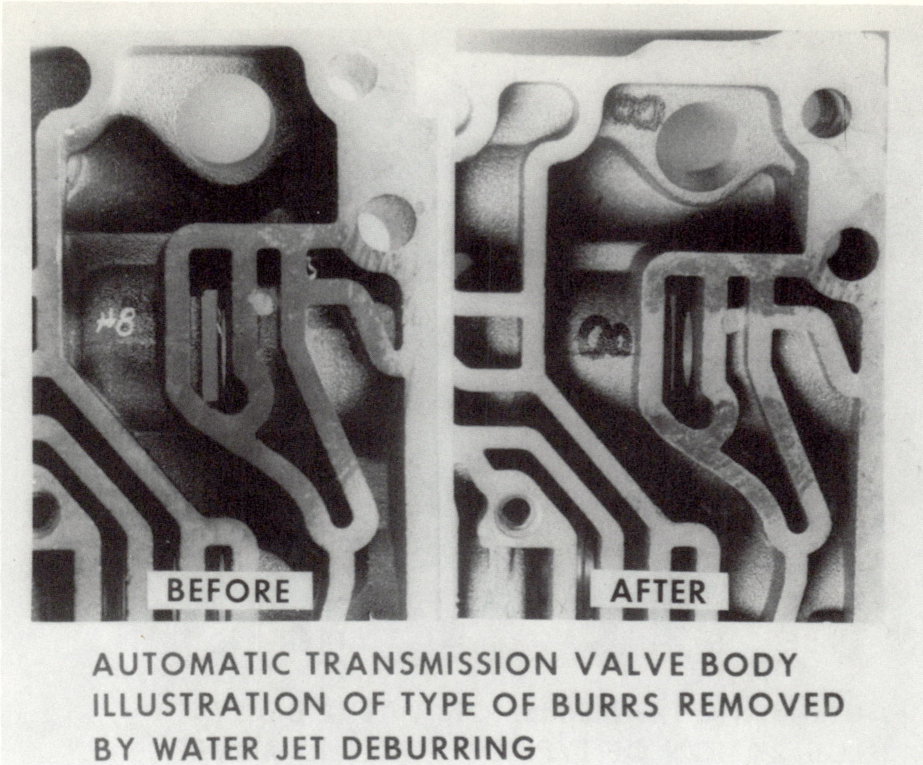

FIGURE 13-4: AUTOMATIC TRANSMISSION VALVE BODY ILLUSTRATION OF TYPE OF BURRS REMOVED BY WATER JET DEBURRING

FIGURE 13-5: AUTOMATIC TRANSMISSION VALVE BODY ILLUSTRATION OF TYPE OF BURRS REMOVED BY WATER JET DEBURRING

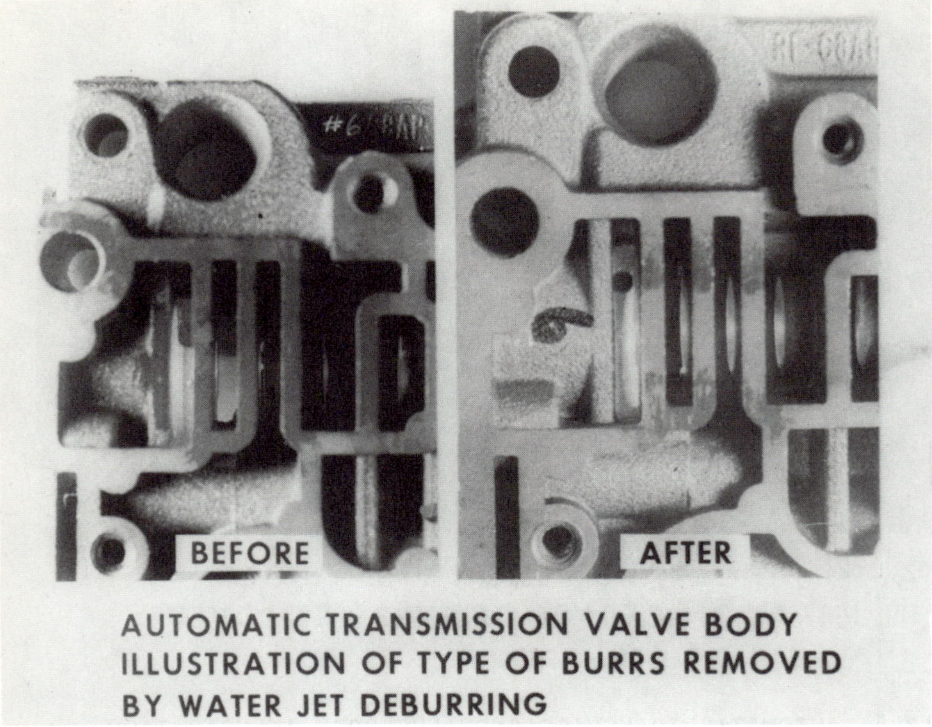

FIGURE 13-6: AUTOMATIC TRANSMISSION VALVE BODY ILLUSTRATION OF TYPE OF BURRS REMOVED BY WATER JET DEBURRING

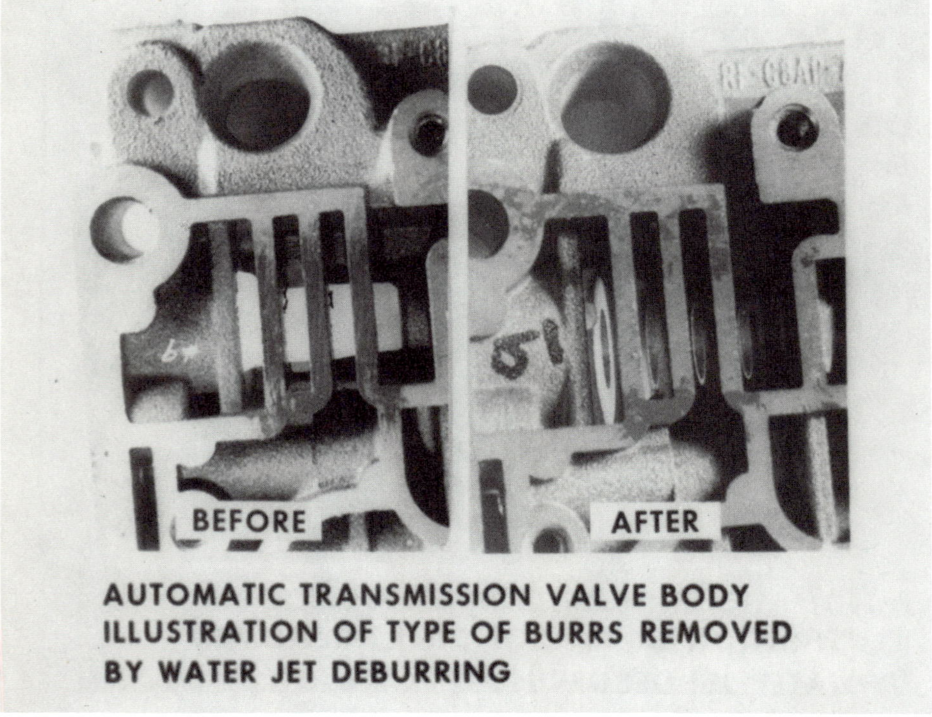

FIGURE 13-7: AUTOMATIC TRANSMISSION VALVE BODY ILLUSTRATION OF TYPE OF BURRS REMOVED BY WATER JET DEBURRING

FIGURE 13-8: AUTOMATIC TRANSMISSION VALVE BODY ILLUSTRATION OF TYPE OF BURRS REMOVED BY WATER JET DEBURRING

FIGURE 13-9: AUTOMATIC TRANSMISSION VALVE BODY ILLUSTRATION OF TYPE OF BURRS REMOVED BY WATER JET DEBURRING

CHAPTER 14

BRUSH DEBURRING

While brushes have been used for at least 35 years for deburring, many potential users often overlook this highly flexible, inexpensive technique. This chapter describes some of the capabilities of this process. Examples of brush styles, applications, and recent research are described.

Introduction

The technology of deburring with power brushes has increased to a point where the economics of their use has resulted in substantial savings of both time and money in almost every type of industry. It might still, however, be difficult to decide when to use brushes and when to use some other deburring technique. What then is the rationale for the use of brushes? What are the reasons that brushes have gained such widespread acceptance in the past in so many deburring operations? Specifically, if brushing has any inherent advantages, what are they? Why should you use a brush to remove a burr? Some of the reasons are:

1. Innovative new concepts of brush types and configurations including abrasive impregnated nylon filaments; elastomer bonded steel wire brushes; nonwoven nylon brushes containing silicon carbide, aluminum oxide, flint or talc; and multiple-strand crimped wire brushes have given new dimensions to deburring.

2. Unlike some of the other techniques, power brushing can be selective to an edge so that it removes burrs without changing the dimensionsl stability of the two intersecting machined surfaces. Nor will power brushes create secondary burrs formed by two sharp edges on the machined surfaces.

3. Fragmented metal or sharp edges can be completely removed. No vistage of a burr is left. Good control of the radii produced on the edge deburred is a secondary advantage.

4. When used with suitable equipment, a deburring power brush will produce uniformly finished parts.

5. The fact that a brush is flexible with a controlled surface permits it to conform and follow the contours of irregular surfaces. Relatively inaccessible surfaces can be reached because of the flexibility of the protruding filaments.

6. The various configurations of brushes are so extensive in range that we find them deburring the most minute of feather machining and grinding burrs on the smallest of miniaturized elements and

the most precision of components, on up to the heavy burr removal with brushes weighing many hundreds of pounds each.

7. Brushes can be set up with varying degrees of mechanization that ranges from offhand manual deburring with hand held tools such as used on small quantity runs or extra large workpieces to fully automatic, highly sophisticated equipment as required in high production applications.

8. The use of brushes reduces labor cost by minimizing the skill requirements.

9. Components being deburred that will be subjected to cyclic motions (such as gears) will have their fatigue life increased since brushing relieves the high stress concentrations.

10. Brushes operate on power available in all plants--air and electricity.

11. Brushes operate quietly at acceptable noise levels.

12. Brushes fit in with our ecology environmental programs--no carry-over materials to pollute air or water.

Of all mechanical burr removal methods--grinding, skiving, belt and disc sanding, etc.--only brushing, because of the inherent flexibility combined with strong cutting action of power brushes, simultaneously deburrs and edge blends in one operation.

Type and Configuration of Brushes Used in Deburring

A number of primary types and configurations of power brushes are advantageously used in deubrring; namely, radial or wheel, cup, end, tube, wide face, or cylinder. Each of these basic types vary in design and composition, and are otherwise modified to provide the versatility required in the thousands of complex, sophisticated deburring applications. For example depending upon the application, various types and sizes of fill materials are used in deburring. These include tempered and untempered high tensile wire, both carbon steel and stainless; a number of nonferrous alloys such as brass or beryllium copper and, of course, many other more glamorous alloys required because of their heat or corrosion resistant qualities; synthetics, including nylon and polypropylene, animal hairs, such as pig bristle and horse hair; vegetable fibers, especially tampico; abrasive impregnated nylons and the nonwoven synthetic, abrasive filled brushes.

Carbon Steel Wire

Steel wire is probably used in power brushes more than all other fill materials combined, with the tempered rather than the untempered variety being predominant. This steel wire is a high quality material similar to that used in springs. It is generally made of a 0.60 to 0.75 percent carbon steel with tensile strengths ranging from 300,000 psi to 380,000 psi (2068 to 2619 MPa). The untempered steel wires are hard drawn to their highest tensile strength and usually have lower hardness values and tensile strengths than the tempered wire.

Stainless Steel Wire

Some nuclear applications and others involving elevated temperatures or corrosive environments prompt the use of stainless steel wire. Two types are used. The first is hand drawn, bright finished type 302 stainless steel which is drawn to tensile strengths of 300,000 psi to 360,000 psi. This type is usually used in deburring stainless steel, aluminum or other nonferrous materials in an effort to preclude after-rust caused by foreign deposits when carbon steel wires are used.

Type 420 tempered stainless steel, with tensile strengths ranging from 240,000 to 255,000 psi (1655 to 1758 MPa) is also used in deburring brushes. However, type 420 is only 80 percent as corrosion resistant as the type 302. In spite of this deficiency it does have some distinct advantages in many deburring applications. It is vastly superior to untempered stainless in its cutting capabilities, fatigue resistance and abrasion resistance. In some instances, it has a life of seven to ten times that of untempered stainless steel, while doing substantially more cutting.

Type 420 stainless steel is not recommended for use in nuclear applications nor in applications where it is desirable to deburr with brushes constructed of nonmagnetic stainless steel. Type 420 is magnetic at all times and therefore type 302 should be used. Do not make the mistake of checking the wire with a magnet. Wire from all heats of type 302, when drawn to spring temper as used in brushes, can be attracted to a magnet to varying degrees, and this amount of magnetic attraction for a given size of wire in the spring tensile range is a function of the degree of cold working that the wire has undergone. Cold work increases the tensile strength and simultaneously the magnetic permeability of the wire. Only in its fully annealed condition can type 302 stainless steel be considered essentially nonmagnetic because in this state it exhibits permeabilities of less than 1.02.

Tampico

Tampico, a cellular vegetable fiber, is used more widely than any other fiber in deburring brushes and produces a flexible, resilient brushing wheel (Figure 14-1). They are used singly or in multiples to make a wide face deburring brush. It has been found that tampico brushes can be treated and cured with a lacquer for improved capabilities of stiffness and grit retention, and the treatment minimizes the shedding, knifing, and weaving that sometimes occurs at high speeds. Treated brushes can therefore perform an infinitely wider range of deburring and edge blending than can be accomplished with the natural, untreated fiber. The stiffness and tackiness attained by treatment makes it easy to apply grease stick and abrasive compounds used with these brushes.

One of the main reasons treated tampico brushes are used is to increase the life of the part being brushed. Over half of all failure of parts subjected to repeated cyclic stresses are fatigue failures caused by the widening of microscopic cracks that result from stress concentrations. The radii and surface finishes in proximity to the deburred edge can almost totally eliminate the scratches, burrs and sharp corners that cause the cracks and progressive fractures in them.

Abrasive Monofilament

One of the latest innovations in the brush industry is the development of deburring brushes filled with nylon monofilaments in which silicon carbide, aluminum oxide, or pumice grit is permanently encapsulated (Figure 14-2). Between 20 and 40 percent of the abrasive particles by weight are uniformly distributed throughout the filament. Grit sizes range between 800 and 80, with the coarse grist encapsulated in relatively large diameter filaments and the fine size disbursed in small diameter filaments. Usually the monofilament diameter is between 0.014 inch and 0.050 inch (0.36 and 1.27 mm) diameter, depending upon the size of grit used.

Abrasive filament deburring brushes can be used either wet or dry. Experience seems to indicate some advantages of wet use when water, mineral oil, or water soluble oil is used as the cooling medium. Grease sticks have also been used efficiently. Generally these brushes can be used in all of the applications for which treated tampico brushes are used. Their advantage lies in the fact that the need for compounds and slurries is eliminated. Such compounds however can be used with abrasive filament brushes to accelerate the cutting action. The abrasive filament brushes can also be used in applications for which the tackiness of treated tampico brushes would be objectionable, such as the deburring of wood, leather, plastics, cork, rubber, ceramics, and glass.

Nonwoven Abrasives

Abrasive grains are disbursed uniformly throughout a low density, open type, nonwoven web of synthetic fibers and held there with a binding adhesive. This produces a cushioned abrasive that is ideal for many deburring jobs (Figure 14-3). The abrasives used are aluminum oxide, silicon carbide, flint, and garnet. Although the grit sizes vary from 80 to 800, they are not usually designated numerically. Instead the grades used are coarse, medium, fine, very fine, super fine, and ultra fine.

The nonwoven abrasives can be used as a substitute for any of the tampico or abrasive filament deburring brushes. Its use precludes the need for grease sticks or abrasive compounds. It also frequently replaces setup wheels, steel wool, and abrasive flap wheels. It is waterproof and chemically resistant and can be used wet or dry. It is nonloading and highly flexible. Nonwoven abrasives very efficiently remove light burrs, however, they should not be used on large burrs. The scope of burr removal capabilities is substantially enhanced by the fact that nonwoven abrasives are available in a number of different densities (degrees of hardness).

This wide selection of densities, coupled with extensive combinations of grit grades and sizes, and many brush configurations, means that most light burrs can be removed with the nonwoven abrasives.

Brush Selection

Recognizing the significant number of variables available in brushes, one with a deburring problem must apply some basic technical considerations to his choice of brushes. Six basic considerations influence one's choice. These factors are:

1. Type and modulus of elasticity of the fill material.
2. Diameter or cross-sectional area of the individual filaments.
3. Trim length (the length of the filaments protruding from the retaining elements).
4. Density of the fill material.
5. Speed of rotation.
6. Filament configuration (crimped wire, knot or twisted tuft, straight wire, or multiple strand crimped).

Each of these six factors contributes to the physical characteristic known as stiffness. The difference in stiffness of filament materials that differ in modulus of elasticity is frequently self-evident. For example, a steel wire would be stiffer than a similarly sized filament of nylon. Obvious, too, is the fact that for two filaments that are identical in every respect except diameter, the largest diameter will be the stiffest. Likewise, for two filaments identical except for length, the shortest will be the stiffest. It is also true that the mass of filament ends at the periphery of a brush has a feel of stiffness dependent upon the density of the fill material. The more compact density (larger number of filaments) will be the stiffest. The compactness permits the individual filaments to support each other; in effect, to permit them to act as a family of filaments instead of individual filaments. Centrifugal force is a less obvious factor that permits brush filaments to resist lateral displacement and to feel stiffer. The faster a brush rotates, the stiffer it will be. The last basic factor of brush design affecting stiffness is the configuration of the filaments used. Every power driven brushing tool constructed of metallic filaments uses one of four basic forms; namely, crimped, straight, knot or twisted tuft, or multiple strand crimped.

Straight Wire

Least effective in deburring and consequently not frequently used except for very light deburring is straight wire construction. In this type of brush, tufts of straight wire in radial pattern are protruding from holes in a circular hub. The protruding ends of the filament wire being relatively unsupported by other wire filaments have a high degree of flexibility with a minimum amount of filament entanglement.

Crimped Wire

Crimped wire construction as illustrated in Figure 14-4 is very effective in deburring. In this type of construction the filament wires have been crimped to produce a multiple plane displacement with the wavelength (sometimes called frequency) and amplitude held to within 0.002 to 0.004 inch (50.8 to 101.6 μm) tolerance.

Knot or Twisted Tuft

Knot or twisted tuft brushes are shown in Figure 14-5. This type of brush is the most effective for heavy deburring. In this type of construction tufts of straight wire formed in hairpin shape around a retaining member are then twisted into a cable with a helix angle of approximately 45 degrees. The standard twist has the tuft twisted for two-thirds of its length; however, in applications requiring maximum impact such as extremely heavy burr removal the tuft is twisted for its entire length. Multiple rows of tufts, instead of a single row, are sometimes used to give additional stiffness for the most severe deburring operations (Figure 14-6).

Multiple Strand Crimped

Deburring in environments that tend to induce premature fatigue is sometimes best done with brushes constructed with multiple strand crimped wire. In this construction, groups of wires ranging from 5 to 25 are stranded together into a steel core usually ranging from 0.020 inch to 0.050 inch (0.5 to 1.3 mm) in cable diameter. This core or cable is then crimped in multiple planes just as conventional crimped wire. This permits the brush to have filaments that have the stiffness of a coarser wire with the fatigue strength and multiple point advantages of a fine wire brush (Figure 14-7).

Deburring Heavy Burrs

Most burrs with a heavy burr base thickness must be removed by brushes with very substantial cutting action and minimal impact action. These characteristics are found in radial type brushes with short trim. Because they are short trim and densely filled, they give maximum cutting action. They can be furnished either in crimped wire or knotted types, with wire sizes of 0.0118 inch to 0.020 inch (0.3 to 0.5 mm) being the most used range. Knot type brushes in diameters of 12 inch to 15 inch (305 to 381 mm) are extensively used in the automotive and aircraft industries for the removal of burrs from gears and spline bores. The deburring occurs immediately after hobbing or spline broaching just before shaving. These short-trim brushes impart the proper radii and adequately blend the edges of gears while at the same time removing all the minute fragments of metal that frequently are left when other techniques of burr removal are previously employed. One distinct advantage of this type of brush is its capability of removing the burr from an edge without changing the dimensional characteristics of the two intersecting surfaces. Short-trim radial brushes are also used extensively for heavy-duty deburring and edge blending of stampings, metal parts, pipe and tube ends, expanded metal and flash removal in the rubber industry. They are designed for efficient, fast, low cost performance on automatic, semiautomatic, and manual machines. In some types of heavy burr removal it is necessary to impart a circular action such as in the deburring of drilled holes in condenser tube plates. Knot type cup brush configurations lend themselves to this type of work. To preclude side flare and reduce the trim length, a bridle or band is used on the brush which adds a new dimension of stiffness (Figure 14-8). This not only deburrs the holes in the plates but also puts a radius around the periphery on a uniform basis without any evidence of hole elongation.

Deburring Light Burrs

Light and feather burrs are usually removed with brushes that have a medium or long trim. These are available in both knotted and crimped wire construction. Each type has its own sphere of excellence. The knotted brush, which usually is used singly instead of in multiples, is a high impact tool. It is radically influenced in performance by the speeds at which it operates. At low speeds the brush is very flexible and soft acting. As surface speed is increased it becomes stiffer until at the high speeds it becomes extremely hard and consequently is very fast cutting. Normally used are 4500 to 6000 SFPM (22.8 to 30.5 m/s), depending on the stiffness required. Wire sizes of 0.0118 inch to 0.023 inch (0.3 to 0.6 mm) are common. Deburring rubber flash from molded products can be accomplished with knotted, medium to long trim brushes by using substantially higher surface speeds. Up to 12,000 SFPM (61 m/s) are frequently used. To attain these surface speeds, 12 inch to 15 inch (305 to 381 mm) diameter brushes are used.

Crimped wire, medium to long trim brushes are used rather universally for many types of light deburring operations. The fact that they are not very stiff permits them to follow and conform to contours of the irregular surfaces being deburred. Relatively inaccessible places can be reached because of the flexibility of their filaments. Feather machining and grinding burrs can be removed from edges and light flash on rubber and plastic can be taken off with these wheels. Generally, smaller diameter wires are used with 0.006 inch to 0.014 inch (0.15 to 0.36 mm) being common. Surface speeds used are 4500 to 6500 SFPM (22.9 to 33.0 m/s). Long trim, crimped wire brushes of small diameter, because of the inherent softness and flexibility of their construction, are not construed to be efficient deburring tools. Light burrs are sometimes removed with brushes of approximately 0.005 inch (0.13 mm) diameter wire to which a deburring compound has been applied. This give a cutting capability to the brush that it normally would not have by itself. Removing burrs from cylindrical surfaces is sometimes accomplished with centerless brushes (Figure 14-9). These brushes used on centerless grinders are normally 20 inch (508 mm) diameter and 5 inch or 6 inch (127 or 152 mm) brush face. These larger radii brushes have a wire fill (0.005 inch to 0.010 inch diameter), or a fill of treated tampico, abrasive impregnated plastic, or nonwoven abrasive web. When the fill is wire, these tools remove grinding burrs and produce a minimum edge of 0.005 inch. Finishing work can be done dry or (preferably) with normal grinding coolant. These fine wire fill brushes, working with conventional grinding coolant, remove feather burrs and improve surface finishes on such parts as control valves for automatic transmissions. It is important to remember the following facts about centerless brushes:

1. They will not remove metal from a cylindrical surface. Parts must therefore be ground to size before brushing.

2. Greatest improvement in finish (for unbrushed parts) is in the 25 to 30 microinch (0.64 to 0.76 μm) MRS range. A part with a 25 microinch finish can be brushed rapidly to deburr and simultaneously obtain a 10 to 15 microinch (0.25 to 0.38 μm) finish; a part with a 10 microinch finish will have a 4 to 6 microinch (0.1 to 0.15 μm) finish after brushing.

3. Centerless brushing follows centerless grinding principles with the following additions: accuracy in pressure and adjustment is not critical with brushes and a machine is no longer needed to obtain the required finish making it possible to locate the part below the centerline.

The last few years there has been a tendency to replace the tampico centerless brushes with ones constructed with abrasive monofilament. They are especially effective in deburring perforated steel tubing. No compounds are required with a consequent savings since no cleaning is required.

Wire Filled, Elastomer Bonded Brushes

Elastomer bonded, wire filled brushes have been proven to be extremely fast cutting with maximum operator safety. Because the bonding material supports the wire to the very tip, the brush provides a maximum amount of cutting action. By periodically reversing the brush, the operator has a tool that maintains its maximum cutting efficiency throughout its life. Unlike a conventional nonbonded wire filled brush, there is no loss of wire through fatigue. Bonded wire filled brushes are made in radial, cup style, flared end and straight end designs (Figure 14-10), permitting them to replace conventional wire filled products. These brushes are effective for many deburring operations because they maintain a uniform face throughout their life and perform at maximum efficiency. Because elastomer wire filled brushes are so fast cutting, lower surface speeds can be used. Examples of the types of jobs that these brushes can perform are as follow: deburring fine pitch gears, powdered metal components, aluminum die castings, copper tube ends and metal stampings. They also are used for edge deburring on steel strip and sheet, and are used extensively for removing burrs from key slots and other hard to reach parts. In deburring ductile materials, radii are produced without producing a secondary rolled condition. Usual wire sizes are 0.0118 inch to 0.014 inch (0.3 to 0.36 mm) diameter and 5500 to 6500 SFPM (28 to 33 m/s) are appropriate speeds. Two factors should be taken into consideration when elastomer bonded brushes are used for deburring. First, periodic dressing of the brush face will produce a brush life ten times as long as that of conventional nonbonded wheels thus keeping end-of-service costs for a given deburring operation much lower.

Second, because the cutting action is so fast, brush adjustment is important; the part must be held precisely if the brush is to be used efficiently.

Nonwoven Abrasives

Nonwoven abrasives have in many instances replaced treated tampico, abrasive monofilament, and fine wire brushes in deburring applications. Since they can be furnished in various densities and grades of hardness, and are filled with a large assortment of grit sizes (in aluminum oxide and silicon carbide), the potential of deburring applications is vast. Very light deburring can be accomplished with the uncompressed material, but the harder brushes are best for deburring purposes. They are used to deburr grind lines in all manner of metal components, appliances, and automobile parts. This includes stampings and machined parts. Many types of plastics also are deburred. Edge finishing on precision parts is usual. Hole deburring on printed circuit boards, done both wet and dry, is one of the largest users of nonwoven abrasives. Even wood items are defuzzed and finished with them. Removing the parting lines on jet blades and deburring and radiusing gears and machine parts are jobs now done with nonwoven abrasives. Generally, the surface speeds used with the material are 5000 to 6500 SFPM (25 to 33 m/s). One should be admonished not to use nonwoven abrasives on heavy burrs--only light feather burrs.

Guidelines

If one applies the basic brush technology and fundamental information given so far, then he is in a position of rather easily using the brush selection chart (Figure 14-11). Note that it is in two parts. Using the basic considerations already noted in this chapter, one makes a tentative selection from Chart 1. This chart is divided into halves based on the surface finish requirements after deburring. If they must be less than 30 microinch, use the left hand half of Chart 1. If greater than 30 microinch can be tolerated, use the right half. Note that both sides have been arranged in a similar manner with brushes that are most flexible being on the outside and sequentially being less flexible as you approach the center of the chart. Proportionate relationships exist on finishing abilities of the brushes and also on their cutting characteristics. Each of the types of brushes are appropriately listed in similar sequential order and are easily related to the listed major brush characteristics. Listed also are major brush applications in order of general usage. Having selected a brush based on Chart 1, it should be tried and observations made on its performance. Based on the observed results, refinements may be made in brush specifications or operating conditions by applying the information in Chart 11.

Applications

Gear and Sprocket Deburring

High production and automated equipment used on primary gear operations

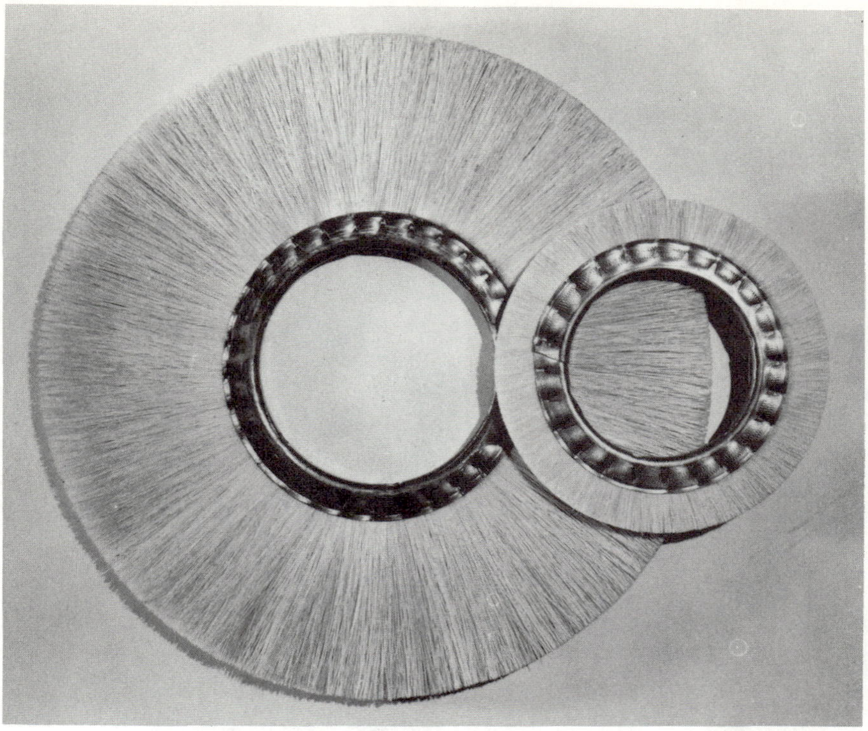

FIGURE 14-1: TAMPICO BRUSHES

FIGURE 14-2: ABRASIVE FILLED NYLON BRUSHES

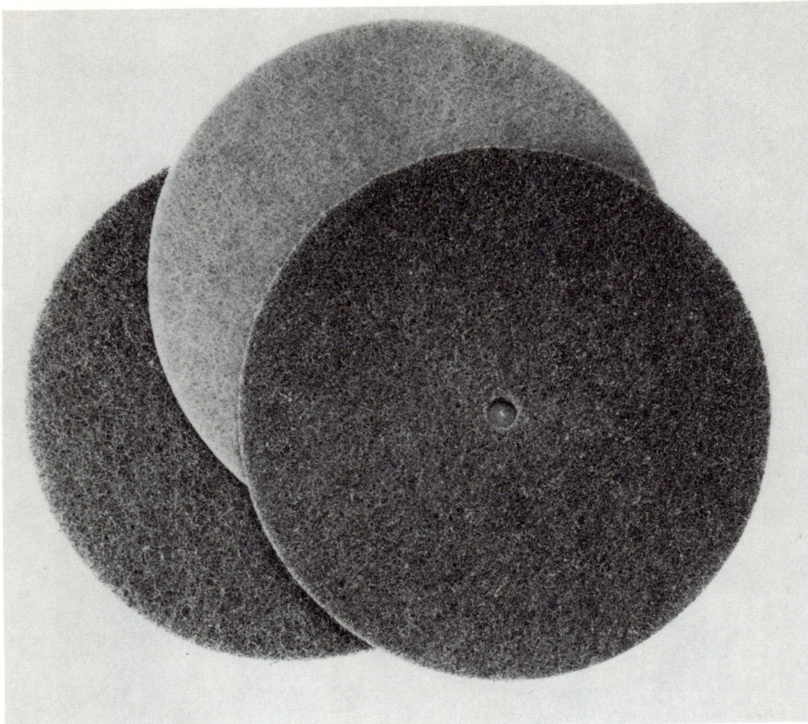

FIGURE 14-3: NONWOVEN ABRASIVE

FIGURE 14-4: CRIMPED WIRE BRUSHES

FIGURE 14-5a: KNOTTED BRUSH

FIGURE 14-5b: TWISTED TUFT BRUSH

FIGURE 14-6a: TUFTED END BRUSH

FIGURE 14-6b: MULTIPLE ROW TUFTED END BRUSH

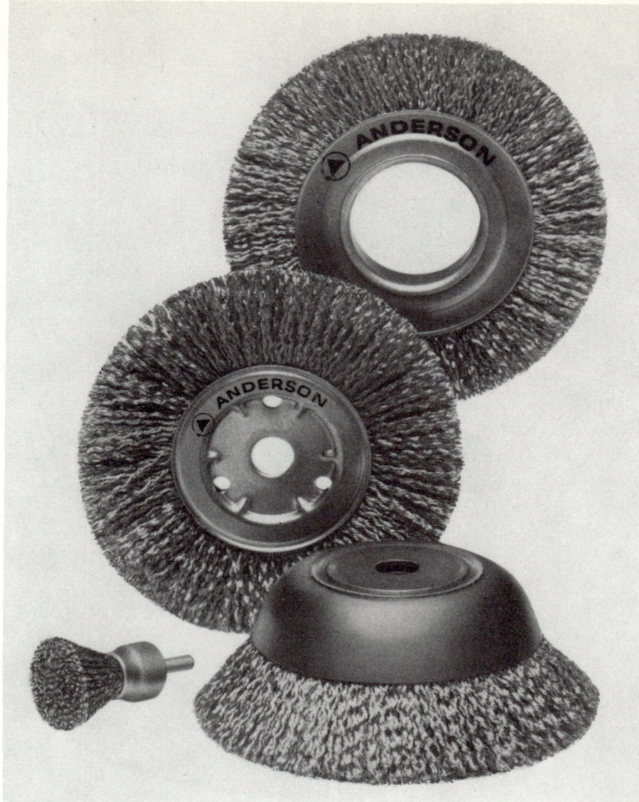

FIGURE 14-7: MULTIPLE STRAND CRIMPED BRUSHES

FIGURE 14-8: END BRUSH WITH BRIDLE

FIGURE 14-9: CENTERLESS BRUSHES

FIGURE 14-10a: BONDED WIRE FILLED BRUSH

FIGURE 14-10b: FILLED BRUSH

TWO STEP METHOD TO SELECT POWER BRUSHES

STEP I: MAKE TENTATIVE SELECTION FROM CHART I
STEP II: ADJUST FIRST SELECTION AS NECESSARY BASED ON CHART II

CHART NO. I

STEP I

FOR JOBS REQUIRING SURFACE FINISH LESS THAN 30 MICRO INCHES		FOR JOBS ALLOWING SURFACE FINISH TO EXCEED 30 MICRO INCHES	
BRUSH FLEXIBILITY: HIGH ← → LOW		BRUSH FLEXIBILITY: LOW ← → HIGH	
BRUSH FINISH ABILITY: LOW MICRO INCH ← → MEDIUM		BRUSH FINISHING ABILITY: FINE ← → COARSE	
BRUSHING ACTION STRENGTH: LOW ← → FAST		BRUSHING ACTION STRENGTH: MEDIUM FAST ← → FAST	

Brush	MAJOR BRUSH CHARACTERISTICS	MAJOR BRUSH APPLICATIONS LISTED IN ORDER OF GENERAL USAGE
SYNTHETIC FILAMENTS OR NATURAL FIBERS LIKE TAMPICO	HIGH BRUSH FLEXIBILITY — LOW BRUSHING ACTION STRENGTH — GENERALLY USED WITHOUT FILAMENT TREATMENT OR ABRASIVE COMPOUND	• CLEANING WET AND DRY • APPLICATOR TO APPLY OILS, COATING COMPOUND, POWDERS, ETC. • FINISHING LIGHT BUFFING & POLISHING • EDGE BLENDING
TREAT FILAMENT BRUSHES LIKE TAMPICO FIBER	MEDIUM FLEXIBLE FAST CUTTING — USED WITH ABRASIVE COMPOUND OF A TACKY COMPOSITION. PRODUCES LOW MICRO INCH SURFACE FINISH	• EDGE BLENDING • SMALL BURR REMOVAL • LIGHT SCALE REMOVAL • SURFACE BLENDING • SURFACE FINISH FOR APPEARANCE • CLEANING
ABRASIVE FILLED FILAMENT BRUSHES (ANDERLON)	MEDIUM FLEXIBLE FAST CUTTING — USUALLY USED WITHOUT ABRASIVE COMPOUND SINCE ABRASIVE IS IMBEDDED IN THE FILAMENT — MAY BE USED WET OR DRY — AVAILABLE WITH ALUMINUM OXIDE OR SILICON CARBIDE GRIT.	• EDGE BLENDING • SMALL BURR REMOVAL • CLEANING • LIGHT SCALE REMOVAL • SURFACE FINISH FOR APPEARANCE • SURFACE BLENDING
FINE WIRE BRUSHES USED WITH ABRASIVE COMPOUNDS	FOR EFFICIENT USE OF WIRE FILLED BRUSHES WITH COMPOUND SELECT BRUSHES HAVING HIGH DENSITY AND WIRE SIZES .008 DIAMETER OR FINER.	• MEDIUM SIZE BURR REMOVAL • LIGHT SCALE REMOVAL • EDGE BLENDING • SURFACE FINISH FOR APPEARANCE
ELASTOMER BONDED BRUSH ←→ (ANDERBOND)	LOW FLEX — RIGID WHEEL WITH FAST BRUSHING ACTION AND CONTROLLED BRUSH FACE	• BURR REMOVAL • EDGE BLENDING • SCALE REMOVAL • ROUGHENING • CONFINED AREAS • CLEANING
HIGH DENSITY REGULAR CRIMPED-WIRE OR ANDERLITZ BRUSHES	SUPER DENSE, LOW FLEX, FOR HIGH PRODUCTION USAGE ON AUTOMATIC OR SEMI-AUTOMATIC BRUSHING EQUIPMENT	• HEAVY BURR REMOVAL • HEAVY SCALE REMOVAL • EDGE BLENDING • SURFACE FINISHING • GENERAL PURPOSE CLEANING • SURFACE REFINING
REGULAR CRIMPED-WIRE OR ANDERLITZ BRUSHES	WIDE FACE, DENSE FILL, FOR FAST CUTTING AND FINE FINISH. BEST GENERAL PURPOSE WHEEL	• BURR REMOVAL • GENERAL PURPOSE CLEANING • SURFACE FINISHING • EDGE BLENDING • HEAT SCALE REMOVAL • SURFACE REFINING
REGULAR CRIMPED-WIRE OR ANDERLITZ BRUSHES	NARROW FACE, MEDIUM DENSE, USED WHERE MEDIUM BRUSH FLEXIBILITY IS REQUIRED	• GENERAL PURPOSE CLEANING • SURFACE FINISH FOR APPEARANCE • SURFACE ROUGHENING • EDGE BLENDING • SURFACE REFINING • HEAT SCALE REMOVAL
HIGH-COMPLIMENT KNOT BRUSHES	MAXIMUM DENSITY MEDIUM FLEX FOR RUGGED PRODUCTION JOBS. ESPECIALLY SUITED FOR AUTOMATIC OR SEMI-AUTOMATIC MACHINE APPLICATION	• HEAVY SCALE REMOVAL • HEAVY BURR REMOVAL • HEAVY CLEANING RUST, PAINT, RUBBER, PLASTICS • SURFACE ROUGHENING • EDGE BLENDING
STANDARD KNOT BRUSHES	NARROW FACE HIGHLY FLEXIBLE BRUSH. USED WHERE SURFACE IMPACT ACTION IS NECESSARY	• WELD SPLATTER AND SCALE REMOVAL • GENERAL PURPOSE CLEANING • SURFACE ROUGHENING • EDGE BLENDING • BURR REMOVAL • SURFACE FINISHING

CHART NO. II

STEP II

AFTER TESTING BRUSH FIRST SELECTED BASED UPON CHART I — THE FOLLOWING REFINEMENTS MAY BE MADE IN BRUSH SPECIFICATIONS OR OPERATING CONDITIONS

OBSERVED RESULTS

BRUSH WORKS TOO SLOWLY	*BRUSH WORKS TOO FAST*	*FINER OR LOWER MICRO INCH SURFACE REQUIRED*	*SURFACE FINISH IS TOO SMOOTH AND LUSTROUS*	*BRUSHING ACTION NOT SUFFICIENTLY UNIFORM*	*ACTION OF BRUSH PEENS BURR TO ADJACENT SURFACE*
• INCREASE BRUSH OUTSIDE DIAMETER • INCREASE BRUSH SPEED • DECREASE TRIM LENGTH THEREBY INCREASING THE BRUSH FILL DENSITY • INCREASE FILAMENT DIAMETER • CHANGE BRUSH TO HEAVIER DUTY TYPE	• REDUCE TRIM LENGTH • DECREASE FILAMENT DIAMETER • REDUCE BRUSH OUTSIDE DIAMETER • REDUCE BRUSH SPEED • SELECT BRUSH TYPES HAVING LOWER BRUSHING ACTION STRENGTH	• DECREASE TRIM LENGTH THEREBY INCREASING BRUSH FILL DENSITY • DECREASE FILAMENT DIAMETER • SELECT BRUSH CAPABLE OF PRODUCING LOWER MICRO INCH SURFACE FINISHES • USE AUXILARY ABRASIVE COMPOUNDS OR OIL BASE LUBRICANTS WITH BRUSH • INCREASE BRUSH SPEED	• INCREASE TRIM LENGTH • INCREASE FILAMENT DIAMETER • SELECT A BRUSH TYPE CAPABLE OF PRODUCING COARSER FINISHES • REDUCE BRUSH SPEED	• INCREASE TRIM LENGTH • CHANGE BRUSH TYPE FROM SECTIONAL TO HELICAL CONSTRUCTION • SELECT SUITABLE BRUSHING MACHINERY • OSCILLATE BRUSH TO AVOID "TRACKING"	• DECREASE TRIM LENGTH • SELECT HIGH DENSITY CRIMPED WIRE WHEEL • FOR DUCTILE METALS CHANGE FROM WIRE FILLED BRUSH TO NON-METALLIC FILLED BRUSHES
BRUSH FACE LOADS WITH MATERIAL BEING REMOVED	*BRUSH DEVELOPS EXCESSIVE HEAT*	*BETTER CONTROL OF BRUSHING PRESSURE*	*BRUSH WIRE FILAMENT ENDS BECOME DULL*	*IMPROVE BRUSH DYNAMIC BALANCE CONDITION*	*BRUSH FACE WEAR UNEVEN OR GROOVING*
• INCREASE BRUSH TRIM LENGTH • INCREASE BRUSH FILAMENT DIAMETER • IF WET APPLICATION APPLY SOLUTION THROUGH BRUSH ARBOR AND DISPERSE OUTWARD THROUGH BRUSH FACE • WHERE CLEANING SOLUTIONS ARE USED WITH BRUSH MODIFY CLEANER FOR BETTER SOLUBILITY • CHANGE BRUSH FILL SPECIFICATIONS	• USE INTERNAL COOLING OF BRUSH AND BRUSH ARBOR • REDUCE BRUSHING PRESSURE • SELECT HEAVIER DUTY BRUSH • INCREASE BRUSH TRIM LENGTH • REDUCE BRUSH PERIPHERAL SPEED • USE EXTERNAL COOLING OF BRUSH	• INCREASE BRUSH TRIM LENGTH • INSTALL METERS TO MEASURE BRUSHING LOADS • WHEN PART SIZE VARIES DESIGN PRESSURE RELIEF IN PART HOLDING FIXTURE • WITH AUTOMATIC BRUSHING EQUIPMENT ALLOW BRUSH TO FLOAT OR PROVIDE AUTOMATIC PRESSURE SENSING AND PRESSURE ADJUSTING	• REVERSE ROTATIONAL DIRECTION OF BRUSH TO AUTOMATICALLY RECONDITION • DRESS THE BRUSH FACE WITH AN ABRASIVE STONE (WITH THE BRUSH OPERATING IN A DIRECTION REVERSE OF NORMAL USE) • SELECT A HEAVIER DUTY BRUSH • KEEP BRUSHING HEAT AS LOW AS PRACTICAL	• WHEN WIDE FACED BRUSH UNITS ARE MADE UP OF THIN SECTIONS, STATIC BALANCE EACH SECTION AND STAGGER HEAVY SIDES • CHANGE FROM SECTIONAL BRUSH ASSEMBLY TO A HELICALLY FORMED BRUSH • DYNAMICAL BALANCE BRUSH ARBOR AND BRUSH ASSEMBLY • REDUCE BRUSH SPEED • CHECK ENTIRE BRUSH DRIVE TRAIN FOR IMBALANCE	• IF POSSIBLE HAVE PART CONTACT BRUSH SO THAT THE FULL FACE OF BRUSH IS ON THE PART • CONSIDER OSCILLATING BRUSH OR PART • PROVIDE A MEANS FOR PERIODICALLY DRESSING TH THE BRUSH FACE • AVOID EXCESSIVE BRUSHING PRESSURE • CHANGE FROM SECTIONAL BRUSH ASSEMBLY TO A HELICALLY FORMED BRUSH PART

FIGURE 14-11: TWO STEP METHOD TO SELECT POWER BRUSHES

cause heavier burr conditions. These burrs develop as a result of excessive tool feed speeds and the use of tools beyond their most efficient cutting point.

The advantages of using wire brushes for gear deburring in preference to other tools are many. When used with good finishing equipment the proper type of brush will quickly remove burrs and produce uniformly finished parts. This is in contrast to fragmented metal or sharp edges often left on parts as metal deburring tools dull or fixed abrasives become loaded or dull. In addition to removing burrs, powered brushes with their flexible action blend sharp edges and produce radii at the junctures of intersecting surfaces without changing critical dimensions of machined surfaces.

For a great majority of gear deburring jobs, radial short trim wire brushes (knot type) are recommended, and gears are brushed in the green. Fill wire can be any of several diameters, however, a 14 inch (356 mm) O.D. brush with .0118 inch (0.3 mm) special analysis steel wire is most widely used (Figure 14-5a). Although it develops relatively little impact action, its short trim and dense construction give the brush a high degree of cutting action. At 6500 sfpm (33 m/s) this brush is extremely fast cutting.

In gear deburring, correct positioning of the brushes is very important. By spacing brushes to work on both sides of the gear face, the size of the edge break and uniformity of blend can be controlled. Brushes can produce a uniform break around the entire tooth profile or can be positioned to break the sides of the tooth more than the root. Conversely, the root can have a larger break than the sides. There are few deburring methods where this can be accomplished and result in uniformity of tooth profile part after part.

For spur gears, brushes are spaced and contact the tooth profile on the gear's centerline (Figure 14-12). With this setup, brushes can be reversed thereby keeping the wire points at their maximum cutting efficiency.

Brushing spline bores requires a different setup. Here the brushes must be located offcenter, shown in Figure 14-13. The brushes used generally do not have to be reversed to maintain their high degree of cutting effectiveness.

When helical gears are brushed it is sometimes necessary to favor the acute side of the gear tooth to develop a generous radius. Such a radius is required because subsequent shaving of the gear will otherwise produce an extremely sharp edge. The correct setup for helical gears is shown in Figure 14-14.

Another plus with brushing is that the fillet on the gear tooth can be broken thus relieving this high stress concentration area. This is particularly essential on hardened gears. Many gear manufacturers find less failures and greatly reduced replacement problems after brush finishing. Easier assembly and safer handling are added benefits of this method.

Elastomer bonded wire brushes are used for deburring fine pitch gears. These brushes have minimum flexibility on the brush face and remove burrs

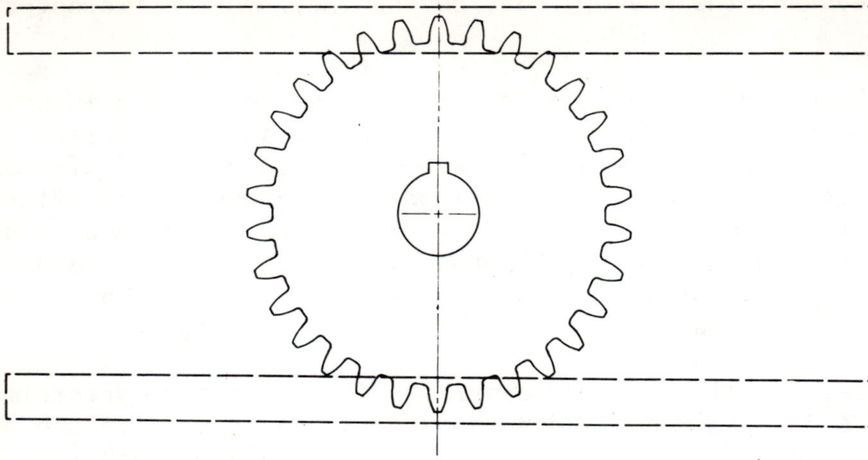

FIGURE 14-12: BRUSH POSITION FOR DEBURRING SPUR GEARS

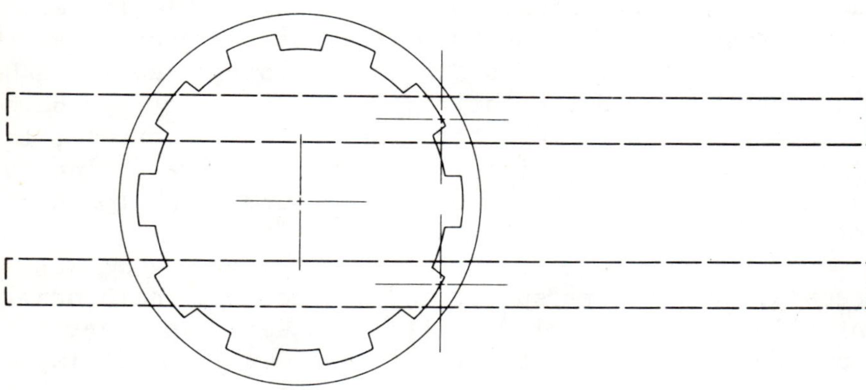

FIGURE 14-13: BRUSHING SPLINES

without leaving a secondary roll. Bonded brushes are also very effective on powdered metal and aluminum diecast parts. At maximum efficiency, speeds of these brushes range from 2600 to 6000 sfpm (13.2 to 30.5 m/s). Because this type has such fast cutting action, surface speeds can be lower than those for conventional wire brushes.

The finishing of hardened pump gears requires treated tampico brushes used with a grease base deburring compound. The same type of gear positioning in relation to the brushes applied for these gears.

Deburring Cylindrical Parts

Cylindrical parts are most effectively deburred on centerless grinding equipment. Large 20 inch (508 mm) O.D. brushes (5 or 6 inch (127 or 152mm) face width) filled with a variety of materials, working with conventional grinding coolant, remove feather burrs and improve micro finish on hydraulic valves and spools, pistons and similar parts. Altering part size generally is not a problem as the brush will attack the edges and improve the finish but not change the finished dimension. An old surplus grinder can often be utilized as the brush is flexible enough to compensate for any wear in the machine. Many manufacturers set the brushing machine in line with the parts grinding machine. The cylindrical parts can be ground to size then fed in line to the second machine for deburring by the same operator.

A typical application is shown in Figure 14-15. Here a centerless brush filled with .005 inch (0.13 mm) steel wire deburrs the grooves in extremely close tolerance rifle pistons without disturbing critical dimensions. The pistons are inserted between guide rails, driven by a regulating wheel, and automatically moved through the machine for brushing at a rate of two seconds per piece.

Brushes filled with treated tampico fiber or cord material, used in conjunction with grease base polishing compound, are capable of producing extremely fine finishes in the range of 2 through 7 microinch (0.05 through 0.18 µm). Brushes filled with abrasive impregnated nylon filaments are recommended for fast, thorough burr removal without compound and subsequent part cleaning. Figure 14-16 shows a setup for deburring perforated steel tubing.

End Deburring--Tube, Pipe and Rod

Power brushes are widely used for deburring ends of tubes, pipe, solid rod, and tubing of all sizes and shapes. A variety of methods are used, ranging from automatic machines built especially for this purpose to simple off-hand operations involving a bench motor or polishing lathe.

Figure 14-17 shows an automatic setup for deburring the ends of steel underfloor ducts after the cutoff operation. Duct sections of various shapes are conveyed through a pair of brushing heads, each made up of five 12 inch (305 mm) O.D. knot type wire brushes. Both duct ends are deburred simultaneously--quickly and efficiently--and edges are uniformly blended.

Another highly successful application is shown in Figure 14-18. Here end burrs are removed from rough cutoff lengths of copper tubing by two heads of 10 inch (254 mm) O.D. elastomer bonded steel wire brushes. The equipment utilizes a ferris wheel arrangement to index the tubes past the brushing heads. Tubes are rotated through the brushing cycle by endless belts. The machine is adjustable for various tube lengths and diameters. For 1 inch (25.4 mm) diameter parts the production rate is 3300 pieces per hour.

An ingenious method for deburring and cleaning the ends of copper tubing after cutting to length is shown in Figure 14-19. Three small diameter radial brushes filled with crimped steel wire are fixed (do not revolve individually) on a metal backing plate. A wire tube type brush is

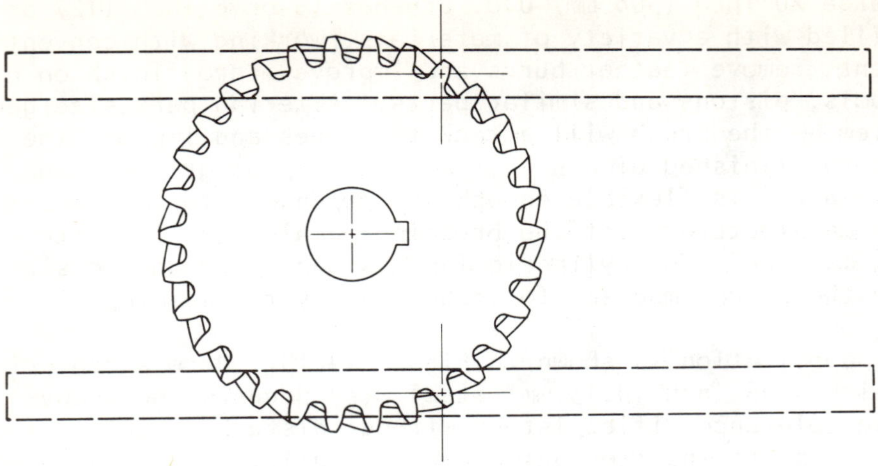

FIGURE 14-14: BRUSH POSITION FOR DEBURRING HELICAL GEARS

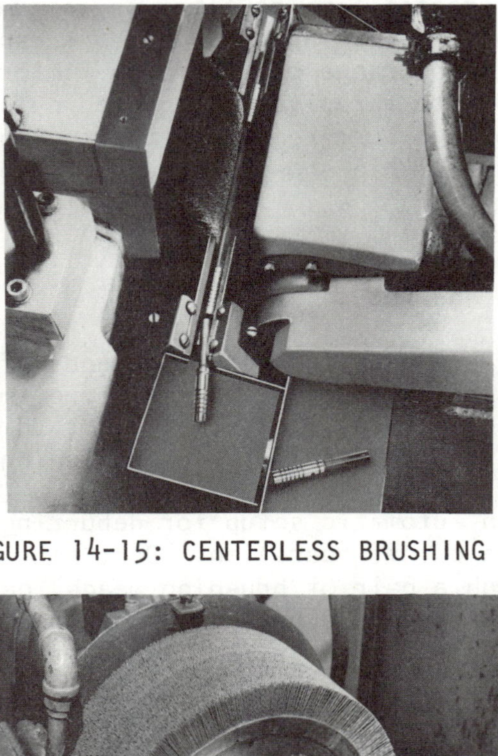

FIGURE 14-15: CENTERLESS BRUSHING

FIGURE 14-16: DEBURRING PERFORATED STEEL TUBING

FIGURE 14-17: DEBURRING ENDS

FIGURE 14-18: DEBURRING TUBE ENDS

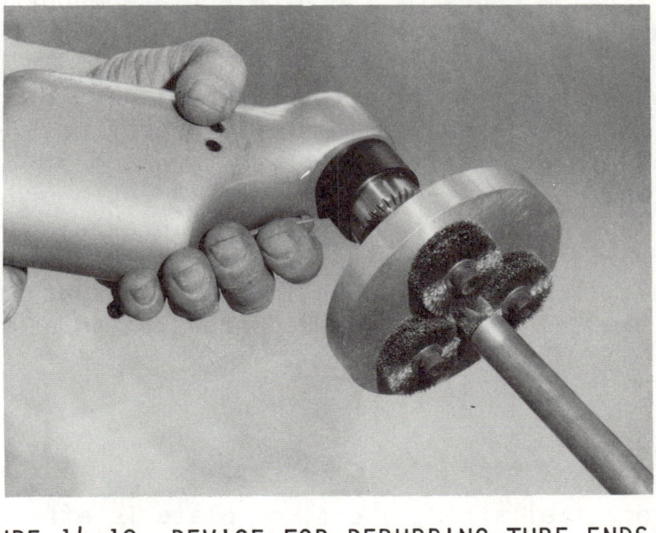

FIGURE 14-19: DEVICE FOR DEBURRING TUBE ENDS

mounted in the center. The tube brush is inserted into the tubing and the plate fixture rotated by a portable tool. End burrs are removed and both inside and outside tubing surfaces are cleaned. Tubes are ready for soldering or brazing. This same type backing plate and brush assembly can be mounted on a polishing lathe or bench motor to handle long lengths of pipe or tubing. Best results can be obtained if a simple angle iron fixture is used as a steady rest and to present the tubing to the brush cluster in the proper position.

Deburring Internal and Threaded Areas

Small power brushes in a wide selection of sizes, shapes, and fill materials are available for deburring and cleaning recessed areas where space limitation is a factor. These are designed for rotary, end, or side action and are readily adaptable to use on portable tools, drill presses, bench grinders, lathes, and special equipment.

The small radial brushes range from 1 to 3 inch (25 to 76 mm) in diameter and are available in wire sizes from .0025 to .014 inch (0.06 to 0.36 mm). Fill materials include steel and stainless steel wire, nylon, tampico, abrasive impregnated nylon filaments, and elastomer bonded steel wire. Figure 14-20 shows a 1 inch (25 mm) diameter brush filled with .005 inch (0.12 mm) nylon deburring a small D.C. armature. The brush is driven at 8000 rpm by a 1/8 HP motor and removes burrs and other foreign particles from the narrow commutator slots. In Figure 14-21 an elastomer bonded wire brush mounted on a drill press cleans internal surfaces and deburrs intersecting holes in a machined casting.

End brushes are designed with a shank that is integral with the cup. The fill material extends from the cup. Generally used with portable tools or drill presses, they are available in a great many types, sizes, and fill materials. In Figure 14-22, an elastomer bonded wire brush with flared face, mounted on a portable air tool, removes machining burrs from internal cavities of an aluminum alloy casting. Figure 14-23 depicts a knot type wire end brush. This is inserted in the pipe end and expands when the power is turned on. Inside edges of the pipe end are deburred when the brush is withdrawn and outer edges then deburred.

Side action brushes are effective for internal deburring and also for removing burrs from threaded areas. Diameters range from 5/32 to 1-1/4 inch (4 to 32 mm) and are available in a broad range of styles and fill materials, metallic and nonmetallic. In Figure 14-24, a brush with wire side tufts feathers and removes sharp areas from internal threads and also removes chips left from the threading operation. In Figure 14-25, a square trimmed brush filled with abrasive impregnated synthetic webbing cleans and deburrs threaded and counterbored areas of an aluminum casting in one operation.

Deburring Flat Shapes

Edge deburring of steel strip, sheet, or plate is effectively done

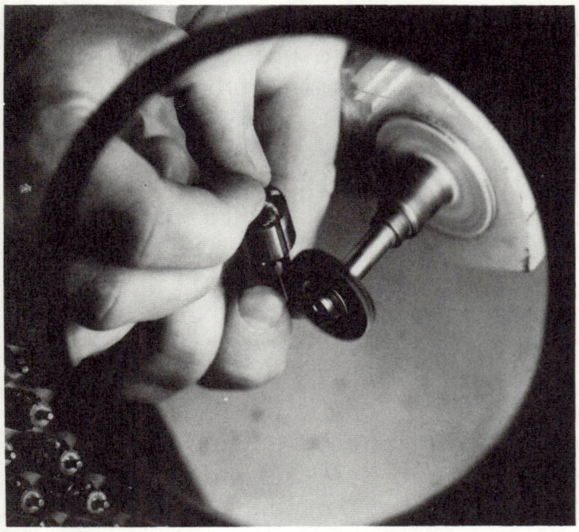

FIGURE 14-20: DEBURRING WITH MINIATURE BRUSH

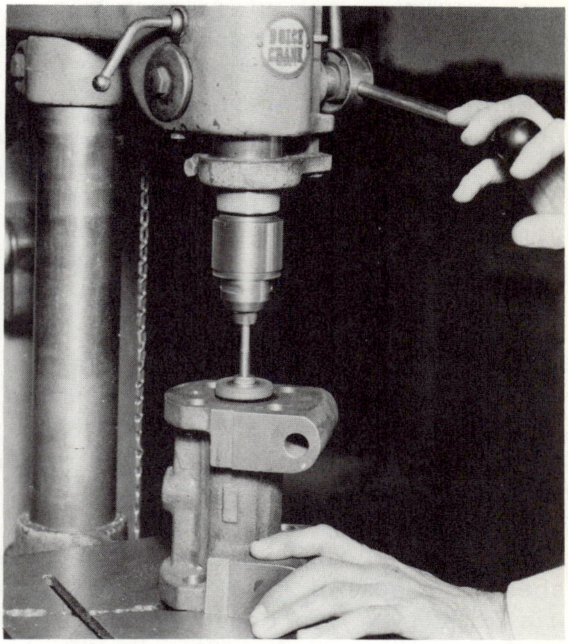

FIGURE 14-21: DEBURRING HOLES ON A DRILL PRESS

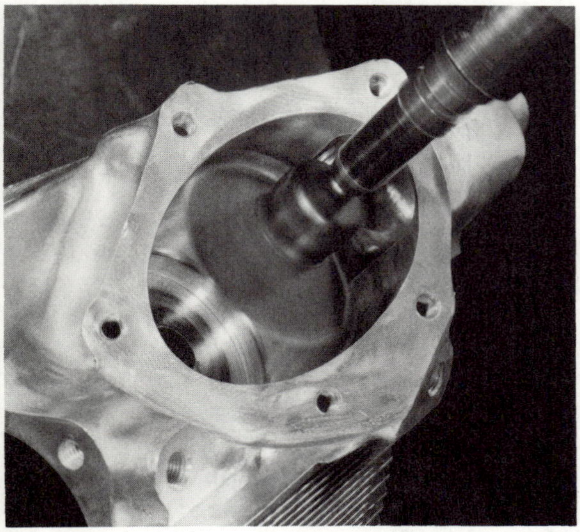

FIGURE 14-22: ELASTOMER BONDED FLAIR FACE BRUSH

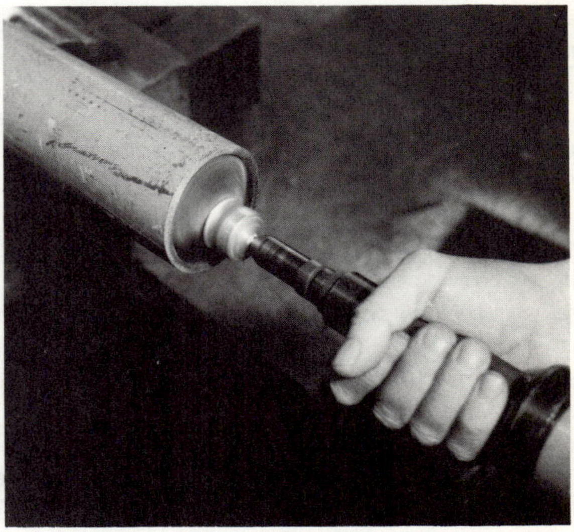

FIGURE 14-23: KNOT BRUSH FOR END DEBURRING

by elastomer bonded steel wire radial brushes, either off-hand using portable tools, or passed under motor mounted brush heads. An important characteristic of this brush is the perpetually short trim with each wire point firmly supported at the brush face. This produces extremely fast cutting action, maximum work efficiency with minimum pressure, and positive control over area of brush contact.

Many manufacturers have replaced grinding wheels with abrasive impregnated nylon filament brushes for finishing and cleaning edges of strip steel (pail stock) prior to welding. Abrasive brushes are quiet in operation (meet OSHA standards), conform to slight distortions in strip, give longer service at less cost.

Expanded metal is another form of sheet that can be thoroughly deburred by power brushing. An extremely simple setup can be assembled to convey the expanded sheet under a wide face brush roll made up of individual knot type steel wire radial brushes mounted in multiple on a steel shaft. The flexible brush tips reach into the uneven surfaces to deburr and edge blend. resulting in a smoothly finished product ready for decorative or other use.

For certain applications, the circular brushing action of cup type brushes is required to obtain desired results. Figures 14-26 and 14-27 show a knot type cup brush with steel bridle deburring thousands of drilled holes in a steam condenser support plate.

This particular setup involves a special machine having two motor driven brushes arranged for lateral and vertical travel positioned to deburr both sides of the steel plate at the same time. The cup brushes remove all burrs and produce a uniform radius around the periphery of each hole without elongating the hole area. For smaller plates, a radial drill press and weld positioning table with the same type cup brush will do an excellent job.

Equipment for Power Brushing

A great many power brushing jobs are effectively done with simple equipment--air and electric tools, drill presses, bench motors, and polishing lathes.

For high production work, it is essential that the correct machines be chosen for driving the brushes. Basic requirements are that the machines have ample horsepower, proper size shafts, and good bearings. Also important is flexibility of machine deisgn so that part design or machining changes which could alter the deburring or finishing problem will not obsolete the equipment.

Horsepower requirements differ greatly from job to job. For example, a deburring operation with a 14 inch (356 mm) diameter narrow face radial brush uses far less power than that needed for wide face brush-and-water cleaning of strip metal. Because of the wide varience in power consumption and lack of basic information available to designers, special equip-

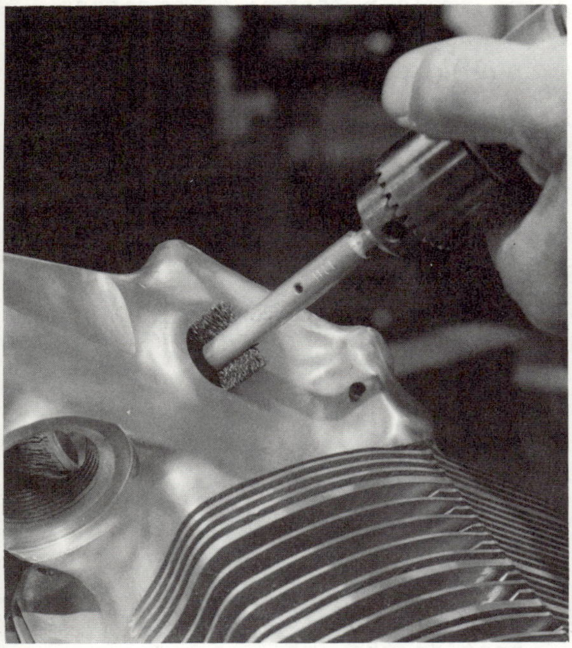

FIGURE 14-24: SIDE ACTION BRUSH

FIGURE 14-25: SQUARE TRIM BRUSH

FIGURE 14-26: DEBURRING CONDENSER SUPPORT PLATE

FIGURE 14-27: CONDENSER SUPPORT PLATE

ment often is not adequate. This has resulted in the design and manufacture of standard brushing equipment by a number of machine builders.

The standard machines are capable of high production and are extremely flexible. They are equipped with motors having ample horsepower, ammeters to measure brushing pressure, switches that reverse the direction of brush rotation to insure wire points with maximum cutting efficiency, and adjustment mechanism for controlling part dwell time. On many machines, automatic feed to compensate for brush wear is available.

Standard units range from manual load to completely automated machines. In all cases, the equipment has been designed to operate effectively with the brushing tool selected for the specific job. With this approach, gear finishing no longer must be regarded as a secondary operation. Today it can be integrated inline following hobbing and preceding shaving.

The best machine for a given job? The answer to this question depends on the application which should be analyzed as follows:

1. Determine job requirements
   a. What is to be done: deburring, producing radii, etc.?
   b. What is the production rate?
   c. How many surfaces must the brush contact?
   d. Must equipment be run separately or integrated inline?
2. Select the type of brush that will best do the job.
3. Determine horsepower requirements.
4. Select standard equipment whenever possible which will best accommodate all of the above factors.

Brush Deburring Research

Components of many precision mechanisms need near-sharp edges to be reliable. These edges must also be burr free to prevent jamming of the mechanisms. When these requirements are joined with the accelerating demands of miniaturization we find parts with the following characteristics:

- Tolerances of 0.0002 to 0.0005 inch (5.1 to 12.7 μm)
- Surface finishes of 8-16 RMS
- Maximum allowable edge radii of 0.002 inch (51 μm)
- Minute size, 0.020 inch (.51 mm) diameter holes for example
- 100 percent burr removal when viewed under 10-30X magnification

When these requirements are combined with short lead times, tough stainless steel workpiece materials, low build quantities, and three dimensional contours, deburring becomes a major processing limitation. Each part must be studied closely to determine which deburring processes might be applicable and how the process can be altered to maintain the rigid restraints imposed by design.

Unlike most manufacturing processes, the capabilities and limitations of the deburring processes have never really been defined. In a review of some 1300 articles and reports on deburring and edge radiusing, less than

50 provide any data on burr properties, edge radiusing, or deburring limitations. The data that is provided is not sufficient to establish any adequate processing guidelines. As a result of this lack of information and the need to assure complete burr removal on high precision (0.0002 - 0.0010 inch tolerance) parts many companies have relied extensively on hand deburring to produce the necessary edge quality. Unfortunately hand deburring is time consuming, expensive, and in itself inherently operator variable.

Test Objectives

In an attempt to provide some quantitative information on deburring with brushes a brief study was initiated. The purpose of the study was twofold:

- Determine the size of burrs which can be removed while maintaining the part tolerances.
- Determine quantitatively how brushing variables affect the deburring/edge radiusing process.

The minute size and contoured geometry of the gears, pawls, latches, shafts, cams, and threads dictated that only miniature brushes could be used. Thus only brushes of 1 inch (25.4 mm) or 1.375 inch (34.9 mm) diameter were used for this study. The brush bristles included brass stainless steel, nylon with impregnated abrasive, and string with impregnated abrasive. The workpiece materials included 303Se and 17-4 PH stainless steel, beryllium copper, 6061-T6 and 7075-T6 aluminum. The 17-4 PH stainless steel was in the H900 condition while the beryllium copper was in the HT condition.

While hand deburring operators have utilized small brushes for many years, it has been used as a secondary deburring tool. The primary deburring tool is typically a small knife or file and the brush just radiuses an essentially burr free edge. The emphasis of this study was to use these brushes as the primary deburring tool. Under appropriate conditions brushing not only should be faster than the two stage deburring process, but it also would help eliminate the accidental gouging of edges which occurs with knives.

To provide burrs of varying sizes some workpieces were ground and others were face milled. The grinding burrs were approximately 0.0005 inch (12.7 μm) thick. The face milling burrs were 0.002 inch thick by 0.0025 inch long (51 by 64 μm) in stainless steel and 0.0034 inch thick by 0.0011 inch long (86 by 279 μm) in aluminum. As previously mentioned the desired final edge condition was a burr free edge with a radii of 0.002 inch (50.8 μm) or less.

Results of Brush Deburring Tests on Grinding Burrs

Grinding burrs can readily be removed using these small brushes and 5000 rpm, 1/12 horsepower bench motors. The resulting edge radii can be as large as 0.0186 inch (0.47 mm) on aluminum (Table 14-1). By appropriate choices of brushes and technique, however, radii of 0.003 inch (76 μm) maximum can be maintained on aluminum in the burr removal operation (Table 14-2). A 0.002 inch (51 μm) maximum radius can also be maintained on 17-4 PH stainless steel. A statistical analysis of variance indicates that edge radii are affected by feedrate past the brush, number of passes made past the brush, and type of brush as well as workpiece material. In each case there is a unique combination of factors which produces the minimum edge radius. As a general rule, edge radii were roughly proportional to workpiece hardness.

The radii shown in Table 14-1 result when the effects of brush feedrate and number of passes in front of the brush are ignored. It represents an average value of radius resulting when these two variables are not controlled. The data in Table 14-2 represents the maximum radius produced at the optimum combination of feedrate and number of passes. Optimum as used here again indicates the conditions which produce the smallest edge radius.

The data in Tables 14-1 and 14-2 were obtained by feeding the workpiece past the brush at feedrates of 1.0 and 2.0 inch per minute (ipm) (0.67 and 1.34 mm/s).* Half the specimen made one pass across the brush, half made four passes. Parts were brushed with one of four brushes at 3000 rpm. This is considerably below the spindle speed recommended for these brushes but is compatible with in-house practice and equipment.

Four passes across the brush typically produced a larger radius than a single pass. The faster feedrate which results in a shorter time in front of the brush typically resulted in smaller radii (Figures 14-28 through 14-32). The stainless steel specimens were an exception to this effect, however--a faster feedrate typically resulted in a larger radius (Figures 14-29 and 14-30), an anomaly not readily explainable. The average radius (r) and the standard deviation (σ) resulting at each condition tested are indicated in Figures 14-29 through 14-33. The standard deviation in edge radius typically ranged from 0.0006 to 0.0012 inch (15.2 to 30.4 μm).

The string brush with impregnated silicon carbide produced the smallest radii. These radii typically varied from 0.002 to 0.004 inch (51 to 102 μm) in stainless steel. Typical radii produced by the other three brushes ranged from 0.004 to 0.008 inch (102 to 208 μm). The exceedingly short life of the string brushes (four to eight edges) and their high cost make them impractical for most applications.

Results of Brush Deburring Milling Burrs

Attempts to use the 1.0 and 1.375 inch (25.4 and 34.9 mm) diameter brushes were largely unsuccessful in removing milling burrs at 3000 rpm. Although the stainless steel (A1) and the brass (A5) brushes removed the aluminum burrs, the adjacent surfaces were severely eroded. In many cases

---

*For this study a contact length (interference between brush bristles and workpiece) of 0.125 inch (3.18 mm) was used for the stainless steel brushes, and 0.25 inch (6.35 mm) for all other brushes. For optimum brush life, the bristles should just touch the workpiece edge.

TABLE 14-1: TYPICAL RESULTS OF BRUSH DEBURRING OF GRINDING BURR

Brush	Workpiece Material									
	17-4 PH		303 Se		BeCu		7075-T6		6061-T6	
	(Inch)	(μm)	(Inch)	(μm)	(Inch)	(μm)	(Inch)	(μm)	(Inch)	(μm)
Stainless Steel	0.0064	163	0.0064	163	0.0085	216	0.0174	442	0.0186	472
Nylon with SiC	0.0055	140	0.0064	163	0.0053	135	0.0070	178	0.0082	208
Nylon with AL_2O_3	0.0064	163	0.0058	147	0.0079	201	0.0062	157	0.0066	168
String with SiC	0.0035	89	0.0035	89						
Typical Material Hardness (Brinell)	415		226		187		150		95	

Values shown are edge radii
Missing data indicates conditions not tested.

TABLE 14-2: BEST CASE RESULTS OF BRUSH DEBURRING A GRINDING BURR

Brush	Workpiece Material									
	17-4 PH		303 Se		BeCu		7075-T6		6061-T6	
	(Inch)	(μm)	(Inch)	(μm)	(Inch)	(μm)	(Inch)	(μm)	(Inch)	(μm)
Stainless Steel	0.0067	170	0.0060	152	0.0030	76	0.0225	572	0.0104	264
Nylon with SiC	0.0055	140	0.0037	94	0.0031	79	0.0050	127	0.0049	124
Nylon with AL_2O_3	0.0030	76	0.0036	91	0.0068	173	0.0017	43	0.0026	66
String with SiC	0.0020	51	0.0036	91						

Values shown represent edge radii in inches which can be maintained on 99.7 percent of the edges finished at optimum conditions.
Missing data indicates conditions not tested.

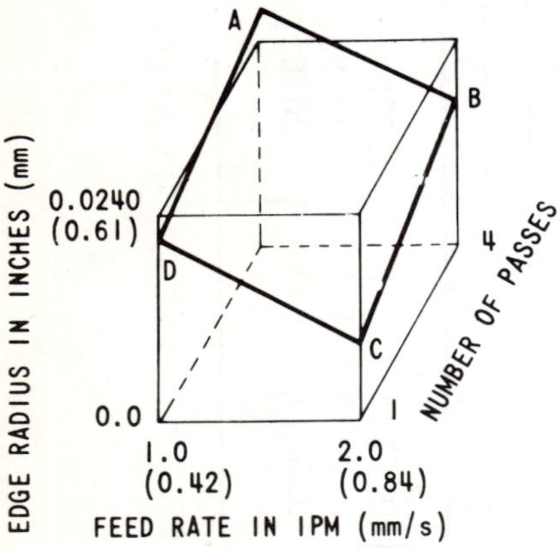

STAINLESS STEEL BRUSH

A - r = 0.0276 (0.701 mm)
σ = 0.0057 (144.8 μm)
B - r = 0.0173 (0.439 mm)
σ = 0.0044 (111.8 μm)
C - r = 0.0089 (0.226 mm)
σ = 0.0005 (12.7 μm)
D - r = 0.0207 (0.526 mm)
σ = 0.0014 (30.5 μm)

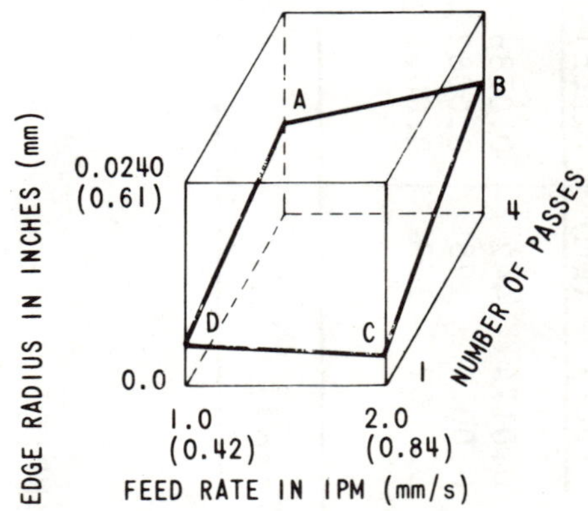

NYLON BRUSH WITH SiC

A - r = 0.0100 (0.254 mm)
σ = 0.0010 (25.4 μm)
B - r = 0.0150 (0.381 mm)
σ = 0.0050 (127.0 μm)
C - r = 0.0035 (0.089 mm)
σ = 0.0007 (17.8 μm)
D - r = 0.0043 (0.109 mm)
σ = 0.0002 (5.0 μm)

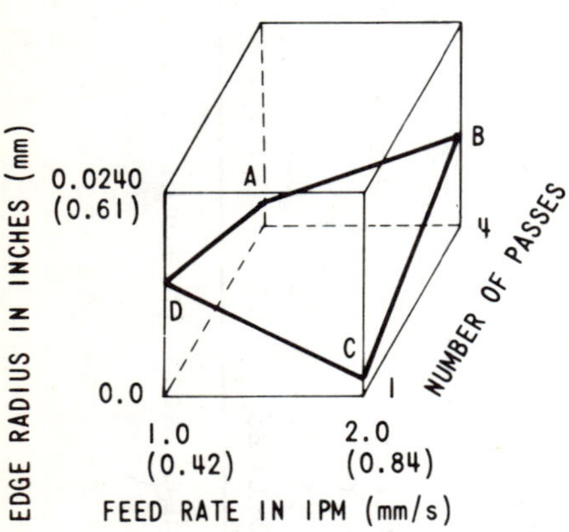

NYLON BRUSH WITH AL_2O_3

A - r = 0.0021 (0.053 mm)
σ = 0.0004 '101.6 μm)
B - r = 0.0107 (0.272 mm)
σ = 0.0012 (30.5 μm)
C - r = 0.0023 (0.058 mm)
σ = 0.0001 (2.5 μm)
D - r = 0.0113 (0.287 mm)
σ = 0.0012 (30.5 μm)

FIGURE 14-28: BRUSH DEBURRING RESULTS ON BERYLLIUM COPPER

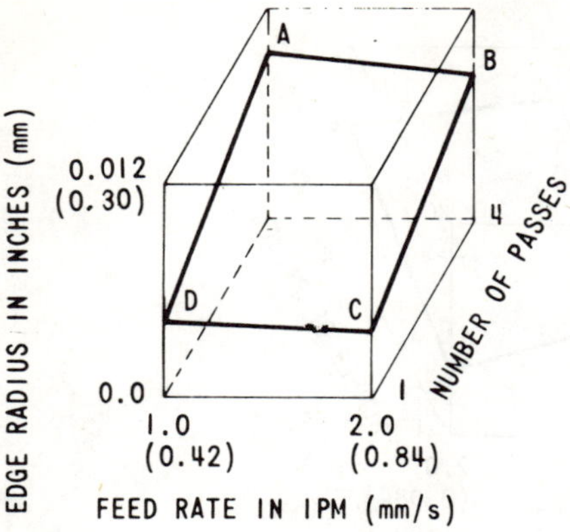

STAINLESS STEEL BRUSH

A - r = 0.0095 (0.241 mm)
σ = 0.0006 (15.2 μm)
B - r = 0.0082 (0.208 mm)
σ = 0.0016 (40.6 μm)
C - r = 0.0036 (0.91 mm)
σ = 0.0011 (27.9 μm)
D - r = 0.0045 (0.114 mm)
σ = 0.0012 (30.5 μm)

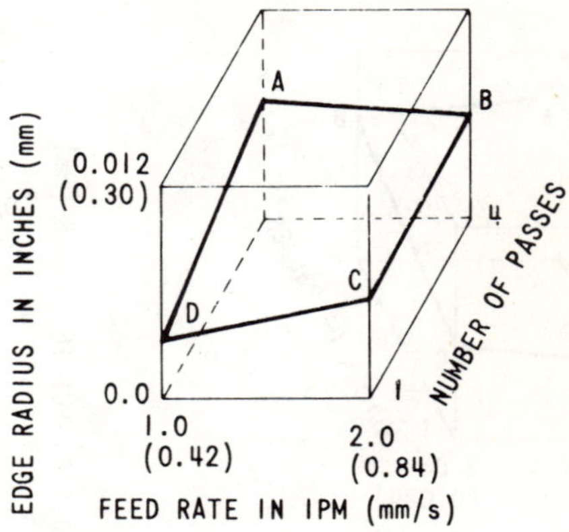

NYLON BRUSH WITH SiC

A - r = 0.0068 (0.173 mm)
σ = 0.0006 (15.2 μm)
B - r = 0.0060 (0.152 mm)
σ = 0.0009 (22.9 μm)
C - r = 0.0059 (0.150 mm)
σ = 0.0037 (94.0 μm)
D - r = 0.0034 (0.086 mm)
σ = 0.0007 (17.8 μm)

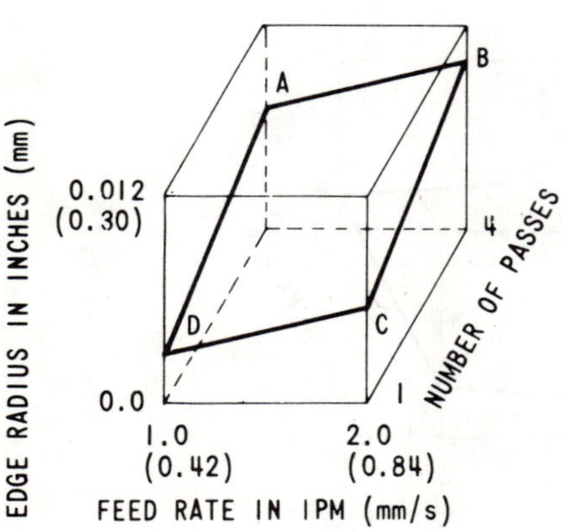

NYLON BRUSH WITH AL_2O_3

A - r = 0.0072 (0.183 mm)
σ = 0.0011 (27.9 μm)
B - r = 0.0101 (0.257 mm)
σ = 0.0026 (66.0 μm)
C - r = 0.0055 (0.140 mm)
σ = 0.0021 (53.3 μm)
D - r = 0.0027 (0.069 mm)
σ = 0.0001 (2.5 μm)

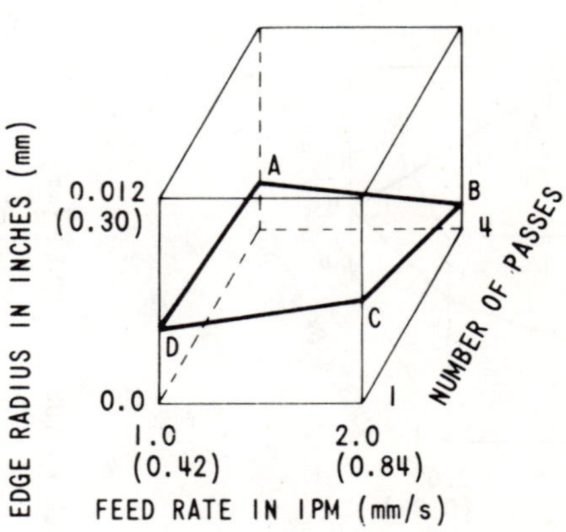

STRING BRUSH WITH SiC

A - r = 0.0028 (0.071 mm)
σ = 0.0006 (15.2 μm)
B - r = 0.0014 (0.035 mm)
σ = 0.0002 (5.1 μm)
C - r = 0.0056 (0.142 mm)
σ = 0.0012 (30.5 μm)
D - r = 0.0042 (0.11 mm)
σ = 0.0005 (12.7 μm)

FIGURE 14-29: BRUSH DEBURRING RESULTS ON 17-4 PH STAINLESS STEEL

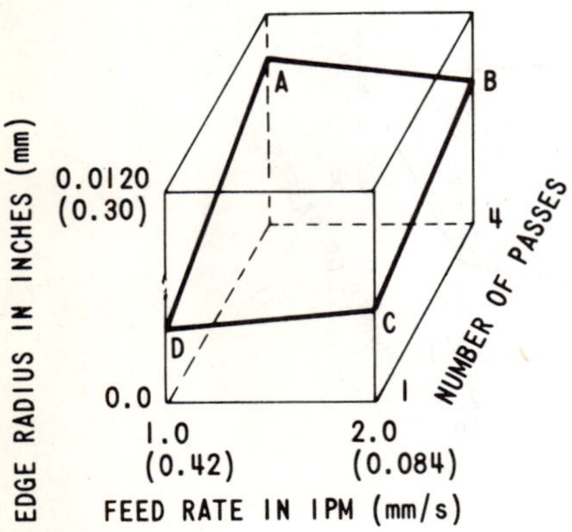

STAINLESS STEEL BRUSH

A - r = 0.0088 (0.224 mm)
σ = 0.0013 (33.0 μm)
B - r = 0.0074 (0.188 mm)
σ = 0.0002 (5.1 μm)
C - r = 0.0050 (0.127 mm)
σ = 0.0012 (30.5 μm)
D - r = 0.0042 (0.107 mm)
σ = 0.0006 (15.2 μm)

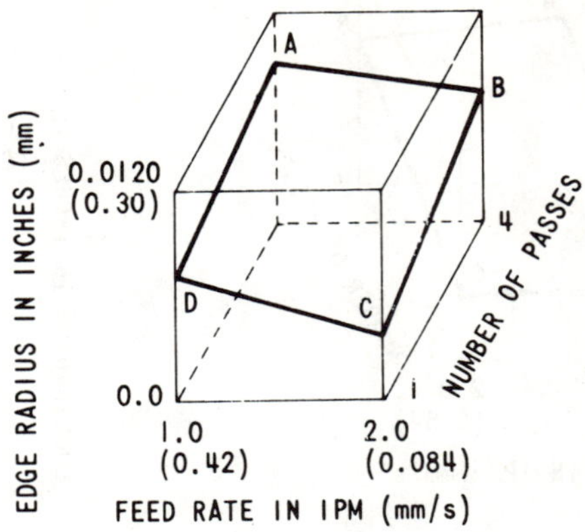

NYLON BRUSH WITH SiC

A - r = 0.0083 (0.211 mm)
σ = 0.0004 (10.2 μm)
B - r = 0.0064 (0.163 mm)
σ = 0.0007 (17.8 μm)
C - r = 0.0037 (0.094 mm)
σ = 0
D - r = 0.0070 (0.178 mm)
σ = 0.0002 (5.1 μm)

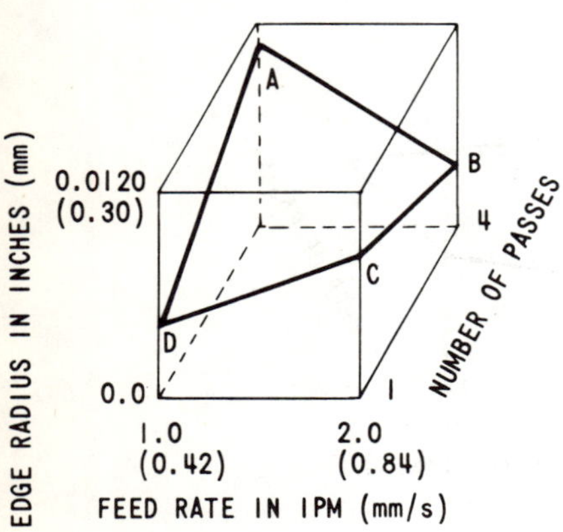

NYLON WITH AL_2O_3 BRUSH

A - r = 0.0080 (0.203 mm)
σ = 0.0005 (12.7 μm)
B - r = 0.0024 (0.061 mm)
σ = 0.0004 (10.2 μm)
C - r = 0.0083 (0.211 mm)
σ = 0.0018 (45.7 μm)
D - r = 0.0044 (0.116 mm)
σ = 0.0006 (15.2 μm)

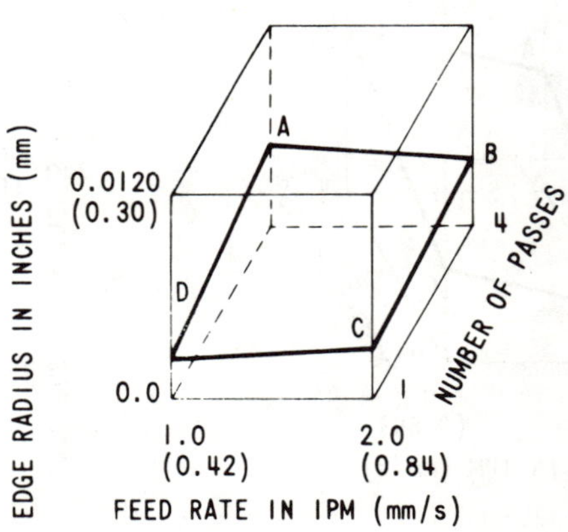

STRING BRUSH WITH SiC

A - r = 0.0048 (0.122 mm)
σ = 0.0003 (7.6 μm)
B - r = 0.0040 (0.102 mm)
σ = 0.0001 (2.5 μm)
C - r = 0.0027 (0.069 mm)
σ = 0.00030 (7.6 μm)
D - r = 0.0023 (0.058 mm)
σ = 0.0009 (22.9 μm)

FIGURE 14-30: BRUSH DEBURRING RESULTS ON 303SE STAINLESS STEEL

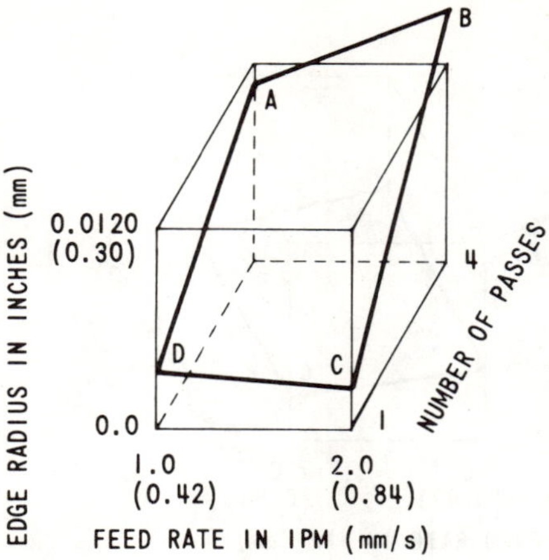

STAINLESS STEEL BRUSH

A - r = 0.0113 (0.287 mm)
σ = 0.0012 (30.5 μm)
B - r = 0.0168 (0.426 mm)
σ = 0.0010 (25.4 μm)
C - r = 0.0027 (0.069 mm)
σ = 0.0001 (2.5 μm)
D - r = 0.0053 (0.084 mm)
σ = 0.0002 (5 μm)

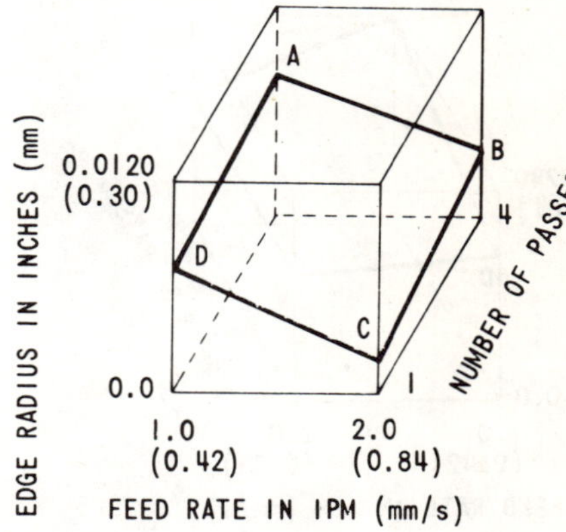

NYLON BRUSH WITH SiC

A - r = 0.0080 (0.203 mm)
σ = 0.0020 (50.8 μm)
B - r = 0.0039 (0.099 mm)
σ = 0.0005 (12.7 μm)
C - r = 0.0022 (0.056 mm)
σ = 0.0003 (7.6 μm)
D - r = 0.0069 (0.175 mm)
σ = 0.0009 (22.9 μm)

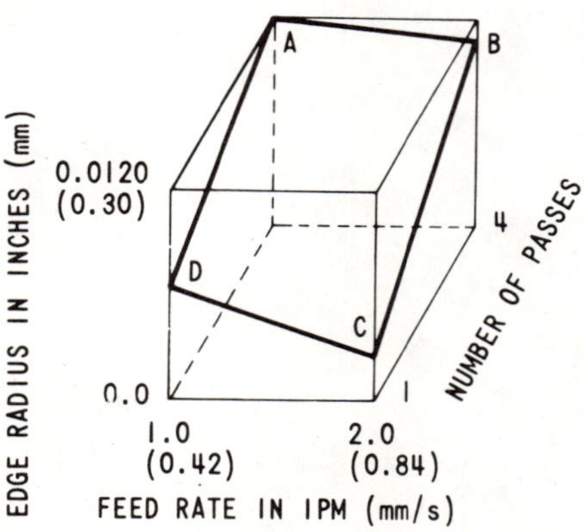

NYLON BRUSH WITH AL_2O_3

A - r = 0.0120 (0.305 mm)
σ = 0.0005 (12.7 μm)
B - r = 0.0107 (0.272 mm)
σ = 0.0012 (30.5 μm)
C - r = 0.0036 (0.066 mm)
σ = 0.0014 (35.6 μm)
D - r = 0.0063 (0.160 mm)
σ = 0.0017 (43.2 μm)

FIGURE 14-31: BRUSH DEBURRING RESULTS ON 7075-T6 ALUMINUM

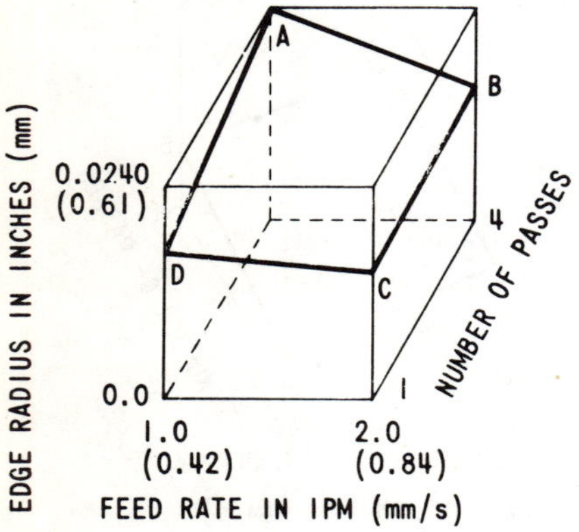

STAINLESS STEEL BRUSH

A - r = 0.0242 (0.615 mm)
σ = 0.0019 (48.6 μm)
B - r = 0.0153 (0.389 mm)
σ = 0.0037 (94.0 μm)
C - r = 0.0141 (0.358 mm)
σ = 0.0028 (71.1 μm)
D - r = 0.0161 (0.409 mm)
σ = 0.0011 (27.9 μm)

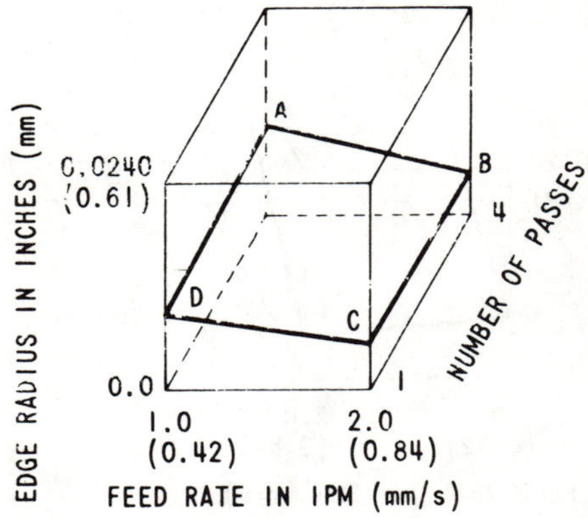

NYLON BRUSH WITH SiC

A - r = 0.0101 (0.257 mm)
σ = 0.0018 (45.7 μm)
B - r = 0.0048 (0.122 mm)
σ = 0.0002 (5.1 μm)
C - r = 0.0047 (0.119 mm)
σ = 0.0001 (2.5 μm)
D - r = 0.0083 (0.211 mm)
σ = 0.0004 (10.2 μm)

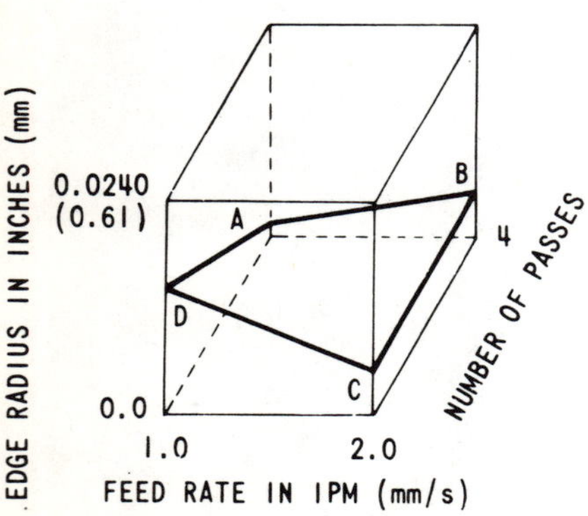

NYLON BRUSH WITH AL_2O_3

A - r = 0.0011 (0.028 mm)
σ = 0.0002 (5.1 μm)
B - r = 0.0050 (0.127 mm)
σ = 0.0007 (17.8 μm)
C - r = 0.0046 (0.117 mm)
σ = 0.0004 (10.1 μm)
D - r = 0.0042 (0.107 mm)
σ = 0.0032 (81.3 μm)

FIGURE 14-32: BRUSH DEBURRING RESULTS ON 6061-T6 ALUMINUM

the metal brushes caused enough plastic flow of the workpiece that a new burr was produced on the part. The use of an abrasive slurry in conjunction with the brushes increased the amount of burr removal but did not totally remove them.

By using spindle speeds of 25,000 - 45,000 rpm (1/2 HP motor) these brushes were successful in removing the milling burrs and maintaining small radii (Table 14-3). Although the nylon bristle brushes produced small radii at these speeds brush life was too short to be economical. At speeds of 40,000 rpm, the nylon bristles melted on contact with the workpiece when too much pressure was applied; in some cases nylon solidified on the part. Because of material melting, the very high speeds cannot be recommended at this time. Speeds of 10,000 to 20,000 rpm did provide better deburring, however, than the 3000 to 5000 rpm's typically used. This is in contrast to previously published information which indicated that speeds above 6000 rpm do not improve cutting action.

At 10,000 to 20,000 rpm not all burrs were removed from the workpieces. Based on the variation in burr sizes produced by milling the burrs probably ranged from 0.001 to 0.003 inch (25 to 76 μm) thick. It would appear therefore that burrs thicker than 0.002 inch (51 μm) would not be economically removed by these small brushes. Burrs thinner than 0.002 inch can be removed by using speeds of 10,000 to 20,000 rpm. The resulting radii will typically be 0.004 inch (101.6 μm) or less in stainless steel.

Future Work

These tests represent a first attempt at defining the capability of these brushes. The use of a controllable brushing force (as opposed to a controlled interference length) may allow smaller radii to be maintained. Providing a repeatable burr would also help lower the controllable edge radius since the variability of the process influences the edge radius

TABLE 14-3: HIGH SPEED BRUSH DEBURRING RESULTS FOR MILLING BURRS

Brush Code	Brush Description	Spindle Speed (rpm)	Number of Passes	Workpiece Material	Average (Inch)	Resulting Radius (μm)
A2	Nylon w/ SiC	45,000	1	303Se*	0.0012	(30)
A2	Nylon w/ SiC	45,000	4	303Se*	0.0008	(20)
A2	Nylon w/ SiC	45,000	1	6061-T651**	0.0019	(48)
A3	Nylon w/ AL_2O_3	45,000	2	6061-T651**	0.0094	(237)
A3	Nylon w/ AL_2O_3	45,000	2	6061-T651**	0.0026	(65)
A3	Nylon w/ AL_2O_3	45,000	3	6061-T651**	0.0027	(69)
A3	Nylon w/ AL_2O_3	45,000	4	6061-T651**	0.0052	(132)

*Stainless steel

**Aluminum

which can be assured.

To establish a complete picture of the capabilities of these miniature brushes several additional tests are required. Some of the additional factors which need to be considered include:

. Effect of brush face width
. Effect of bristle length
. Effect of bristle diameter
. Effect of bristle "sharpness"

Each of these factors needs to be evaluated for different burr thicknesses, workpiece materials, spindle speeds, and brushing force.

Summary

With over 500 combinations of sizes, shapes, and materials available brush deburring is flexible and capable of handling a wide range of work conditions. It is one of the few processes which can remove burrs without producing other burrs. The existence of guidelines for usage such as presented in this chapter should help both the novice and the long time user of brushes find better conditions faster.

References

[1] James E. Bateman, "Capabilities and Limitations of Brush Deburring," SME Paper MR75-826, 1975.

[2] Richard F. West, "Deburring with Power Brushes," SME Paper MR75-476, 1975.

Code Number	Wheel Brush Type	Description
A1	Stainless Steel Wire	0.003 inch (76 µm) bristles by 1 inch (25.4 mm) outside by 0.125 inch (3.2 mm) face width
A2	Nylon Bristles	0.014 inch (356 µm) bristles by 1.375 inch (34.9 mm) outside diameter by 0.25 inch (6.4 mm) face width. Impregnated with 600 grit SiC
A3	Nylon Bristles	0.018 inch (475 µm) bristles by 1.375 inch (34.9 mm) outside diameter by 0.25 inch (6.4 mm) face width. Impregnated with 500 grit AL_2O_3
A4	String	0.015 inch (381 µm) bristles by 1.375 inch (34.9 mm) outside diameter by 0.25 inch (6.4 mm) face width. Impregnated with AL_2O_3
A5	Brass Wire	0.014 inch (356 µm) bristles by 1 inch (25.4 mm) outside diameter by 0.25 inch (6.4 mm) face width

TABLE 14A-1: BRUSHES USED IN RESEARCH STUDY

CHAPTER 15

THERMAL ENERGY METHOD

The Thermal Energy Method of deburring was introduced to industry in 1971. Since that time it has found acceptance in several diverse industries. This chapter will discuss the capabilities and limitations of the process as well as its operation.

Operation

The SURF/TRAN Thermal Energy Method or TEM, as it is called, is probably the fastest of all the known methods of deburring. The actual time to remove even the largest burrs is generally less than 20 milliseconds. TEM works by utilizing two distinct fires. Fires always require: (1) a fuel, (2) oxygen, and (3) heat. If one of these three elements is missing, a fire cannot start and once started, if one of these elements is removed, the fire goes out.

The Thermal Energy Method starts by placing a manufactured part with burrs into a chamber, closing the chamber, and then pressurizing the chamber with a fuel gas and oxygen mixture. The first fire is started by adding heat in the form of a spark from a spark plug. This causes the combustible mixture to ignite. Unlike automobile engines where the ignition causes a piston to move, all of the energy of the fuel must be converted into heat. The heat hits everything in sight. It hits the walls of the water cooled chamber where it is carried away. It hits the main body of the part itself tending to warm it. It hits the burrs themselves regardless of their location. The heat hitting both sides of the burr soon meets itself in the middle and seeks a way to conduct itself into the main body of the part. The only way out for the heat is through the root of the burr into the part itself. Since burrs are created by the constant feed of cutting tools, the route through the root of the burr is relatively narrow--restricting the flow of heat. This restriction causes the heat to build up within the burr until it reaches it automatic ignition temperature.

After the first fire utilizing the fuel gas and oxygen goes out because all of the fuel gas is consumed, the second fire starts using the heat of the initial detonation, excess oxygen which we have placed in the chamber, and for fuel, the material from which the component is made. This second fire, the burning of the burrs, will continue until one of the three essential elements for a fire is lost. This will obviously never be the fuel unless the entire part is consumed, and as stated before, an excess of oxygen exists in the chamber by design. The second fire goes out only after it reaches the corner or surface of the part itself by the root of the burr. At this point, the heat is no longer restricted by the narrowness of the burr but has a suddenly wide path in which to flow. It rushes into the component itself lowering the temperature of the flame and the flame goes out.

Figure 15-1 illustrates some high speed photographs (5000 frames/second) of the process. These were taken through a quartz window of a thermal chamber by the Gas Dynamics Laboratory of the University of Michigan. These photos clearly show the initial burst of heat caused by the fire of the gas fuel explosion which in the films lasts two milliseconds. The second fire consumes the burr down to the root in the sequence lasting an additional 18 milliseconds.

These infrared photos show portions of the burr being oxidized. In each case the metal combines with oxygen to form the oxide of the metal that is present. This oxide then settles on the surface of the part. The oxides can be removed by means of a liquid solvent if desired. In the case of steels that are later heat treated, cleaning can be omitted; in the cases of zinc which is dichromated, aluminum which is anodized, or brass which is brite dipped, extra cleaning is not necessary; and in the case of inexpensive components sometimes the oxides are simply left on the parts.

The burr is not removed by vibration, wear, or the shock wave which accompanies ignition. The burr is actually burned off.

Capabilities

It is thus easy to understand that TEM can work on all combustible materials not only including ferrous and nonferrous metals, but also thermoplastics and rubber. It should now be apparent that one would have difficulty in doing a thermosetting plastic which would not burn but simply get harder as heat is applied and one would be correct in assuming that the process requires more energy to deburr corrosion resistant metals than unalloyed metals. The fact that the second fire goes out due to heat flowing into the component would logically mean that materials which are better heat conductors would have the fire go out prior to poor conductors. Thus it is easier to obtain an edge break or radius on a poor conductor such as steel as opposed to brass which is a relatively good conductor.

What does this mean with regard to the size of the burr which can be removed and the edge breaking which can be performed? The TEM process has instantly removed fingernail size burrs as thick as .020 inch under ideal conditions. Ideal means that the remainder of the component has no sections thinner than 15 times the thickness of the largest burr and that it is capable of sustaining the detonation without damage. Usually of course burrs are .005 inch to .010 inch (0.127 to 0.25 mm) in thickness. Burr length is not a factor in TEM. It is the thickness of the burr root that is critical.

Having described the process which surrounds burrs with gas regardless of where they are and burns them down to the root, one may ask why should not the Thermal Energy Method replace all known forms of deburring? An immediate problem is highlighted in an attempt to define the properties of a burr. Broadly speaking, it is something one wants to get rid of lest it cut somebody, restrict mechanical motion or fluid flow, or fall off and cause electrical short circuits or mechanical damage. Maybe one just wants to get rid of it because its appearance proclaims "bad workmanship." In any event,

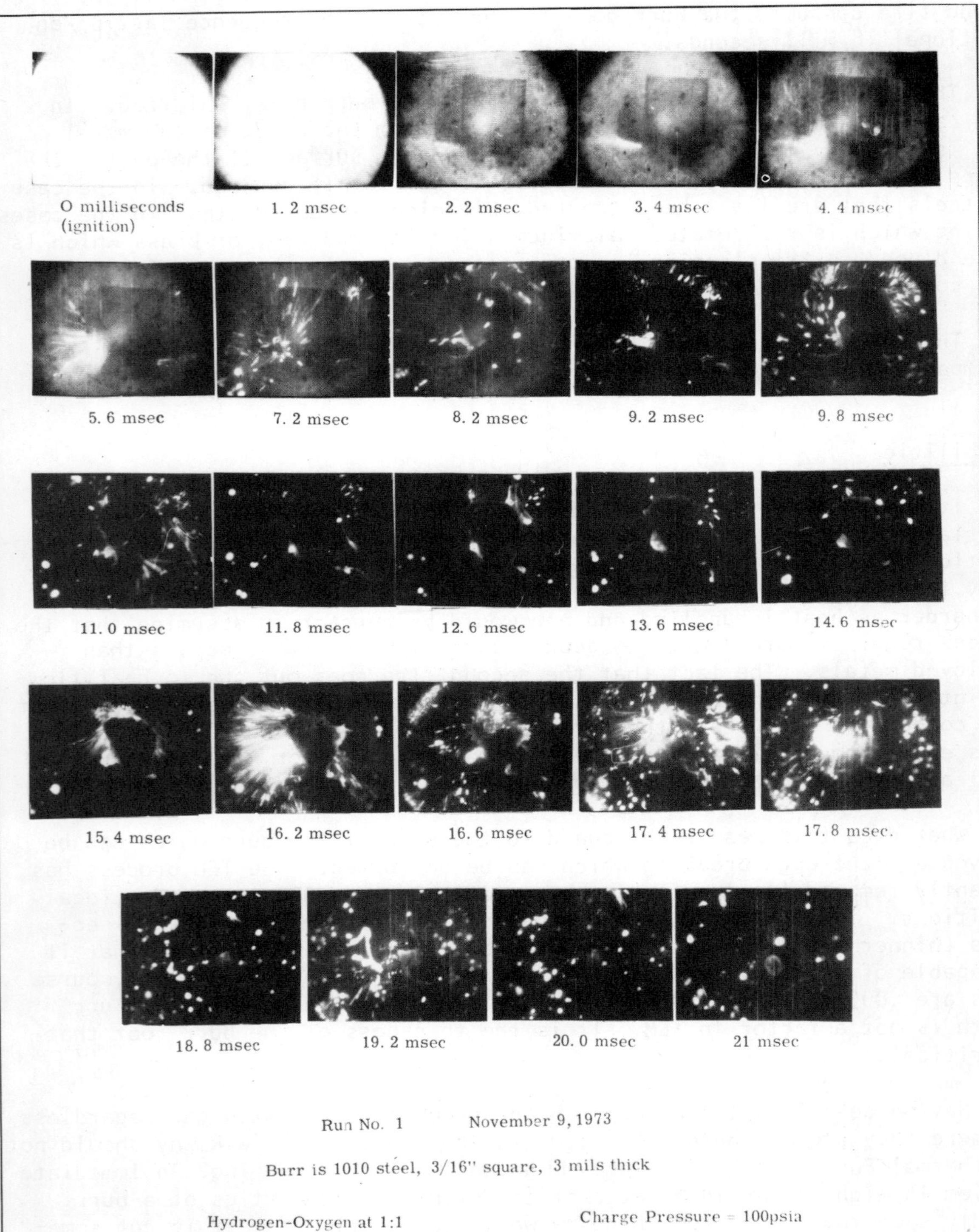

FIGURE 15-1: IGNITION OF SIMULATED BURR

it is a man-made decision as to where a protrusion on a component starts being bad and something which must be eliminated.

Many of the deburring processes actually remove flash formed in a casting or die casting process rather than burrs. Flash has the characteristic (as shown in Figure 15-2) of being wider at its root than at its tip. This is distinctly different from a burr produced by normal sharp cutting tools with a fixed feed per revolution. The two fires story could be repeated for this example of flash and the tip of the flash would light after auto ignition, but the point at which the heat flows into the main body of the part is different in flash than in burrs. As much as we would like the process to follow the man-made decision of burning down to a flat surface on the part, the process follows the laws of physics and the fire goes out as the heat drains into the part leaving a nicely rounded lump (Figure 15-3).

When extremely dull tools are used in machining, burrs are created which are extremely similar in their characteristics to flash. Figure 15-4 shows an example from using an extremely dull drill. The drill has essentially pushed the metal originally in the center of the hole toward the outside forming a pushed-out situation in which the root of the protrusion is definitely wider than its tip. Normal low energy level detonations will reduce these push-out burrs to various degrees of roundness depending on the heat conductivity of the metal being used. Figure 15-5 shows typical results of different energy levels with different materials.

The TEM process has put radii on steel from .002 to .060 inch (0.05 to 1.5 mm), and on aluminum from .002 to .010 inch (0.05 to 0.25 mm). The size of the radius can only partially be influenced by the controls available to the process. The major determinants are the workpiece material and the size of the burrs. In Thermal Deburring the radius achieved is in direct relationship to the size of the burrs. One will get larger radii where there are burrs on various edges and virtually no radius on edges where there was virtually no burr. This is the opposite of what one might expect, however, it is explainable if one considers the burrs as kindling to start a fire. The radii achieved will be consistent within ± .002 inch (±0.05 mm) for a given material with the same size burr. In general, a sufficient radius will be put on all edges so as to prevent hand cuts from the typical ragged edges left on untreated parts. A part which is properly placed in a TEM chamber can, both inside and outside, meet that elusive specification "remove all burrs--break all sharp edges."

The equipment used to perform Thermal Deburring generally averages about 20 seconds in total cycle time for each load of parts with factory units operating as low as nine seconds and the slowest as high as 30 seconds. TEM productivity is measured by the number of components one can deburr at one time. In the average 20 seconds, although one can only deburr one large cast iron pump body, one could deburr four ordinary sized aluminum carburetors, about 30 or 40 brass elbow fittings, or several hundred zinc lock cylinders. Thus rather than measuring part times in minutes, when one gets above 200 parts per load, cycle time per part is in tenths of seconds per part.

The distinguishing feature of all the parts just mentioned is that the

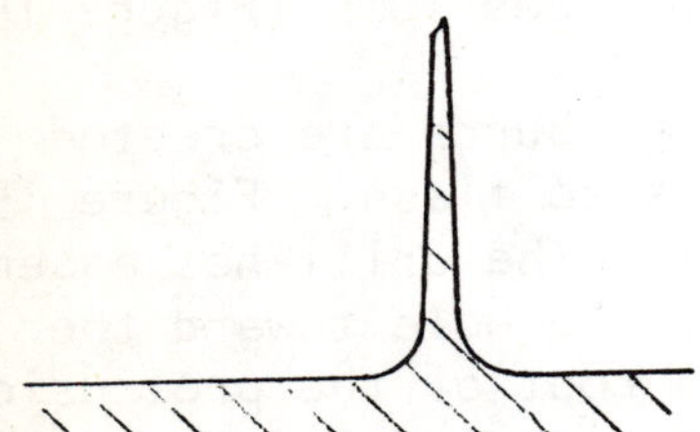

FIGURE 15-2: PARTING LINE FLASH

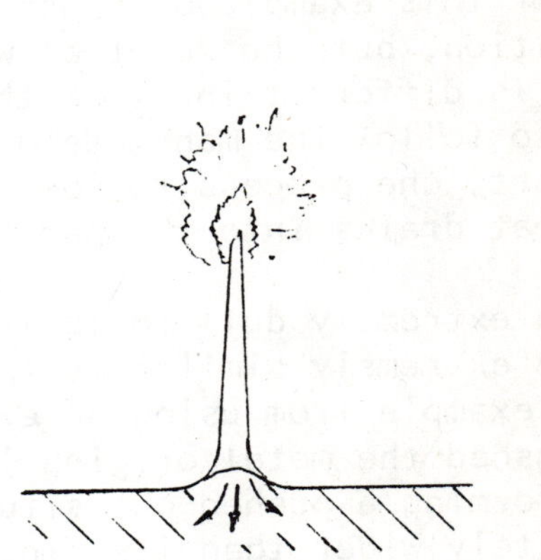

FIGURE 15-3: HEAT FLOW IN BURRS AND FLASH

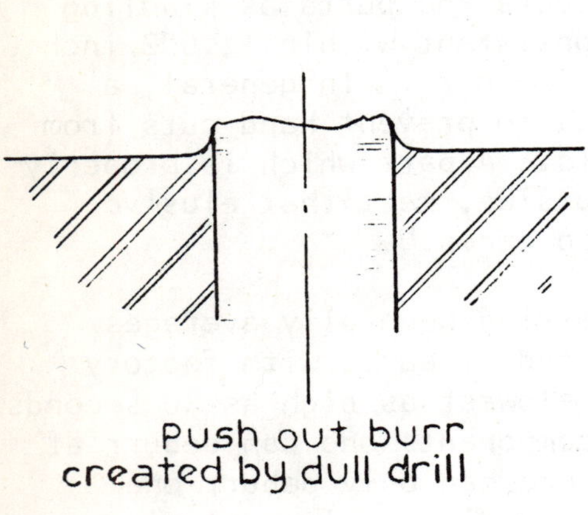

FIGURE 15-4: HEAVY DRILL BURR

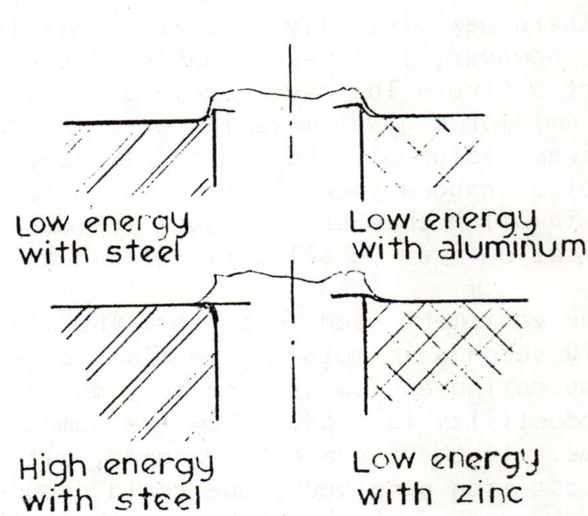

FIGURE 15-5: RESULTS ON DIFFERENT WORKPIECES

deburring problem area is internal. TEM is the fast method of removing burrs which are inside a part, possibly located in a blind hole, an inaccessible groove, or even in locations which are not readily visible. Certainly any pump, valve, metering device, fitting, etc., through which air, water, gas, oil, or any liquid must flow is an ideal candidate for the TEM process. TEM is an excellent way of eliminating the loose debris often found internally in such components. TEM for example is now used by virtually every major carburetor manufacturer in the world as the method of providing consistent quality to meet the new environmental standards. It is gaining in popularity among hydraulic and pneumatic component manufacturers.

The limitations to the process generally result from the effects of the shock and heat waves connected with the detonations to fragile or thin parts. There is a stress relieving characteristic equivalent to dropping the parts from a height of over one foot which is especially noticeable on highly stressed parts such as formed or stamped sheet metal parts. Air hardening steels which have thin sections can be troublesome if it is impossible to shield or heat sink the thin sections and if the part is not to be heat treated later. On rare occasions, customers will not accept the recast layer under a removed burr. The process is also not recommended for materials which have been previously hardened in that the surfaces occasionally develop heat check cracks.

Some papers list as a side effect the fact that the process converts burrs into oxides since this requires, in many instances, secondary cleaning operations. In actuality, all of the tumbling and vibrating methods must also clean the parts to assure that there are no bits of the burrs remaining in them and the chemical methods must wash the parts to remove electrolyte or acid. Today a line of liquid solvents are available which will dissolve the oxides which the TEM process forms without reducing the size of the main surfaces. This oxide removal can be easily performed up to three days after the thermal deburring because the oxides which form are not yet married to the part. In other words, the oxides are distributed from the converted burr all over the part as opposed to forming oxides all over the surface. If one anodizes aluminum, dichromates zinc, brite dips brass, or hardens steel, no additional cleaning equipment will be needed along with Thermal Deburring since these processes remove the TEM oxides.

The process can and is used for the removal of flash, however it will not remove parting line flash entirely since the second fire goes out as the flash widens at its root. The results are generally acceptable for functional purposes, however if the beauty of an abrading or buffing action is required, TEM will not fill the bill. On the other hand, one has the assurance that TEM will not abrade large flat surfaces of a part while removing most of the parting line flash.

To discuss the economics of TEM, it is necessary to visualize the equipment used. A typical unit using natural gas is shown in the schematic (Figure 15-6). Basically, the unit is a toggle type press having a C-frame capable of squeezing force of 250 tons (2.24 MN). This force is used to hold the lid on the pressurized chamber. The unit happens to have the chamber upside down and utilizes six lids which are raised up to the chamber. These lids, which we call lower closures, are the items upon which parts

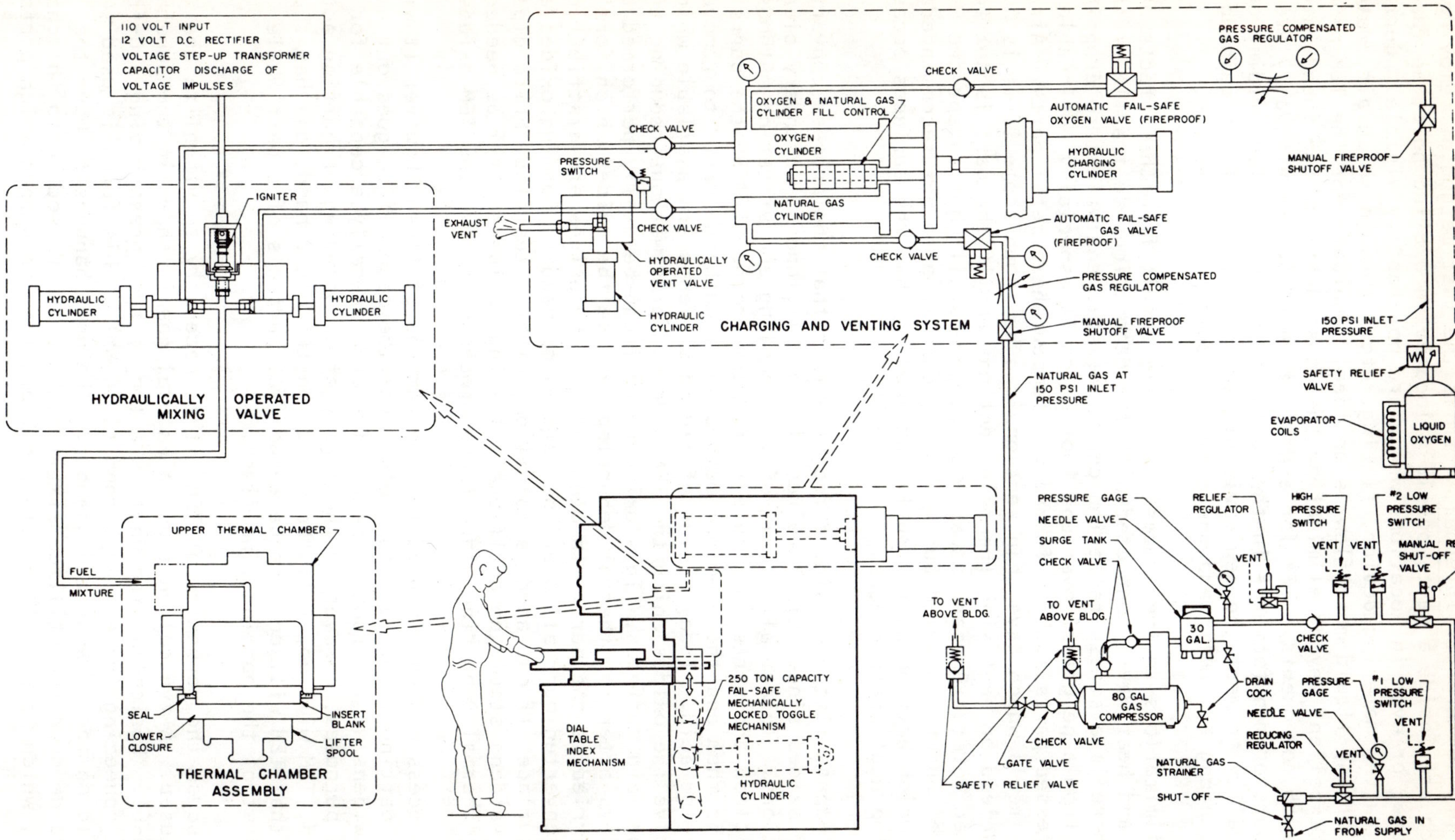

FIGURE 15-6: SCHEMATIC OF T.E.M. SYSTEM

are loaded. Depending on size and amount of automation, units such as these can cost between $60,000 and $250,000.

Customers with large rectangular parts with internal burrs utilize no tooling at all and they simply pay for the fuel costs which, including rental of the tanks from a supplier, average between $.01 and $.015 per detonation.

Customers with many small parts do have to contain them in some sort of basket to prevent them from spilling over the sides and these stainless steel baskets generally cost below $100 each. Customers with parts which have external burrs which are to be removed simultaneously with the internal burrs must raise the parts up in a manner so that the external burrs are "in the gas." This requires some form of pin, post, or nest which, although it must be made of stainless steel, does not require the precision of an ECM electrode, the form fitting of the abrasive putty machines or the complexity of the holders on spindle finishing machines. Some customers for a few thousand dollars have made standard subplates and pin sets capable of handling several parts of varying sizes and shapes.

Just as all the deburring methods utilizing an abrasive wear out various components, various seals for TEM are subjected to wear by the process and must be changed regularly. The chamber and mixing valve seals can be compared to cutting tools on a lathe or milling machine when stressing the need for regular, scheduled changes. Rather than having this changing time classified as downtime, it should be classified as setup time as it is on normal machine tools.

It is difficult to give an average value for the cost of seals, spark plugs, thermocouple tips, etc., for the process because their use is directly proportional to the energy level which is used. For example, a zinc manufacturer would use relatively low energy levels in which a set of such components could last well over 30,000 detonations which, in turn depending on size, could mean 120,000 to 240,000 carburetors. This could be about two cents per detonation or mills per part. On the other hand, a customer deburring a stainless steel part one at a time in a special fixture at a very high energy level could require new seals before 5000 detonations requiring costs in excess of $.05 per workpiece.

Guidelines for Usage

Three rules must be followed for successful Thermal Deburring, and these rules further highlight process limitations.

Rule 1: KEEP THE BURRS IN THE GAS. That sounds simple enough but it is easy to violate the rule. For example, if you have external burrs on the corners of the parts and set the part into the chamber with the burrs touching the chamber, these burrs will be heat sunk by the chamber--not brought up to auto-ignition temperature and therefore not eliminated. In the case of bulk loaded parts which may be run several hundred at a time such as brass fittings, tees, elbows, etc., the burrs are all internal and there are no problems in getting the burrs "in the gas." As soon as the component has some external burrs, however, bulk loading presents the probility that an external burr will touch a mating part and be heatsunk. Although rerunning the components twice, having stirred them between thermal shots, can improve the odds, there will always be come external burrs remaining. Blind or intersecting holes which are filled with cutting oil, coolant, or packed with chips will obviously not allow the gas to surround a burr located at the bottom of such a hole. If only a thin layer of coolant is present, one will be able to utilize the unique capability of this process to eliminate burrs in blind, inaccessible locations.

Just as burrs which are folded over can be hammered into the part by vibratory system, a tightly hammered-in burr will heatsink itself in the thermal process. Users who wanted to tumble large external burrs and use the thermal method for small internal ones complained when the thermal process could not correct the occasional misses of the tumbling system. Obviously the processing should be reversed and the Thermal Deburring be done first.

TEM has been correctly described as the removal of thin sections, therefore,

Rule 2: MAKE SURE THAT WHAT YOU WANT TO KEEP IS DESIGNED TEN TIMES THICKER THAN WHAT YOU WANT TO GET RID OF. Users are often asked what is the maximum size burr that one can remove. They generally will answer, any size, if your component can take it. The ability to take it means that one should not produce burrs with a root of 1 mm thickness if on the same part one has a threaded hole whose thread has a root of .5 mm thickness. Actually threads withstand the thermal process quite well in that they have a shape resembling flash which allows the heat to flow back into the part. However, when the energy level is set to eliminate massive burrs, a fine thread must go. Previously published information that the process works only on thin copper alloy burrs is, therefore, completely false. Steel burrs as large as one's thumbnail and over .020 inch (0.5 mm) are quite easily removed when the remainder of the part is massive.

Rule 3: RESTRAIN THE PART IF REQUIRED. Picture the detonation as the release of energy in all portions of the chamber at once with vectors and ricocheting forces beating on the part on all sides. It is quite natural for components to want to move in the direction of least resistance. If the part is unequally shaped, it will tend to move in one direction or another gaining velocity the further it moves. Utilize a nest on individual parts to stop them after about an eighth inch or so of movement so they do not damage themselves by hitting the sides of the chamber. In the case of bulk loading into a basket, one part tends to support another and no particular fixturing is necessary providing the parts do not have external burrs. Some publications have stated that tooling costs for the Thermal Method exceed liquid hone and ECM tooling costs. On many parts one needs no tooling to deburr. Gas conforms to all components without help being required. Occasionally nests are required to prevent the part from moving or some heatsinks for thin sections designed by a product engineer not familiar with the process.

Most manufacturers of valves, manifolds, and other rectangular pieces of "swiss cheese" type parts use no fixture at all. Manufacturers of irregular parts such as small die castings or brass fittings utilize a stainless steel basket costing about $100 which holds many parts.

The movement caused by the shock wave associated with creating the heat is a definite limitation on the process in that it is roughly the equivalent of dropping the part about a foot. Thus, if the parts have any extremely close tolerances that could be changed by such a blow, they could be eliminated as appropriate for the process. Most stampings and drawn parts are stressed to such a condition that the shock and sudden heat applied tend to relieve them slightly unless they are made of relatively heavy stock (over 1/8 inch thick). Box shaped, thin wall die castings must occasionally be filled with a fixture to avoid inflation or collapse.

This seems an appropriate moment to mention that blanked sheet metal is often vibrated or tumbled to remove the "burr" at the sheared edge. A study of the section of this edge, Figure 15-7, shows that it generally represents the wide root configuration which is resistant to the thermal process. Stampings are unfortunately the segment of the industry that TEM can only rarely help.

Most of the literature available describes the SURF/TRAN TEM with the use of hydrogen and oxygen. Actually very few hydrogen units have been sold in the last two years. Natural gas increased the economy of operating the unit many times in spite of the rising cost of natural gas. Fuel costs are only slightly over $.01 per shot with one shot sometimes being able to deburr hundreds of parts within the chamber. Also important, the maintenance of natural gas units is a lot easier in that leaks occur less frequently with the more viscuous natural gas.

Unlike EDM and ECM, the surface of the deburred part where the burr used to be is affected by the thermal process. Obviously if one were burning the metal, it was in a gaseous form and it had to be in a liquid form before vaporizing, there is a very slight recast layer generally about .00008 inch thick; but sometimes as thick as .001 inch which has had its

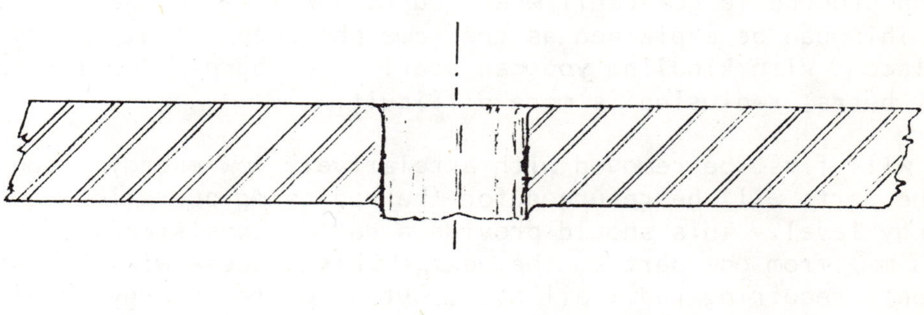

FIGURE 15-7: BURR PRODUCED BY BLANKING

metallurgical properties changed. Generally the layer is so thin that this does not bother anyone and one prominent manufacturer indicates that the layer did not affect the hobbing tools which were shaving gears deburred thermally.

Guidelines for future consideration of the process are as follow:

1. If one can use a satisfactory automated bulk method--stick to it. TEM is still less economical than these methods at this time of writing.

2. The investment required means that one must have three or four people hand deburring in order to justify the investment if only the economics and not added quality are taken into consideration.

3. The largest advantage is the universality and ability to change from one component to another in minutes without tooling. It is only on the first half dozen or so shots that trial and error methods are used to establish the energy level, thereafter, the parameters are noted and one can get consistency.

4. Do not expect TEM to treat wide surfaces as done by practically all other deburring methods. Most wear down the surfaces in some manner leaving a smaller remaining part but one which has improved surface finish. Although TEM may alter the shape of some fragile parts due to stress relieving, it definitely does not remove metal on flat surfaces.

The use of improper solvents for oxide removal initiated improper rumors that TEM causes surface pitting. The only pits which could be caused by TEM would be created by the removal of a thin section covering near surface porosity on a casting.

Future Trends

Units are in development which will be used for thermal energy machining. These will have the capability of reaching energy levels which will consistently radius even a square edge. Current machines function in a manner which produce larger radii where burrs are located than where there are none. This can be explained as the same phenomenon that occurs in your fireplace. With kindling you can start a log burning easily--without kindling or burrs, radiusing is more difficult.

Burrs will first be removed with a relatively low energy detonation and then the parts will be returned for the radiusing operation at the higher energy level. This should provide a radius consistency of ± .002 inch (±0.05 mm) from one part to the next. This process will be suited for components requiring radii all over, but since the energy level is extremely high, will be limited to parts having strength to withstand relatively severe stress relief. Some high strength materials are actually advantageous to the TEM process contrary to the assertions of some articles.

Those articles meant to say that anticorrosion alloys--which are in effect antioxidation alloys--make TEM action more difficult.

Summary

This chapter has tried to answer the questions of how and why the process works without describing the hardware used to make it work. In four short years, there are over 100 of this type of unit in operation. The average 25 second cycle time per shot is apparently as big an advantage as the ability to deburr in otherwise inaccessible places. TEM is in fact still in its infancy, but it already has been shown to be a significant approach to reducing deburring costs.

If components are designed correctly with the TEM process in mind, it will meet realistic deburring requirements for internal and external burrs. In fact the process will offer a design freedom for both product design and processing never yet achieved. It will allow you to make use of the fact that gas goes everywhere, and therefore you can be assured that wherever there are burrs TEM will find them and eliminate them.

References

[1] A. C. Montag, "Thermal Energy Method of Deburring," SME Paper MR75-485, 1975.

[2] A. C. Montag, "Deburring With TEM," SME Paper MR75-835, 1975.

CHAPTER 16

ELECTROCHEMICAL DEBURRING

Operation

An electrochemical cell is formed when two conducting electrodes, one at positive potential and the other at negative potential, are fixed adjacent to each other, separated by a gap through which flows a conductive liquid (electrolyte), usually an aqueous solution of brine. The two electrodes are connected to the output poles of a Direct Current Rectifier operating at low voltage (5-20 volts) (Figure 16-1).

During the passage of current between the two electrodes, metal is eroded from the positive electrode according to the laws of Faraday which state that the mass (weight) of metal removed in a given time is proportional to the quantity of current passed and the chemical equivalent to the material.

Following removal from the component, the metallic products of erosion undergo chemical changes, eventually forming hydroxides of the component material and are evidenced in the electrolyte as a brownish colored sludge.

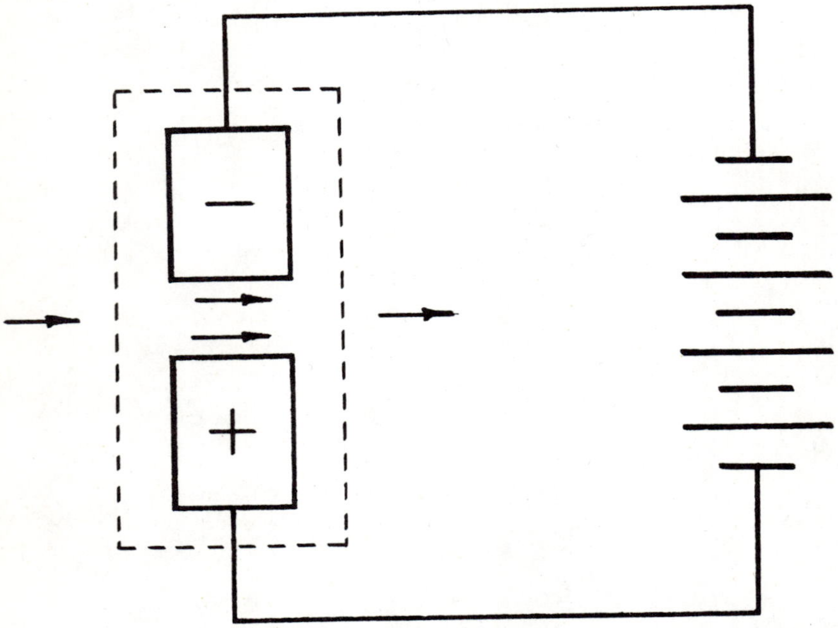

FIGURE 16-1: ELECTROCHEMICAL CELL

An electrochemical deburring (ECD) fixture forms such a cell as described above, however, in order for an electrochemical cell to function efficiently, various ingredients are required. These may be listed as:

1. A Cathode at Negative Potential
   This is the deburring electrode supported in a suitable fixture designed to suit the configuration of the work part in the areas where the burrs appear.

2. An Anode at Positive Potential
   This is the workpiece or component to be machined or deburred, also supported in a suitable fixture.

3. A Direct Current Source
   This is provided from a rectifier unit with established voltage output for greater accuracy.

4. A Conducting Electrolyte
   Usually sodium chloride or nitrate in water, at concentrations between 60 g/liter and 200 g/liter. The electrolyte is contained in a storage tank of about 1000 liter capacity.

5. Temperature Control
   This is an integral part of the electrolyte storage tank, to either heat or cool the solution in order to maintain a stable working temperature, again with a view to greater accuracy in the process.

6. Clarification System
   In order to clean the electrolyte, coarse filters and a centrifuge are used to remove the products of erosion.

7. Protection Devices
   It is advisable to "police" the electrochemical cell gap with electronic circuitry located between the cell and the rectifier supply to protect the tooling and the parts in the event of a short circuit in the gap.

   Gap short circuits may be caused by:

   a. Metallic particles bridging the gap during the time machining power is switched on, e.g., a burr breaking away from the parent component. This is detected by a spark detection unit.

   b. Extra large burrs bridging the gap <u>before</u> the power is switched on. This is detected by a contact sensing unit which causes the cathode to retract and prevents the power switching on.

   c. Metallic debris in the system due to poor system housekeeping.

d. Faulty electrical connections to the fixturing causing intermittent sparking.

Figure 16-2 shows a diagram of a deburring fixture with all the associated equipment outlined above.

Tooling Considerations

Component Study

As with most processes and particularly with ECD and ECM, the expertise is centered around the ingenuity of the tooling design.

A complete study of the manufacturing sequence of the subject component is essential before ECD tooling design can begin.

With most mechanically machined parts, the direction the final burrs are facing is dictated by the last operations to be performed. It is important that the tooling designer be aware of this so that the electrodes are correctly placed in relation to the burrs. The ultimate significance of this statement will become evident in the next paragraphs.

Size of Burrs

How much easier life would be if all burrs were the same size. The fact that burrs do not conform to American Standards is a further consideration in the design of deburring tooling.

The first decision therefore is to either machine the burr away, i.e., total erosion, or sever it from the part by eroding away the root of the burr.

For total erosion, the electrode may be placed above at right angles to the burr, while in order to erode away the burr "root," the electrode may be alongside the burr.

To assist in making the decision, an attempt may be made to classify burrs in terms of their size, their ease of removal, and the technique to be adopted. Such a classification may take the following form for ECD:

Burr Classification for ECD

Class 1	Large, say more than 3 mm long, with thick root section (more tenacious).
Class 2	Large, as above, with thin root section.
Class 3	Small, say less than 1.5 mm, with thick root section.
Class 4	Small, as above, with thin root section.
Class 5	Little more than a ragged, sharp edge.

As indicated in Chapter 4, the size and characteristics of burr formation are influenced by:

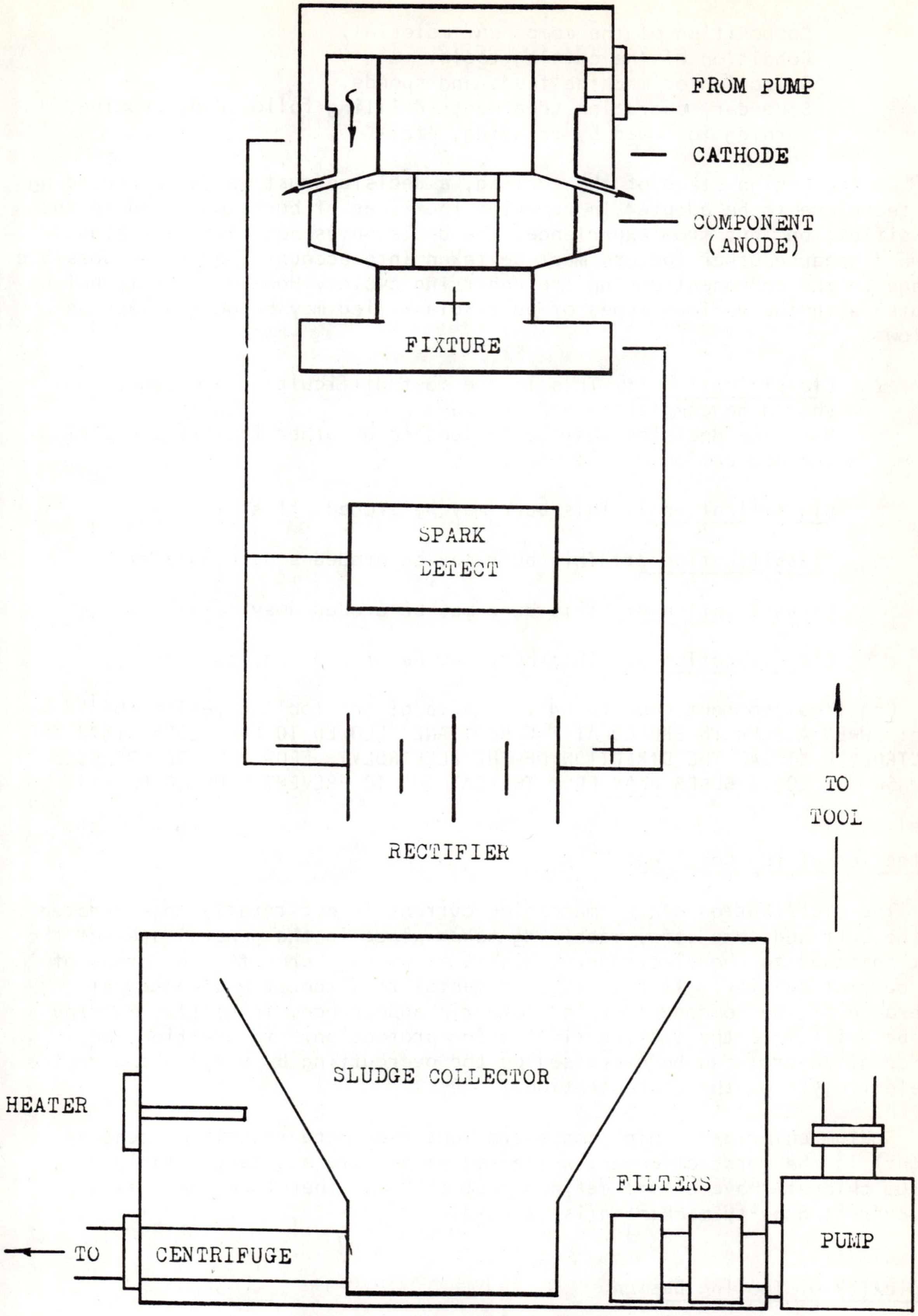

FIGURE 16-2: ECD EQUIPMENT

- Composition of the component material.
- Condition of the cutting tools.
- Selection of machine feeds and speeds.
- Secondary operation treatment (drilling followed by reaming, turning followed by grinding, etc.).

At the design stage of the tooling, a decision must be taken regarding the technique to be adopted in removing the types of burrs outlined in the classification, but from experience, the decision is not always straight forward because other factors must be taken into account concerning possible damage to the component during the deburring cycle. However, the technique adopted with the various types of burrs classified may be generalized as follows:

Classification 1: This is the most difficult type of burr and should be completely eroded away.
N.B. The decision must be influenced by other factors concerning the protection of the part.

Classification 2: This burr may be eroded off at the root.

Classification 3: This burr may be eroded away completely.

Classification 4: This burr may be eroded away completely.

Classification 5: This burr may be eroded away completely.

One final comment concerning this area of the tooling design should be made: WHEN A BURR IS ERODED AT THE ROOT AND ALLOWED TO FALL LOOSE INTO THE ELECTROLYTE STREAM THE DIRECTION OF THE ELECTROLYTE FLOW MUST BE ARRANGED TO WASH THE LOOSE BURRS AWAY FROM THE CATHODE TO PREVENT SHORT CIRCUITS.

Protection of the Component

The distribution of the machining current is not totally concentrated on the burr and some stray machining takes place in the general area of the part adjacent to the electrode. This is known as overcutting or areas of low current density. If this is detrimental to either the dimensional tolerance of the component or its cosmetic appearance, insulative masking may be built into the tooling fixture for protection. In addition, some degree of control can be exercised on the overcutting by varying the choice of electrolyte or the concentration.

Sodium chloride at high concentrations (say more than 15 percent by weight) is the worst offender, while sodium nitrate and particularly sodium chlorate have a more defined "cut off" and therefore the area of low current density machining is reduced.

Complexity of Tooling Design

In general, ECD tooling electrodes are relatively simple in design and do not require prolonged development as do some of the electrodes used in ECD machining. Any good conducting material may be used for the manufacture of the cathode but usually stainless steel, brass, copper, or tungsten copper is used. The latter material has become popular as an electrode material due to its physical toughness which provides protection against damage from rough handling and excellent protection against occasional spark (short circuit) damage.

In applications where the cathode must be insulated away from the workpiece or provision made for machining only in the burr laden areas of the component, insulation materials must be used on either the cathode or component or both.

There are many plastic insulative materials available which are suitable for use in ECD tooling either for the purposes of insulation just mentioned or as materials which may be used for the main construction of ECD or ECM fixtures. To use insulative materials for fixture building should be the rule rather than the exception--the nature of the process dictates this.

In the selection of insulating materials, care should be taken to select those with low water absorption factor. Any electrolyte absorbed by the material may be later troublesome because:

1. Separation of the material laminates may take place as the electrolyte dries and crystallizes during periods when the fixture is not in use causing:

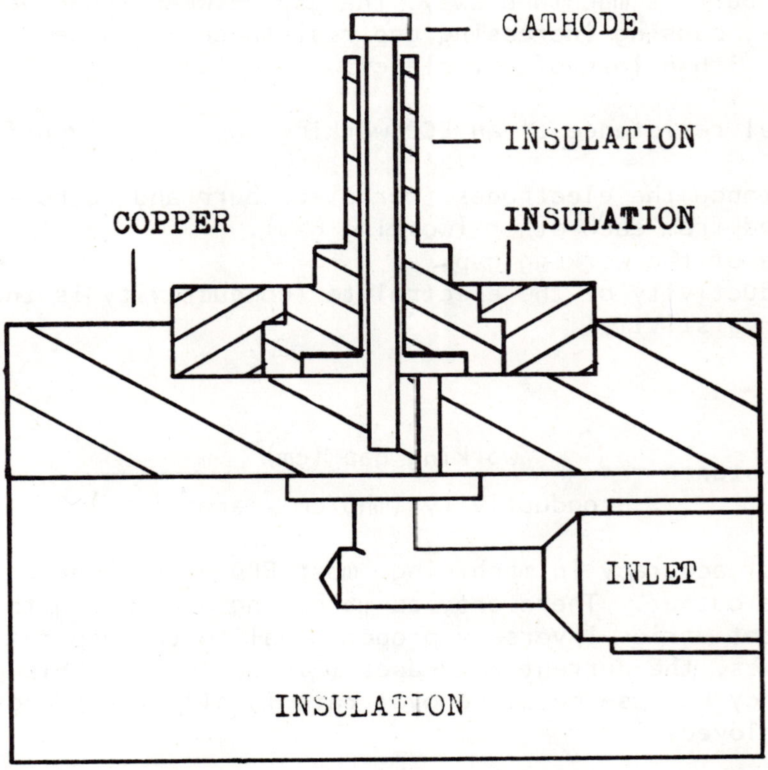

FIGURE 16-3: DEBURRING ELECTRODE

2. Ill fitting mating parts on the fixture which become troublesome to operators, or

3. Current leakage between cathodic and anodic parts on the fixture which may damage the component or fixture.

Insulative materials which are generally unsuitable are laminates using PHENOLIC RESIN as a binding agent. Any laminate manufactured with the use of EPOXIDE RESIN is suitable. There are usually epoxide/fiberglas, epoxide/linen, or epoxide/paper--the former being more suitable than the latter.

Other materials which may be used for insulation are: PVC, ABS, PTFE, nylon, epoxy resin coatings or mouldings, etc. Figure 16-3 shows a typically insulated deburring electrode.

ECD Using Constant Current

From the point of view of process efficiency in ECD, one of the problems has always been that of machining components with large heavy burrs as outlined in burr classification 1 or, to a lesser extent, the burrs classified as 2 or 3.

Essentially ECD tooling works on a fixed gap principle, i.e., the cathode does not feed forward as in more conventional metal cutting but is fixed at some predetermined distance from the face or edge of the part. Therefore, as the burr is machined away, the gap between the electrode and the burr increases, causing increasing gap resistance and a decrease in machining current with a loss of efficiency.

The electrical resistance of an ECD working gap is derived from:

1. The distance the electrodes (workpiece burr and cathode) are separated from each other (working gap).
2. The area of the working gap.
3. The conductivity of the electrolyte (conductivity is the reciprocal of resistivity).

Therefore:

$$\text{Gap Resistance} = \frac{\text{working gap (cm)}}{\text{conductivity (mho/cm)} \times \text{area (cm}^2\text{)}}$$

For reasons of accuracy in machining, most ECD or ECM rectifiers have a constant voltage output. Therefore, as machining proceeds with fixed gap tooling, the current varies inversely proportional to the gap resistance. As the gap increases, the current decreases (Figure 16-4). This promotes a loss in efficiency because metal removal rate is directly proportional to the current employed.

One of the methods of overcoming the problem of decreasing current is to arrange the rectified output to increase the voltage proportional to

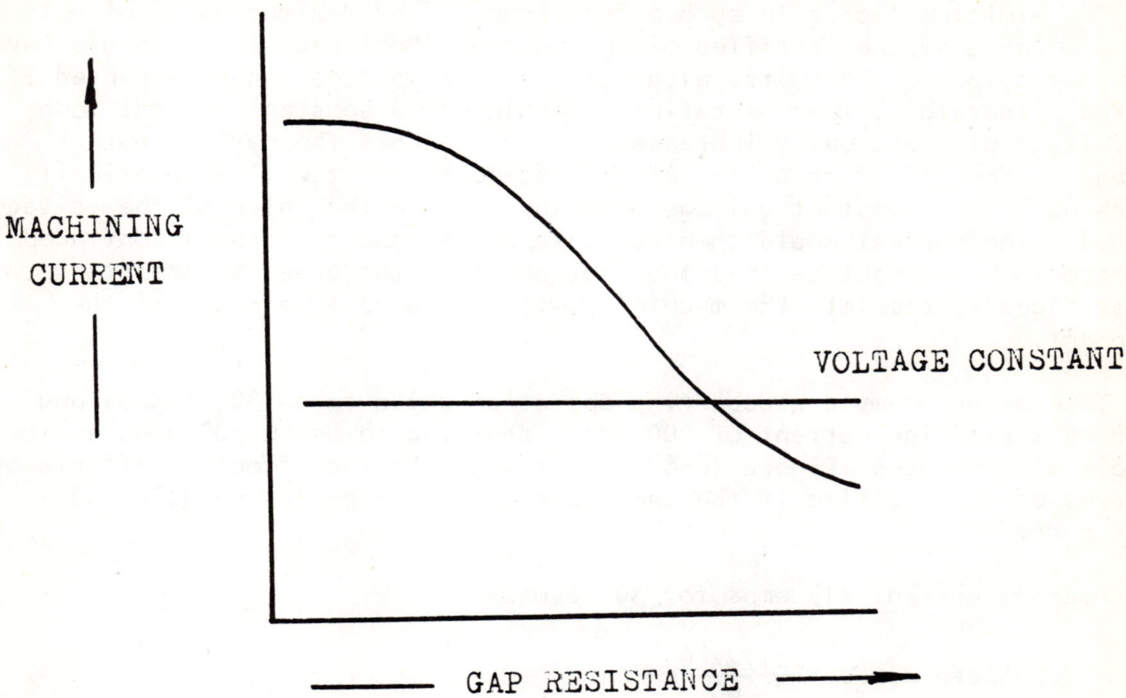

FIGURE 16-4: CURRENT VERSUS RESISTANCE

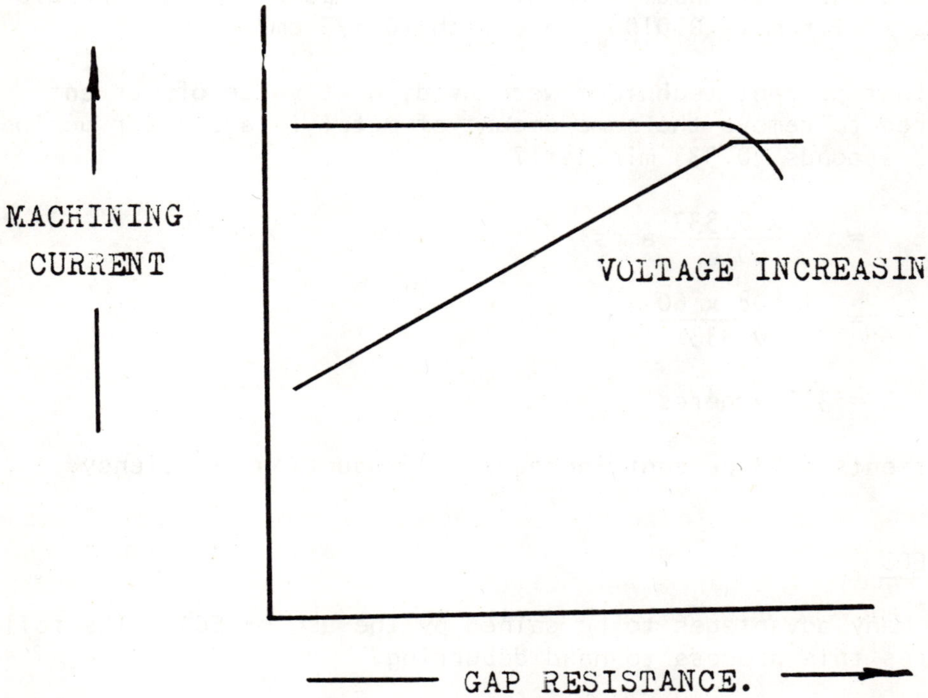

FIGURE 16-5: VOLTAGE VERSUS RESISTANCE

the increase in the machining gap resistance and therefore maintain the current constant (Figure 16-5).

The limiting factor in such a technique is the maximum value of voltage available at the rectified output source. Most rectifiers in use have a maximum output of 24 volts, although a higher voltage can be arranged if desired. Therefore, when operating a machine on a constant current mode, the voltage will gradually increase until it reaches the maximum output voltage of the rectifier unit. At this stage the unit will automatically switch back to a constant voltage mode and sustain the value of the voltage reached. The current would then decrease as the gap resistance continued to increase. In practice this does not occur as the process parameters are arranged to complete the machining within the voltage range of the equipment.

Take as an example a deburring operation which takes 30 seconds and produces a starting current of 400 amps, reducing to below 100 amps as the gap distance changes (Figure 16-6). What would be the effect on efficiency in terms of reduced time if the same operation were performed using constant current?

Average current 217 amps for 30 seconds

$$\text{Ampere Hours} = \frac{217 \times 0.5}{60}$$

$$= 1.808$$

Assuming 10,000 amps removes 1 cubic inch (16.39 cm^3) in one minute, this represents a current consumption of 166.666 ampere hours. Therefore 1.808 amp hours will remove 0.0108 cubic inch (0.177 cm^3).

If a constant current technique were used, what value of current would be required to remove the same amount of metal in a shorter period of time--say 20 seconds (0.333 minutes)?

$$1.808 = \frac{y \times 0.333}{60} \text{ amps}$$

$$y \text{ amps} = \frac{1.808 \times 60}{0.333}$$

$$= 325 \text{ amperes}$$

This represents a 33 percent increase on production efficiency.

Advantages of ECD

There are many advantages to be gained by the use of ECD. The following list compares this process to hand deburring.

ELECTROCHEMICAL DEBURRING	HAND DEBURRING
Provides consistent results	Results are subject to human error and fatigue
Assures service reliability of parts in the field	May not provide consistent results
Can be used following heat treatment	Burrs may be difficult to remove after heat treating
Removes burrs gently, does not break or bend burr off, therefore no damage to part	Operators frequently scratch or nick parts
Has no tool wear	Deburring tools must be sharpened
Unskilled labor can be used	Labor skill is required proportional to part difficulty and precision
Facilitates deburring in normally inaccessible places (internal holes)	A difficult hand operation
Can reduce manpower necessary by a factor of 6	A labor intensive operation

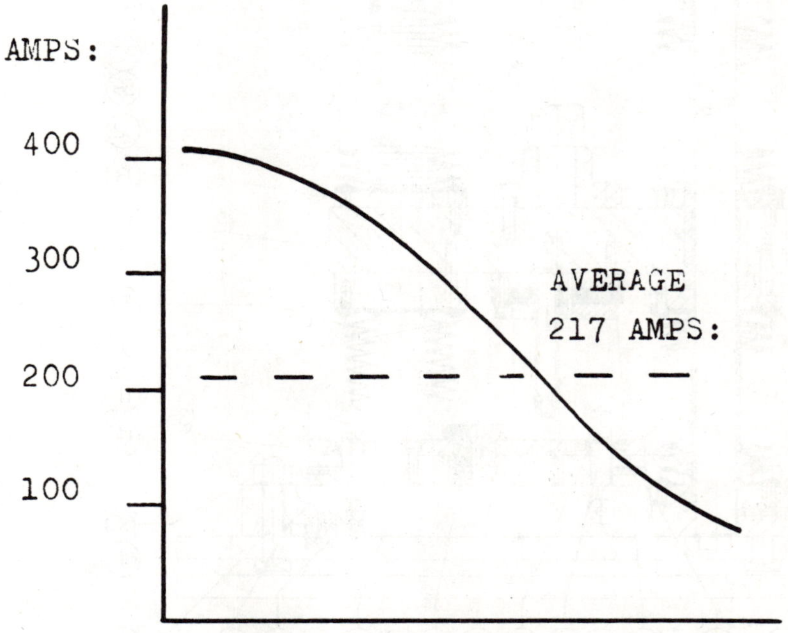

FIGURE 16-6: CURRENT PLOT OVER 30 SECONDS

Summary

Little can be written of an academic nature with respect to ECD and there is no necessity for reams of complex mathematical equations. The technique is practical and all its problems can be solved on the workshop floor.

When the list of ingredients necessary to cause an electrochemical deburring cell to function efficiently was formed, one important ingredient was deliberately omitted and reserved for mention in this conclusion. It is the patience, perseverance, ingenuity, and determination of the tooling development engineer to make it work. Figures 16-7 through 16-9 illustrate tooling used and workpiece results.

As with most high technology processes, a period of some months is required for familirization with the techniques of tooling and development and getting to grips with the nature of the process. Many mechanical engineers exhibit concern with a process which embraces chemicals, electricity, mechanical motion, pumps, filters, and all the paraphernalia of ECD, therefore the manner of man required to complement the process is not mentioned lightly.

References

[1] Eric Morris, "Electrochemical Deburring Using Constant Current," SME Paper MR75-477, 1975.

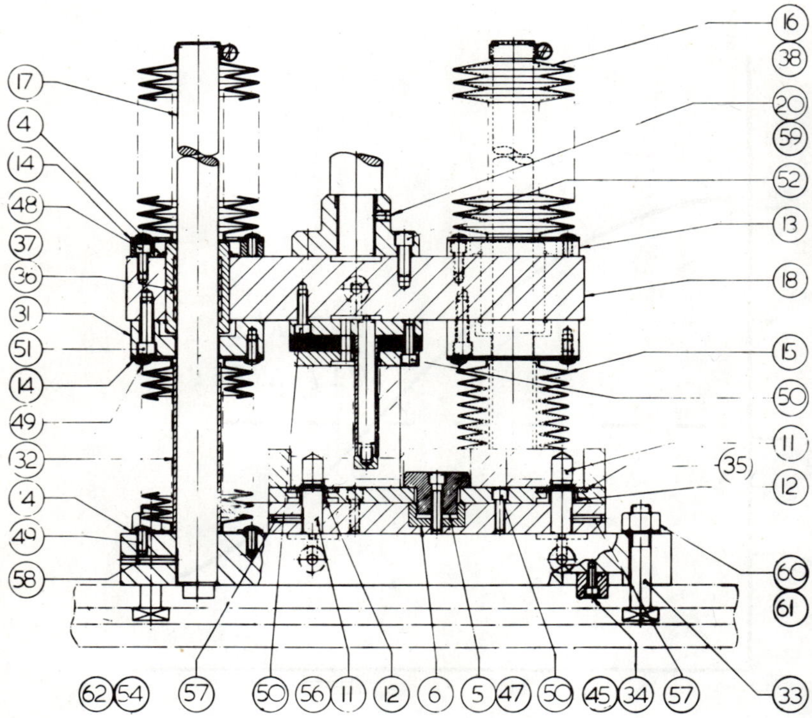

FIGURE 16-7: DEBURRING TOOL FOR PUMP COVERS

FIGURE 16-8: GEARS BEFORE AND AFTER ECD

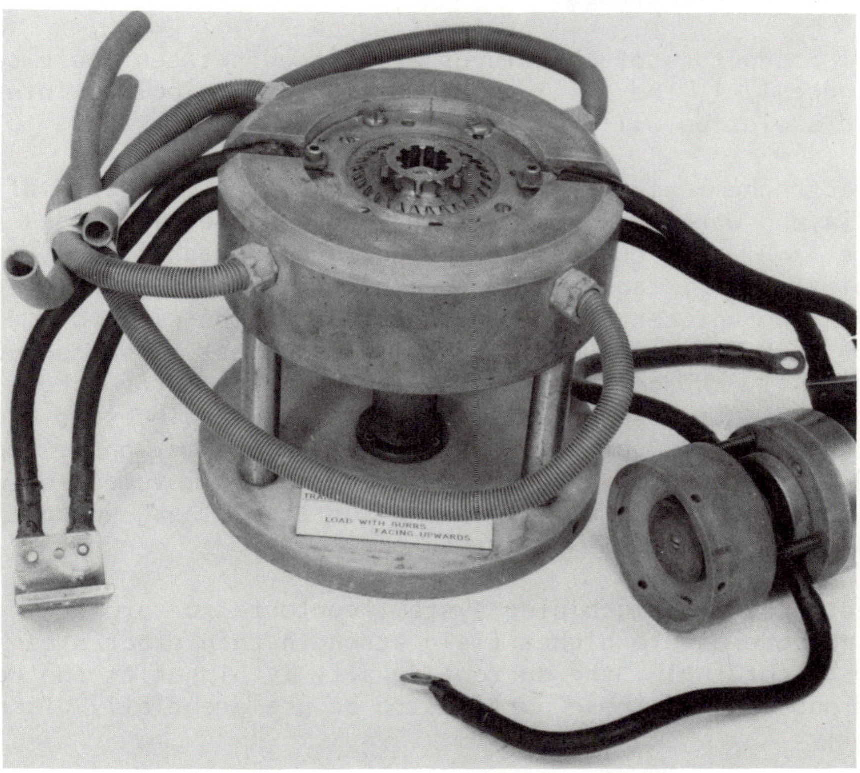

FIGURE 16-9: TOOLING FOR GEARS

CHAPTER 17

ELECTROCHEMICAL BARREL FINISHING

Electrochemical barrel (ECB) finishing is a method of finishing, deburring, and descaling workpiece surfaces by utilizing electrically conductive media as dynamically moving electrodes in an electrolyte. The workpieces are placed into a mass of the conductive media loosely held in the electrolyte and agitated with the mix while an electric current is passed through the moving bodies. Unlike conventional barrel finishing, burr removal is affected electrochemically. This process however requires no fixturing or jigs which makes it different from conventional ECM deburring. Machining efficiency is extremely high and a wide range of applications is expected.

Machining Principles

When an electric potential is applied between a metal conductor and a cathodic electrode in an electrolyte, material is dissolved from the anodic conductor into the electrolyte. The amount of material dissolved is, in accordance with Faraday's Law, proportional to the quantity of electricity delivered between the electrodes.

When metal conductors of workpieces are placed between the electrodes as shown in Figure 17-1, they will each function as a bipolar, intermediate electrode and dissolution will take place at their anodic sides.

In the electrochemical barrel finishing system, a multitude of conductive solid media and workpieces constitute intermediate electrodes as shown in Figure 17-2. The conductive media are composed of a material which is dissolution-resistant and of an electric resistivity higher than that of the metal workpieces but lower than that of the electrolyte. A typical form of such media are graphite balls (spheres). The conductive media act as resistive spacers to prevent the workpieces from short circuiting and as counterelectrodes against the adjoining workpieces to create enhanced machining current paths. Rotary or vibratory motion is imparted to a carrier drum to maintain constant random relative movement between the media and the workpieces so that the entire surface of each workpiece can evenly contact adjoining media.

In an electrochemical machining system, contoured or protruding portions of an electrode have a higher field strength than other areas of the electrode. Accordingly, the current density is higher at the burr and contoured portions causing those portions to be preferentially dissolved electrochemically.

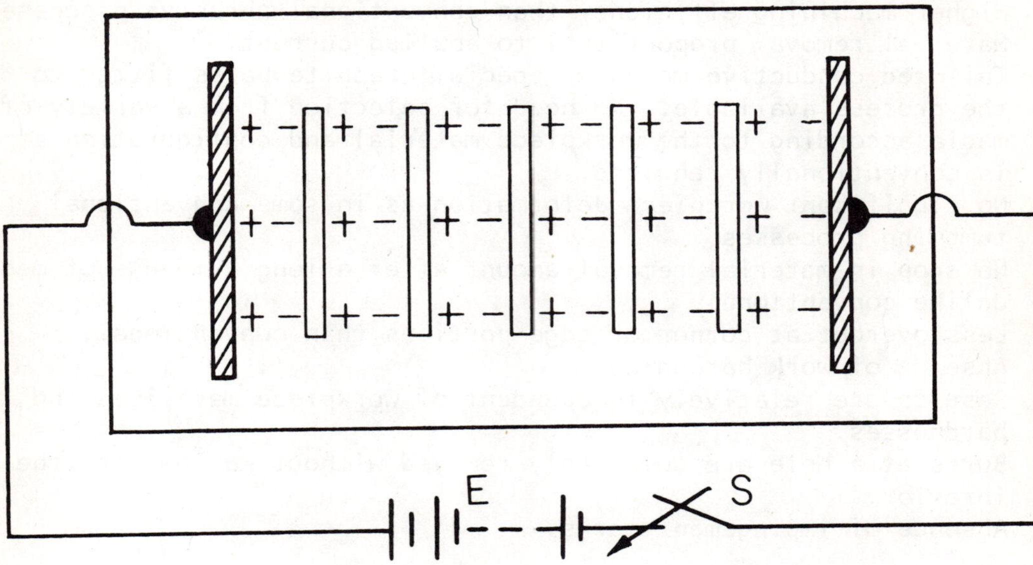

FIGURE 17-1: PRINCIPLE DIAGRAM OF ECB

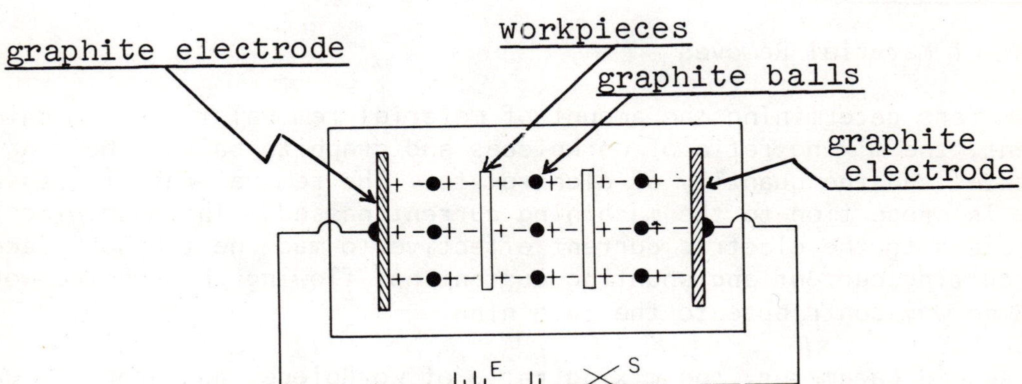

FIGURE 17-2: PRINCIPLE DIAGRAM OF ECB WITH GRAPHITE BALLS

Advantages and Disadvantages of ECB

Advantages of ECB include:

1. Higher machining efficiency than conventional abrasive processes. Material removal proportional to applied current.
2. Tailored conductive media or special graphite balls fitted to the process available. No need for selection from a variety of media according to the workpiece material and configuration as is conventionally required.
3. No significant workpiece deformation as in some conventional tumbling processes.
4. No drop in material removal amount after a long time use of media unlike conventional.
5. Less overcut at corner or edge portions than conventional.
6. Absence of work hardening.
7. Results are relatively independent of workpiece materials and hardnesses.
8. Burrs at a hole are completely removed without moving into the interior.
9. Absence of impingement marks.

Disadvantages of ECB include:

1. Appearance on ferrous workpieces of semiglossy black/brown colored finish surface.
2. Possible roughening of finish surfaces depending on particular materials, particular kinds of electrolyte, and particular machining conditions (current density, electrolyte temperature).
3. When mirror face finish is essential, post-treatment such as chemical polishing is required.

Process Performance

The Amount of Material Removed

Parameters determining the amount of material removal include machining current, the mixing ratio of workpieces and graphite balls, their agitating state, and the quantity of electrolyte. The removal rate increases basically in proportion to the machining current passed. The machining current refers to the electric current effective to machine the workpieces. Short circuiting current and shunting current not flowing through the workpieces in no way contribute to the machining.

The second parameter, the mixing ratio of workpieces and graphite balls, is extremely important as illustrated in Figure 17-3. In this figure, the apparent or bulk volume ratio of workpieces to graphite balls is plotted along the abscissa, and both the machining current and the amount of metal removal are plotted along the ordinate. In the graph, the quantity of graphite balls was maintained constant and the quantity of workpieces varied.

When the proportion of workpieces to graphite balls is low, the metal removal is reduced apparently because the shunting current not flowing through the workpieces increases.

When the proportion of workpieces is increased, a maximum amount of metal removal is achieved with the bulk volume ratio around 1:1.

When the proportion of workpieces is further increased, the opportunity for the workpieces to come in direct contact with one another and with the electrodes (fixed) increases. As a consequence of the increase of short-circuiting current, machining efficiency levels down.

It is seen that best results are obtained when the volume ratio ranges between 1:5 and 1:1. In this range, machining condition may be established appropriate to prevent workpieces short-circuiting.

The agitation of workpieces and graphite balls, the third parameter, facilitates the release of ECM gaseous products and the replenishment and repulsion of electrolyte in machining regions to enhance machining efficiency. It is imperative that the relative movement be uniform over the entire mass and continuous (without interruption).

The fourth parameter electrolyte must be present in an amount necessary and sufficient to allow machining. The movement of workpieces and graphite balls and the amount of electrolyte are correlated. In rotary and vibratory systems, the state of agitation may become unsatisfactory unless the amount of electrolyte is properly monitored. In a long time operation the consumption of electrolyte due to vaporization produced by the temperature rise may not be disregarded and then the use of an automatic electrolyte feed device having sensing and control functions may be necessary.

ECB performance under optimum conditions can be expressed by the following formula:

$$W = K \cdot e \cdot I \cdot t$$

where W = metal removal amount (gram), K = constant (2 to 4), e = electrochemical machining unit quantity (gram/ampere • minute), and t = machining time (minute).

In the above, K is determined by the second, third, and fourth of the parameters described. The fact that K is larger than 1 shows that the process involves polarization effects, a point that distinguishes it from usual electrochemical machining.

Surface Roughness and Accuracy

As pointed out earlier, the main mechanism of ECB is workpiece dissolution by electrochemical actions. Accordingly, factors determining the surface roughness and accuracy include the following: the material of workpieces, the kind of electrolyte, the agitation, the mixing ratio, and machining time.

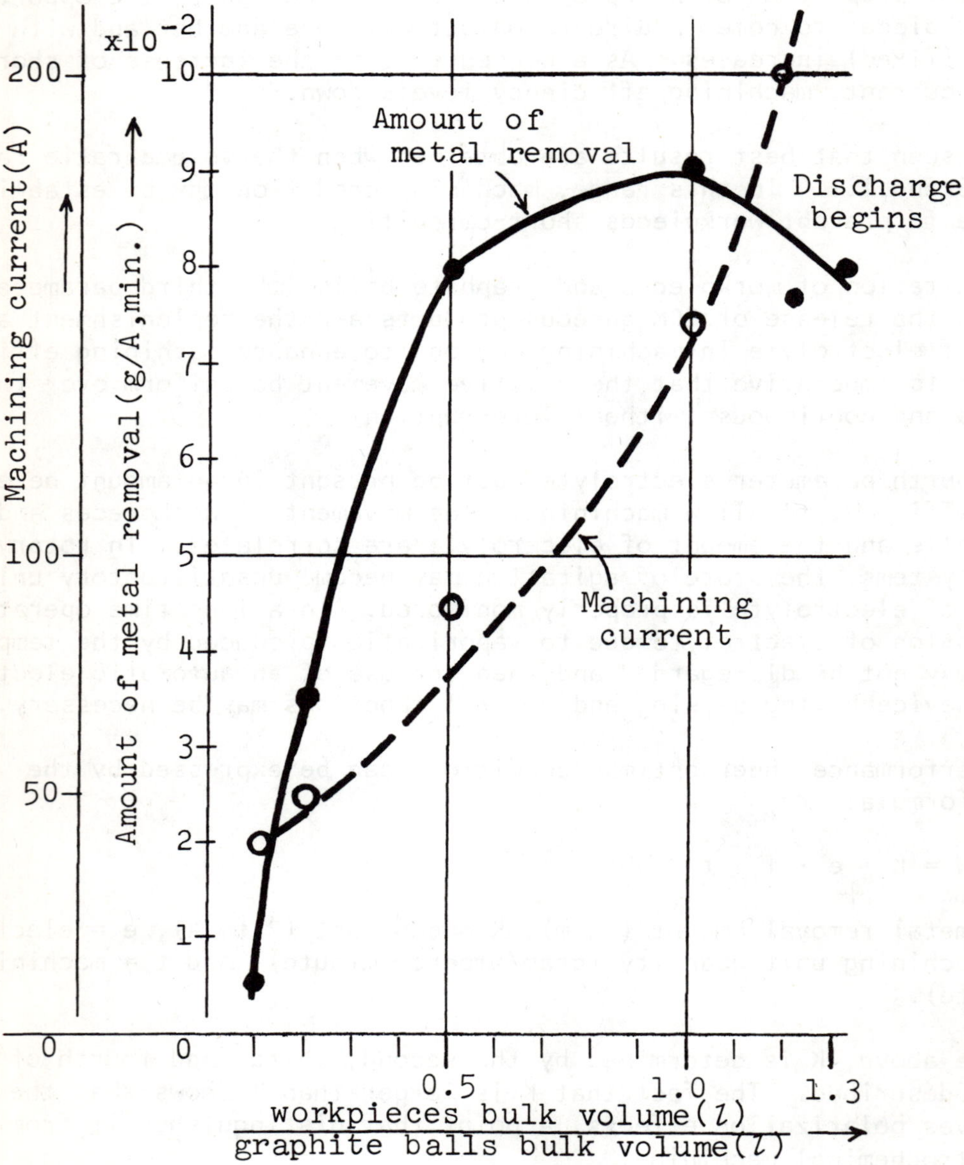

Electrolyte:	H_3PO_4 20 - 30 vol.%aq.
Machining Voltage:	19V
Workpiece:	S45C (AISI 1045 Steel)

FIGURE 17-3: MIX RATIO AND METAL REMOVAL

Dependency of ECB'd surface roughness on materials is shown in Table 17-1 with respect to principle materials. If a fine finish is desired, the roughness may not be satisfactory. By employing chemical polishing as a post-treatment, however, a glossy finish can readily be obtained.

Some alloys have rates of electrochemical dissolution extremely varying from the others. Since cast iron contains such components, the surface of this material becomes rough when subjected to electrochemical dissolution.

TABLE 17-1: MATERIALS AND SURFACE ROUGHNESS

Working Material	Machining Voltage Machining Current	Roughness of Treated Surfaces**
SUS 24***	19V 40 - 44 A	2 - 2.5 μRmax
SUP-1†	19V 28 - 32A	3 - 4
Brass (7-3)	19V 34 - 40A	2 - 2.5
Zinc (diecast)	19V 40 - 44A	5 - 6
Aluminum (diecast) (Si 7.5 - 9%)	19V 36 - 40A	4 - 5
Phosphorus bronze	29V 72A	1.6

*Electrolyte: H_3PO_4 20 - 30 Vol%aq
**1.0 μAA = 40 microinch AA. Finishes expressed as Rmax are not directly translatable to AA however.
***Equivalent to 430 stainless steel
†Equivalent to 1085 steel

Kinds of electrolyte in customary use at the present time include aqueous solutions of H_3PO_4, $NaNO_3$ and NaCl. When fine surface is an important requirement, an optimum electrolyte composition should be used that best suits the particular material of workpieces.

The agitation of workpieces and graphite balls is an important parameter to the workpieces accuracy as well as to the amount of material removal. If there is a lack of the uniformity of agitation or relative random displacement, variations occur in the amount of metal removal over the entire mass of parts and also over the entire surface of an individual part. Especially when a workpiece is flat and long, there may be a difference in material removal brought about between the two ends and workpieces may pile up causing machining inaccuracy. An example of dimensional variations of a typical part before and after ECB is shown in Figure 17-4.

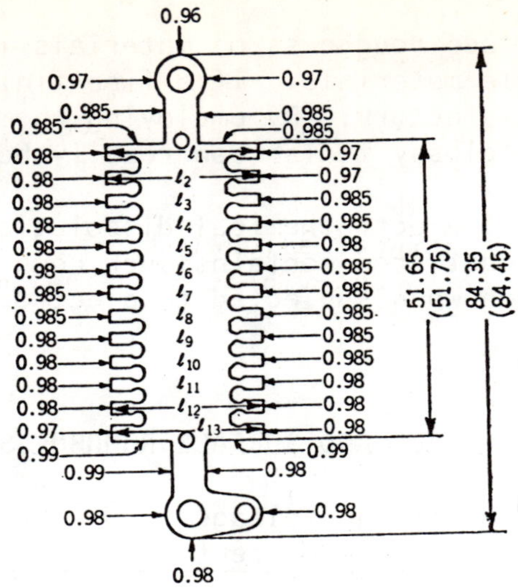

ECB'd parts of electric calculators. Measured variations of thickness, width and length are listed below.

a) Parenthesized are values before ECB (mm).
b) Thickness before ECB is 1 mm.
c) Thickness values after ECB at 1 mm apart from edges.
d) Length and width were measured by calipers and thickness by a micrometer.

Machining voltage: 28V, Machining current: 200A
Machining time: 20 minutes, Electrolyte: 15% NaCl

FIGURE 17-4: DIMENSIONAL VARIATIONS IN A PART

When the proportion of workpieces relative to graphite balls is excessive, workpieces and electrodes tend to short-circuit as mentioned previously. As a result, arcing marks may be produced on the workpieces. Accordingly, when shapes liable to short-circuit are dealt with, it is necessary to limit the bulk volume ratio to 1:5 or so.

It is also necessary to use smaller graphite balls for workpieces of high volume to weight ratio.

Relationship between machining time and the increase of surface roughness is shown in Figure 17-5.

Time required for deburring in general is 15 to 30 minutes.

Wear of Graphite Balls

The required nature of graphite balls is having a specific resistivity of 1 x 10^{-1} to 10^{-4} ohm-cm, shock-resistant and wear-resistant properties.

In machining, graphite balls wear slightly due to electrochemical oxidation and mechanical agitation and vibration. The amount of wear for 1 liter of graphite balls is generally 3 to 4 grams/liter/hour in case of rotary barrel and 10 to 15 grams/liter/hour in case of vibratory barrel.

ECB Equipment

ECB actually consists of five basic loose abrasive approaches. Figure 17-6 illustrates four of these. Figure 17-6a illustrates a rotary ECB machine. Workpieces and graphite balls are contained in a rotary barrel in a volume ratio of 1:5 and 1:1. The barrel is rotated relatively slowly in an electrolyte container to give a uniform agitation to the workpieces and graphite balls in the barrel. A stainless steel cathode is mounted in the barrel while graphite anodes are arranged in the container.

A modification of the rotary ECB system is illustrated in Figure 17-6b. The actual machine is shown in Figure 17-7. The barrel unit, control panel and power supply, and electrolyte cooling unit constitute the equipment.

Figure 17-6c shows diagrammatically a vibratory barrel machine. A pair of graphite electrode plates is mounted inside the barrel receiving workpieces, graphite balls, and electrolyte between them. The use of excessive electrolyte in the barrel impedes the agitation, hinders uniform machining, and reduces machining efficiency. As a countermeasure to the vaporization of machining liquid due to electrically induced temperature rise, it is possible to monitor the electrolyte and automatically replenish the evaporated amount of water to the barrel as shown in Figure 17-6d. The actual form of the vibratory ECB built is shown in Figure 17-8.

A massive part may be ECB'd in the manner diagrammatically shown in Figure 17-6e. The workpiece is mounted to a holder and receives rotary and/or reciprocatory movement to impart random displacement to graphite balls in the electrolyte. The graphite balls act as dynamically moving electrodes to electrochemically deburr the workpiece.

The inner surfaces of massive workpieces in a small number are effectively deburred, descaled, or finished by ECB in the manner shown in Figure 17-6f. A perforated cathode is inserted into the workpiece that is poled anodic and electrolyte is flushed from or into the cathode through the perforations. Graphite balls agitated by the electrolyte uniformly remove dissolution-resistant oxide films to achieve efficient finishing.

Although not shown, centrifugal barrel units can also be made using this principle.

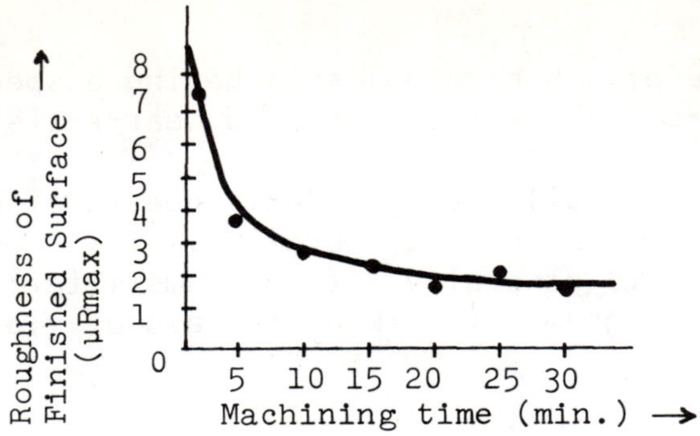

workpieces: brass
machining voltage: 19V
machining current: 40A
electrolyte: H_3PO_4 20 - 30Vol%aq

FIGURE 17-5: MACHINING TIME AND SURFACE ROUGHNESS

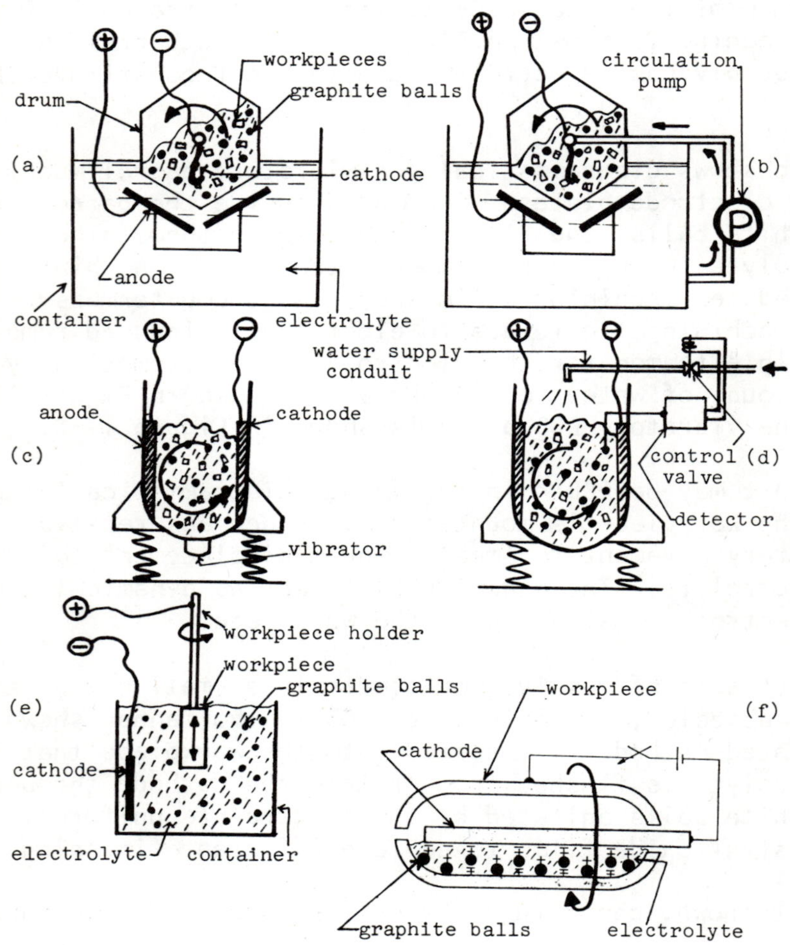

FIGURE 17-6: FORMS OF ECB

Examples of Process Performance

Comparison of Performance

A comparison of performance of ECB with those of other barrel finishing processes is shown in Table 17-2. As seen there, vibratory ECB is competitive with centrifugal barrel finishing for the particular workpieces from which this data was taken.

Applications

Applications in which ECB may be utilized with efficiency include:

1. Deburring of peripheries of machined parts.
2. Deburring of press-formed or ground parts.
3. Descaling of forged or cast parts.
4. Deburring and descaling of massive parts.

Various examples of parts ECB deburred are shown in Figure 17-9.

Figure 17-10 shows a hydraulic accumulator which may be descaled by a modified ECB system (Figure 17-6f). The fixture used to hold the part is also shown. Good results are obtained with this modification in the deburring of massive stainless steel machined parts.

Electrochemical Vibratory Deburring

This process was first described in United States literature in 1969 by DeGroat.[1] At that time, operating time was reportedly reduced by a factor of 10 to 30 over conventional processes then in use.

Two British patents, filed in 1968,[2,3] indicate that phosphoric acid is used as the electrolyte in this process and that silicon carbide or copper coated aluminum oxide can be used as the media. Subsequent United States patents in 1969 and 1970 also cover this process.[4,5]

In a related development, Krapivin[6] connected parts to the positive side of a power supply and followed a similar process. In his study however a physical coupling with the workpiece was apparently used, whereas in the other cases the electrolyte and graphite provided the coupling. Krapivin's results indicated that the process was three to nine times faster than conventional vibratory finishing. Four applications are briefly described in another recent report.[7]

Little information has been published on the burr removal, edge radiusing, surface finish changes, or stock losses resulting from this process. DeGroat, however, indicates that a 0.002 inch thick (50.8 μm) burr was removed from a brass wristwatch housing in 35 seconds without removing noticeable stock from the housing.

Only one commercial facility is known to be running in the United

FIGURE 17-7: ROTARY ECB EQUIPMENT

FIGURE 17-8: VIBRATORY ECB EQUIPMENT

TABLE 17-2: PERFORMANCE COMPARISON

Cost Factor		Rotary Barrel	Vibratory Barrel	Centrifugal Barrel	Vibratory ECB
	(1)	0.25(g/kg/h)	2.5(g/kg/h)	20(g/kg/h)	50(g/kg/h)
	(2)	1	4 - 20	80	200
	(3)	4(g/ℓ/h)	14(g/ℓ/h)	240(g/ℓ/h)	10-15(g/ℓ/h)
	(4)	220(yen/kg)	220(yen/kg)	220(yen/kg)	800(yen/kg)
	(5)	300(yen/kg)	300(yen/kg)	300(yen/kg)	8-20(yen/kg)
	(6)	30ℓ	40ℓ	30ℓ	40ℓ
	(7)	1:2-20	1:3-10	1:1-2	1:1-10
(8)	(9)	4(g/16kg/h)	40(g/16kg/h)	480(g/24kg/h)	1230(g/24kg/h)
	(10)	80(g/20ℓ/h)	420(g/20kg/h)	3600(g/15ℓ/h)	200-300(g/20ℓ/h)
	(11)	17.6(yen/h)	92(yen/h)	790(yen/h)	240(yen/h)
(12)	(13)	4.4 yen	2.3 yen	1.65 yen	0.16 yen
	(14)	16.0 yen	1.6 yen	0.75 yen	2.2 yen
	(15)	20.4 yen	3.9 yen	2.4 yen	2.36 yen

DESCRIPTION OF COST FACTORS

(1) Removal of unit material
(2) Polishing efficiency ratio of unit material
(3) Wear of unit material
(4) Cost of media
(5) Cost of compound
(6) Media quantity into 50 liter barrel
(7) Work/media ratio
(8) 50 liter barrel
(9) Material removal/time
(10) Media wear/time
(11) Media wear cost/time
(12) Material removal cost
(13) Media wear cost/time
(14) Cost of electricity
(15) Partial sum

1 yen = 0.0033 dollars
h = hour
g = gram = 1/454 pounds
kg = kilogram = 2.2 pounds
ℓ = liter = 61 cubic inches

A: Material
B: Machining voltage
C: Machining current
D: Machining time
E: Electrolyte

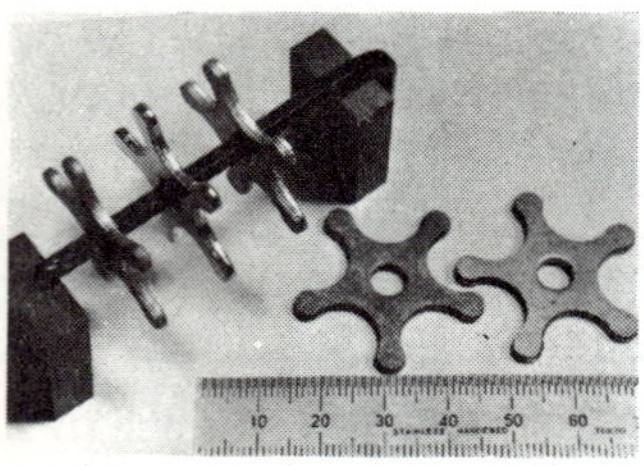

A: Iron
B: 28 volts
C: 68 to 72 amperes
D: 30 min.
E: 15wt%aq. $NaNO_3$

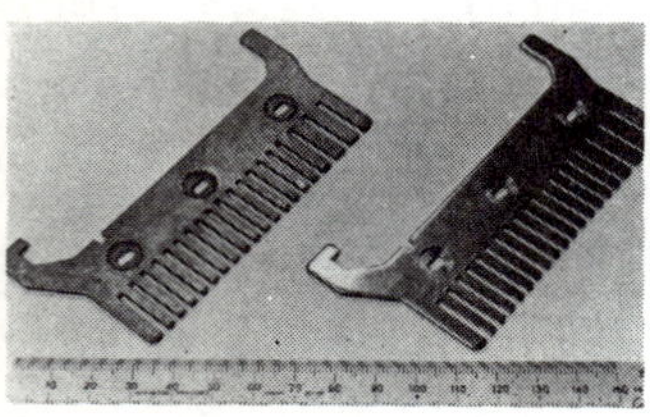

A: Iron
B: 19 volts
C: 40 to 68 amperes
D: 50 min.
E: 20 to 30 vol%aq. H_3PO_4

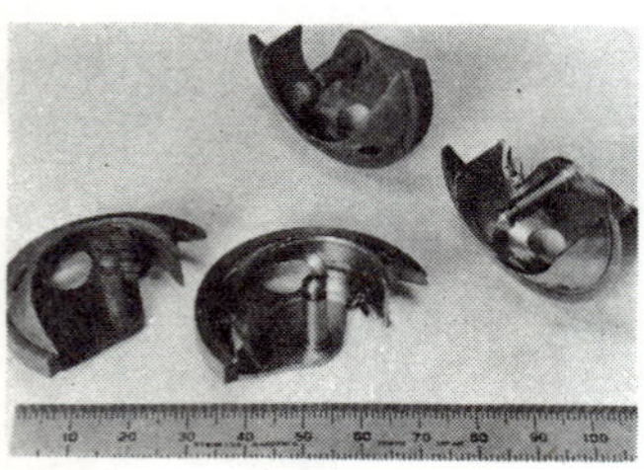

A: Iron
B: 28.5 volts
C: 68 amperes
D: 30 min.
E: 15wt%aq. $NaNO_3$

A: Iron
B: 28.5 volts
C: 44 to 48 amperes
D: 50 min.
E: 20 to 30 vol%aq. H_3PO_4

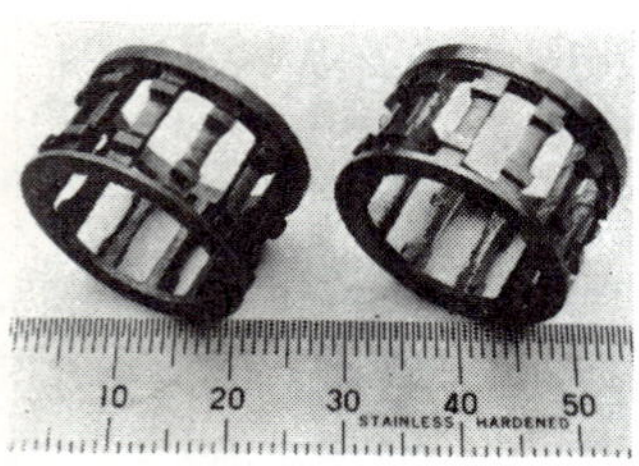

A: Iron B
B: 19 volts
C: 40 to 42 amperes
D: 30 min.
E: 20 to 30 vol%aq. H_3PO_4

FIGURE 17-9a: ECB MACHINING EXAMPLES

A: Material
B: Machining voltage
C: Machining current
D: Machining time
E: Electrolyte

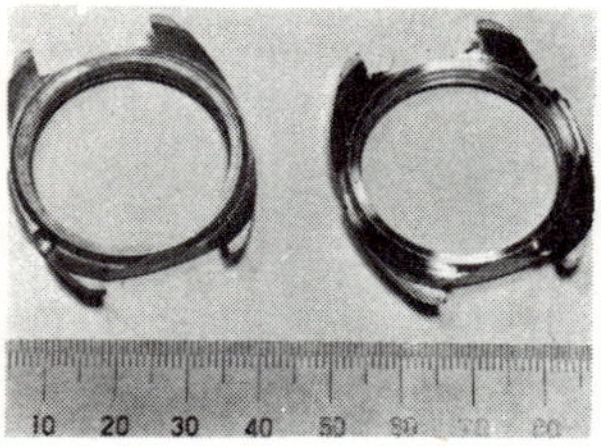

A: SUS-27
B: 19 volts
C: 40 to 56 amperes
D: 60 min.
E: H_3PO_4 + $NiCl_2$aq.

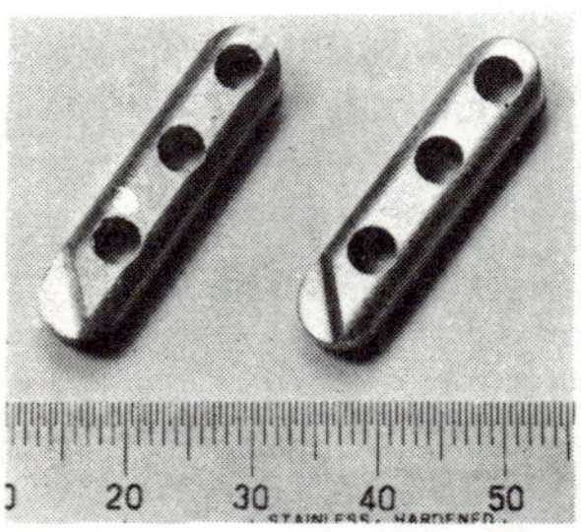

A: Brass
B: 29volts
C: 72 to 96 amperes
D: 30 min.
E: 20 to 30 vol%aq. H_3PO_4

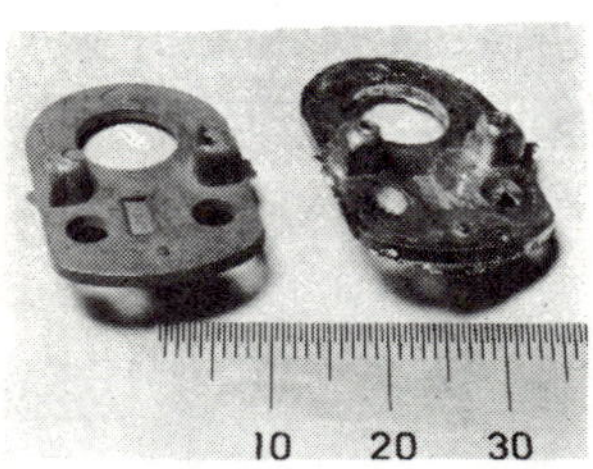

A: Zinc diecast
B: 18 to 19 volts
C: 32 to 34 amperes
D: 20 min.
E: H_3PO_4 + CH_3COOH + H_2O

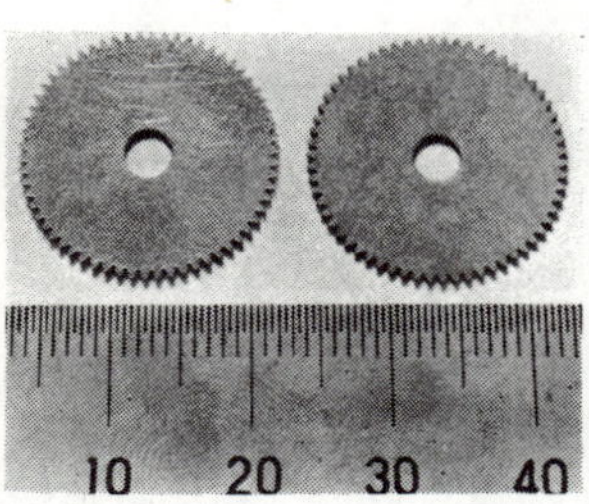

A: Duralumin
B: 19 volts
C: 44 amperes
D: 10 min.
E: 40wt%aq. H_3PO_4

A: Alminum diecast (inc. 10%Si)
B: 39 volts
C: 136 to 168 amperes
D: 30 min.
E: 40wt%aq. H_3PO_4

*SUS-27 = 304 Stainless Steel

FIGURE 17-9b: ECB MACHINING EXAMPLES

States. This application processes millions of small precision parts of alloy steel every week. Other installations exist in Japan, Italy, and apparently in Russia.

References

[1]British Patent Number 1,238,073 - 1968.

[2]British Patent Number 1,194,649 - 1968.

[3]USA Patent Number 3,533,928 - 1969.

[4]USA Patent Number 5,523,834 - 1970.

[5]S. P. Krapivin, "Vibroabrasive Electrochemical Grinding," Vibrats Obrabotka Detale: mashin i priborov (Vibratory Machining of Machine and Instrument Components) Rostov-on-Don, 1972, pp. 80-84 (See Referativnyi Zhurnal--Tekhnolagiia Mashinostroeniia, 1973, 4B 253) (in Russian).

[6]"Elektrochemische Bearbgitung," VDI-RICHTLINIE 3401-2, January, 1972, Beuth Vertrieb GmbH, Berlin, 30.

[7]Akihiko Shimizu, "Electrochemical Barrel Finishing," SME Paper MR75-837, 1975.

FIGURE 17-10: HYDRAULIC ACCUMULATOR AND FIXTURE

CHAPTER 18

ELECTROPOLISH DEBURRING

Introduction

Ten years ago a group of deburring consultants visited a factory in Southern Germany where chrome-nickel-steel filter-cylinders for paper making machines are produced. These cylinders are made of filter plates and have a total surface of more than three square meters (323 sq ft). The perforated plates have conical drill holes every square meter having 300,000 holes. After drilling the plates are faceground in order to remove the burrs of the drill holes. During this working process, however, new burrs are produced inside the drill holes. Formerly all holes had to be deburred inside by hand one by one in order to avoid the stopping up by the paper fibers during the dewatering process of the paper pulp. It took two women two weeks to deburr one single filter plate by hand.

Figure 18-1 shows a conical drill hole of the above mentioned filter plate after facegrinding. The burrs at the edge of the hole are clearly visible as well as the surface roughness on top of the plate and inside the drill hole. One can imagine the tremendous and monotonous work when deburring hundreds of thousands of holes by hand. One can also imagine the cost of labor and the difficulty of finding reliable persons for this work.

This problem was solved by introducing an electropolish deburring process. Since then all the filter plates are deburred in an electropolishing installation as shown in Figure 18-3 where the plates are simultaneously leveled and smoothed. Up to 1,000,000 drill holes are electropolish deburred within 15 minutes and after this treatment every drill hole has exactly the same quality as shown in Figure 18-2. It is evident that no paper fibers will stick to a smooth electropolished surface. Therefore the quality of filter-cylinders manufactured according to this method will be considerably superior to the mechanically deburred cylinders.

One might presume that similar technical problems would be solved by similar technical means all over the world. From experience we know however that there are only a few countries where the advantages and the applicability of electrochemical deburring and polishing processes are appreciated and put into practice. Germany is one of the countries where electropolishing processes are widely known and widely practiced throughout the whole metalworking industry.

This is one of the reasons why the "Deutscher Industrie Normenausschuss" (DIN--German National Bureau of Standards) issued an industrial specification sheet DIN 8590[1] concerning surface treatment by anodic dissolution. Another association, "Verein Deutscher Ingenieure" (VDI--Society of German Engineers), has worked out guiding principles for the users of

FIGURE 18-1: CONICAL DRILL HOLE (0.9 MM DIAMETER) AFTER FACE GRINDING

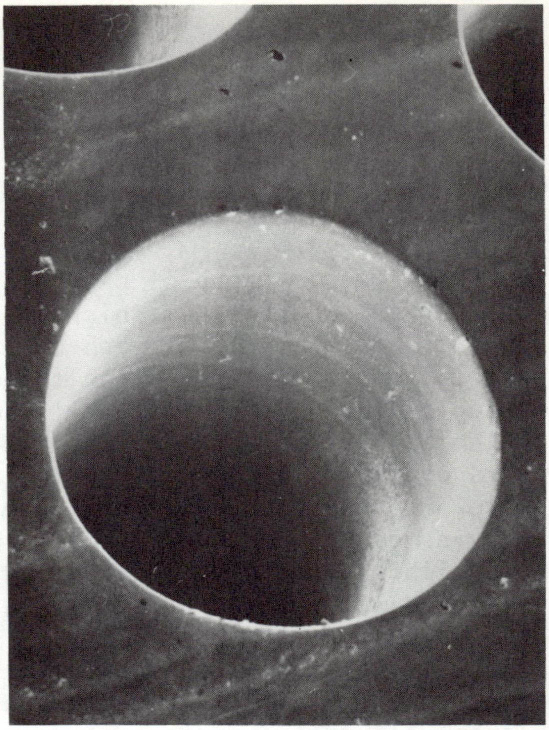

FIGURE 18-2: THE SAME DRILL HOLE AFTER ELECTROPOLISH DEBURRING

FIGURE 18-3: ELECTROPOLISHING PLANT FOR DEBURRING FILTER PLATES MADE OF CHROME-NICKEL-STEEL (TANK VOLUME 4500 LITER)

electropolishing processes and issued in a VDI Handbook No. 3401-2(2). The same association organizes technical courses on electrochemical polishing and deburring.[3]

Principle of Electrochemical Deburring in Electropolishing Plants

It is assumed that the principle of electrochemical surface metal removal as a result of direct current energization within an electrolyte solution is common knowledge. In fact it is more or less the reversal of the plating process: an anode connected component is subjected to selective electrochemical surface attack and as a result of the formation of a viscous layer with a high electric resistance, a preferential removal of the projections out of the base material occurs.

The significant aspect of this method of metal removal is that the microprofile is smoothed and the macroprofile is leveled, all in one single process and quite independently from the original surface roughness.

It is also important to note that the upper material layer is removed without stressing the piece, an effect which was unavoidable in case of mechanical treatment. Consider the mechanical forces arising from turning, milling, drilling, etc., which lead to the destruction of the metal crystals at the surface. Think of the considerable localized overheating of surface layers in the case of grinding or mechanical polishing which may result in coarsening of crystals, crack formations and even chemical reactions within the structure of the surface. Surface inclusions in the form of grease, scale, polishing and grinding compounds, or particles from the cutting tools are inevitable, thereby increasing the danger of contamination and of buildup of localized elements and thereby producing a favorable condition for corrosion attack.

When electropolishing, neither the forementioned defects are produced nor are new stresses introduced. Since the surface layers removed differ in their physical and chemical behavior from the base material, an electropolished surface has the unadulterated properties of the original material. This means that electropolishing can be considered as a surface finish refinement leading on the one hand to an improvement in geometric proportions in the sense of smoothing and/or leveling and producing an appreciable reduction in absolute surface area. On the other hand, there is an improvement of the surface structure in view of the fact that all defects resulting from mechanical or heat treatment are eliminated inasmuch as they are localized within the dissolved layer. Usually the thickness of this layer amounts to a few hundreths of a millimeter.

The above mentioned action of preferential removal of surface projections does not only affect surface irregularities but to a greater extent also sharp corners and edges which are electrochemically dissolved at a more rapid rate than the general surface. Consequently fine burrs adhering to edges as they are produced while stamping, grinding, milling, etc., are removed. This applies also to turned-down burrs, flakes or overlaps over ground or drawn surfaces. This process is described as electrochemical deburring. Since this metal removal takes place simultaneously with electropolishing, the process may be named electropolish deburring. Thereby it is distinguished from ECM deburring, i.e., electrochemical machining.

Which Burrs Can Be Removed By Electropolish Deburring?

First of all, we have to consider the many types of burrs originating from different cutting processes. Figure 18-4 shows various cross sections of burrs.[4] It is easy to understand that the connection of the burr with the base material (burr root) is most decisive for the amount of metal which has to be dissolved electrochemically in order to deburr the edge.

If the burr root is very thick, many Ampere-minutes have to be spent for its removal. Due to the fact that any electrochemical deburring has simultaneously an electropolishing effect, a too heavy dissolution of the overall surface might be the result in this case. Very often this must be avoided for technical reasons--for instance if exacting tolerances have to be observed or if a decrease of the wall thickness is not acceptable. Economic reasons may also be taken into consideration if consumption of electrolyte and electric current becomes excessive.

At any rate it is recommended to examine the burrs in question and the conditions referring to the piece before deciding in favor of electropolish deburring.

In view of the different types of burrs depicted in Figure 18-4, those burrs are immediately recognized which cannot be removed successfully the electrochemical way. This concerns especially the types shown under a, c, and f; while burrs b and d seem to be favorable to the electrochemical process as they show a comparatively thin connection of the burr to the base material.

Figure 18-5 explains the essential determining characteristics for the judgment of a burr: (1) thickness (t) of the burr which should be as little as possible, i.e., normally not exceeding 0.3 mm (0.012 inch); (2) radius (r) of the burr root which also should normally not exceed 0.3 mm. It is interesting to know that the length[1] of a burr is not relevant for the question of being suitably electropolish deburred. Sometimes the burr root has been dissolved before the rest of the burr which then falls off. While disscussing this process, it should be mentioned that hip beads resulting from worn out tools or due to inefficient manufacturing processes naturally limit the applicability of the electrochemical process. Moreover the question may be of interest--whether an edge should become round or remain sharp as it is the case with edges of piston valves.

The unequal dissolution of the surface is to be seen in Figure 18-6 which shows a stamped brass sheet before and after electropolishing. It demonstrates clearly the simultaneous polishing and deburring effect. While the sheet becomes thinner by approximately 0.04 mm (0.0016 inch) the hip bead is reduced to an acceptable size. Of course, the time of treatment is related to the intended use of the piece.

If the burrs are too coarse, mechanical deburring may be employed as

(a) Drilling AlMgSi 0,5
(b) Grinding AlMgSi 0,5
(c) Sawing Carbon Steel
(d) Notching AlMgSi 0,5
(e) Turning Carbon Steel
(f) Shaping Carbon Steel

FIGURE 18-4: CROSS SECTIONS OF BURRS ORIGINATED BY DIFFERENT CUTTING PROCESSES

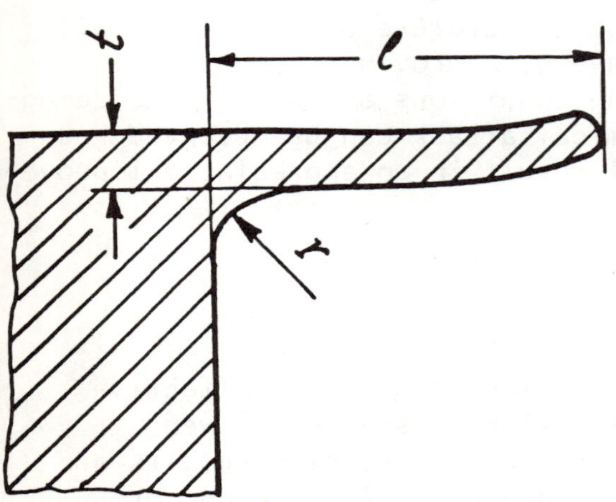

l = Length of the Burr
t = Thickness of the Burr
r = Radius of the Burr Root

FIGURE 18-5: SCHEMATIC CROSS SECTION OF A BURR

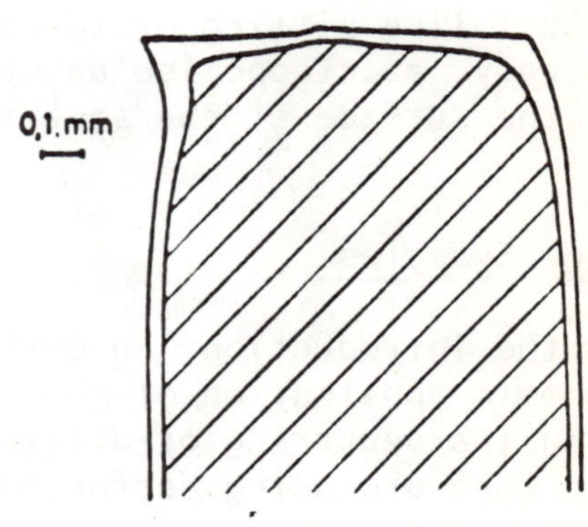

FIGURE 18-6 CROSS SECTION OF A STAMPED BRASS SHEET BEFORE AND AFTER ELECTROPOLISHING

pretreatment. When deburring mechanically for instance by grinding or shaving, new burrs may occur as shown in Figure 18-1 and sometimes instead of one large burr two small burrs are the result. The latter are, however, easily removed by electropolish deburring. This applied also to all broken burrs and splinters which are pressed into the base material.

Electropolish deburring is irreplaceable if "superficial burrs" have to be removed, i.e., turned-down burrs, overlaps or flakes. These particles though having no genuine connection to the base material adhere very tightly to the surface. Their removal by mechanical polishing or brushing processes is practically impossible. In particular when grinding nodular castings, recumbent burrs and overlaps are produced owing to cutting the spherolites on top of the surface. These particles will break off as soon as the surface is submitted to gliding strain and cause wear and tear to the respective machine.[5] Any attempt at deburring this kind of burr mechanically will be in vain because new burrs and overlaps are produced by cutting processes.

When electropolish deburring, it is essential to assure that the burrs are situated within the throwing power of the electrolyte. If they are located at remote places like deep holes or hollows, auxiliary cathodes have to be inserted. If adequate racks and fixtures have once been prepared for this purpose, invariable and constant results will be achieved. It is the reliability of the electropolish deburring process which made this kind of metal finishing so successful.

Finally, it should be added that practically any metal or alloy can be electropolish deburred notwithstanding its physical properties or chemical composition. In Germany and many other European countries more than two dozen special electrolytes, working at room temperature or hot, are available for different materials. They all have been proven by industrial testing. If adequate solutions and working conditions are chosen, no damaging effects like pitting or corrosive chemical attack can be observed. On the contrary, electropolish deburring will result in an appreciable improvement of the surface at the same time.

Deburring Examples

In the introduction, an outstanding example for the most effective and economic application of the electrochemical processes was given regarding the deburring of drilled holes. Similar impressive results are achieved when producing perforated double-faced conductive plates for printed circuits. Hydraulic steering components such as housings or piston valves are electropolish deburred in order to give upmost cleanness to the functional edges and surfaces.

By using precision tools, when cutting, punching, or stamping sheets very fine burrs are produced which are difficult to remove especially when thin walled and flexible components are at stake. If the burrs have an unfavorable position as shown on the punched sieve sheet in Figure 18-7 and 18-8, only electropolish deburring guarantees a reliable smoothness of the slots including an appreciable improvement of the corrosion resistance of the work.

Other examples are stamped ball bearing cages, the cutting edges of which have to be deburred thoroughly in order to avoid wear and tear to the ball bearing. Furthermore components of jet engines made of heat resistant material such as Hastelloy or Nimonic are treated electrochemically for instance in order to deburr perforated combustion chambers, guide plates, or ductings. When electropolishing these parts, they are simultaneously submitted to a most reliable material control revealing the faintest crack and every inhomogenity of the structure.

Burrs arising from grinding or broaching operations are easily electropolish deburred when fine grained grinding tools have been used. A typical deburring example is illustrated in Figure 18-9 showing carbon-steel transport rollers for spinning machines. When grooving the rollers, formation of burrs is inevitable. The finest burrs would however be able to retain textile fibers which on the long run would form a layer around the roller and disturb the operation of the machine. By electropolishing a reliable deburring effect is achieved while the gripping power of the transport rollers is maintained.

Similar results have been obtained with regard to ground gears or pinions, notched shafts and other machinery components such as knitting cylinders or frames and needle bearers of knitting machines, where as a rule the improvement of the surface roughness is appreciable besides the deburring effect.

An interesting example in connection with the continuous production of carbon-steel wire with serrated profile is shown in Figure 18-10. These profiled wires are wound on cylinders which are used for disaggregating textile fibers and for roughening textiles. Electropolish deburring prevents hooking of the fibers on the fine burrs originated by grinding operation of the wire. Figure 18-11 illustrates an electropolishing plant for continuous deburring of profile wire.

Finally, Figure 18-12 demonstrates a good example of electropolish deburring of superficial burrs on nodular cast iron sprocket wheels. Without electropolishing the burrs break off when the wheel is driven by a plastic sprocket belt. The latter is then transferred onto the abrasive belt due to the fact that the broken burrs are pressed into the plastic. As a result the toothing of the sprocket wheel is deformed within a short time. By electropolish deburring, this damage can be avoided and the resistance to wear will be increased. Figure 18-13 illustrates an automatic plant which is designed to deburr 4000 sprocket wheels in two daily shifts. Automatic plants of this kind are working at Ford Motor Company in Germany and England.

<u>Plant and Equipment for Electropolish Deburring</u>

As mentioned before, the electropolish deburring process is more or less the reversal of the plating process. Though the necessary equipment

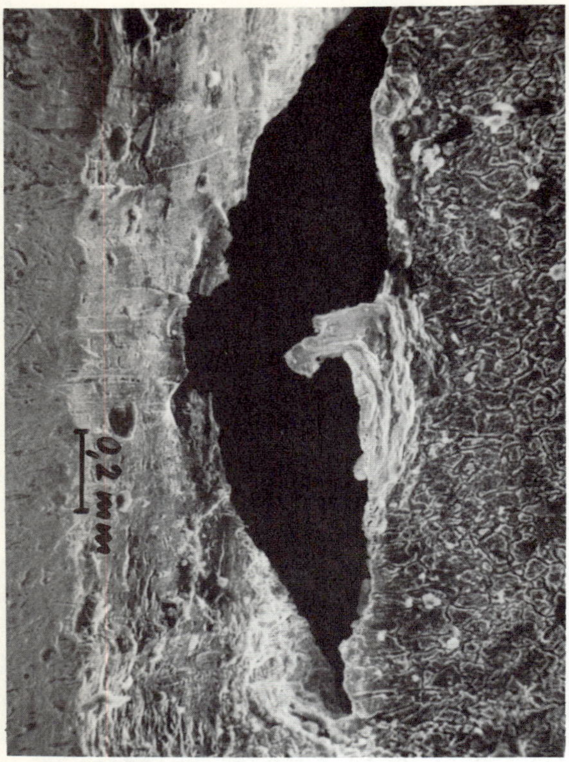

FIGURE 18-7: CORRODED STAINLESS STEEL SIEVE 10 YEARS AFTER PRODUCTION (SLOT = 2.0 MM)

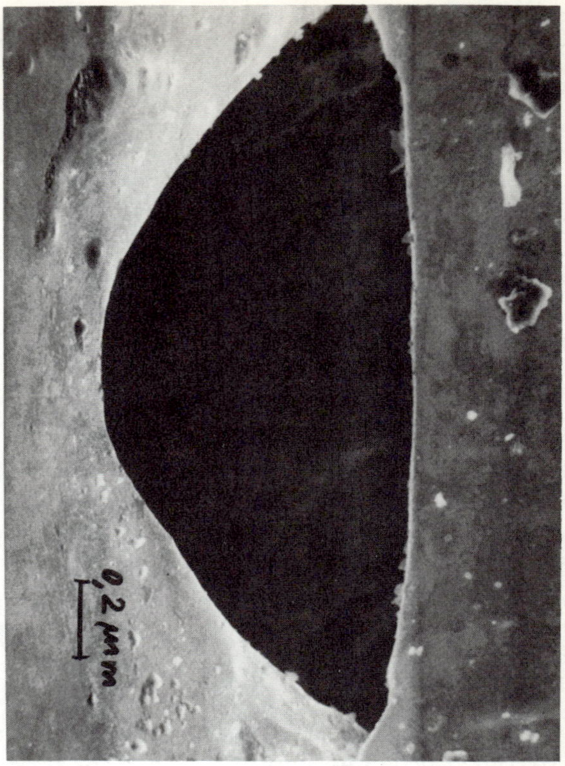

FIGURE 18-8: THE SAME SIEVE 10 YEARS AFTER ELECTROPOLISH DEBURRING

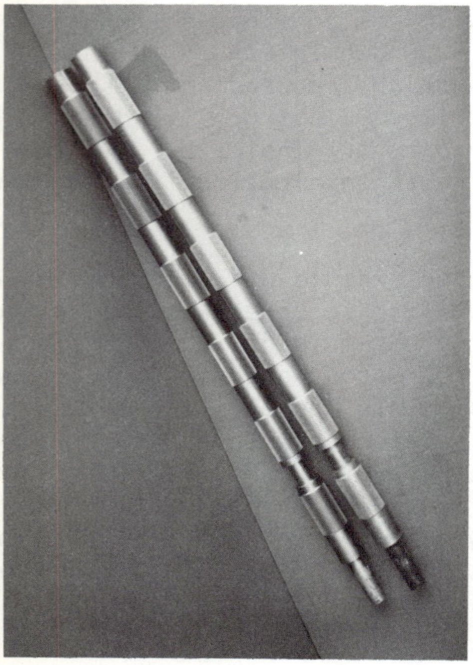

FIGURE 18-9: CARBON-STEEL TRANSPORT ROLLER FOR SPINNING MACHINES BEFORE AND AFTER ELECTROPOLISH DEBURRING

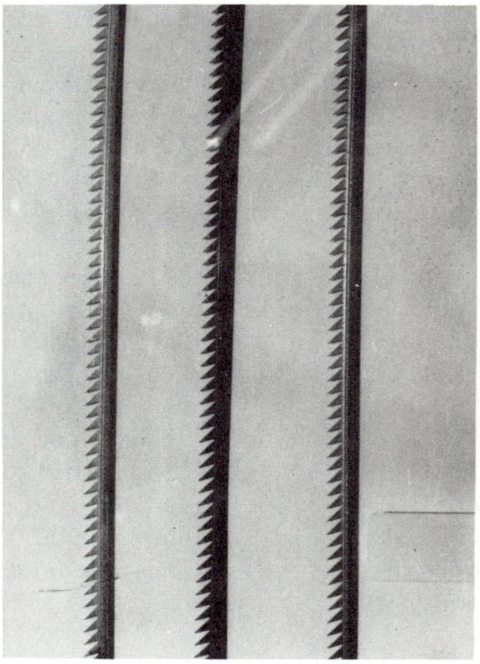

FIGURE 18-10: ELECTROPOLISH DEBURRED CARBON-STEEL WIRE WITH SERRATED PROFILE

is very similar to the plating equipment, special knowledge is required when constructing an electropolishing plant. In this field, a sound technology has been developed in Germany over the years owing to thorough research by specialized companies. German customers with deburring problems are furnished with a complete electrochemical engineering consisting of the delivery of plant plus solution plus technical know-how. They are reluctant to mix their own electrolytes or to try electropolishing on their own--they prefer to be consulted and serviced by specialists guaranteeing them troublefree production and best working conditions.

Today a wide variety of electropolishing plant and equipment is available for all applications. The schematic diagram of Figure 18-14 illustrates the different types of plants, beginning with normal plants for the treatment of jigged parts like those already shown in Figures 18-3 and 18-13[6].

Actually there is much interest in electropolish deburring of bulk components with different types of barrels. Figure 18-15 illustrates a tipping barrel equipment with automatic operational sequence for the batch bulk treatment of components having a total surface up to 80 square decimeter (approximately 7.5 sq ft). This kind of plant for instance is used for electrochemical deburring and polishing of stainless steel screws and is best utilized for a manufacturing program made up of varying components.

Finally a number of special plants have been developed to meet the individual requirements of many different products. Figure 18-11 presents an illustration of a plant for continuous treatment of wire with serrated profile as shown in Figure 18-10. The same equipment may be used for continuous electropolishing of strip or chains in order to electropolish deburr stamped perforations or superficial flakes.

Outlook

In Germany, there is no standstill in further development of electropolish deburring processes since every day brings new deburring problems and new products. New solutions for better alloys are requested as well as improved methods for neutralizing the resulting sludge and rinse water in order to meet the requirements of the water authorities. Therefore, nontoxic electrolytes have just been developed working at low temperature and with low current densities.

Due to the lack of skilled people for mechanical deburring and due to the ever increasing demand for higher surface quality and wear resisting materials, the electrochemical processes will become more and more interesting for the metal finishing industry.

FIGURE 18-11: ELECTROPOLISHING PLANT FOR CONTINUOUS TREATMENT OF WIRE, STRIP, AND CHAINS

FIGURE 18-12: NODULAR CAST IRON SPROCKET WHEEL FOR ENGINE CAM SHAFT

FIGURE 18-13: AUTOMATIC PLANT FOR ELECTROPOLISH DEBURRING OF SPROCKET WHEELS (4000 PIECES IN TWO DAILY SHIFTS)

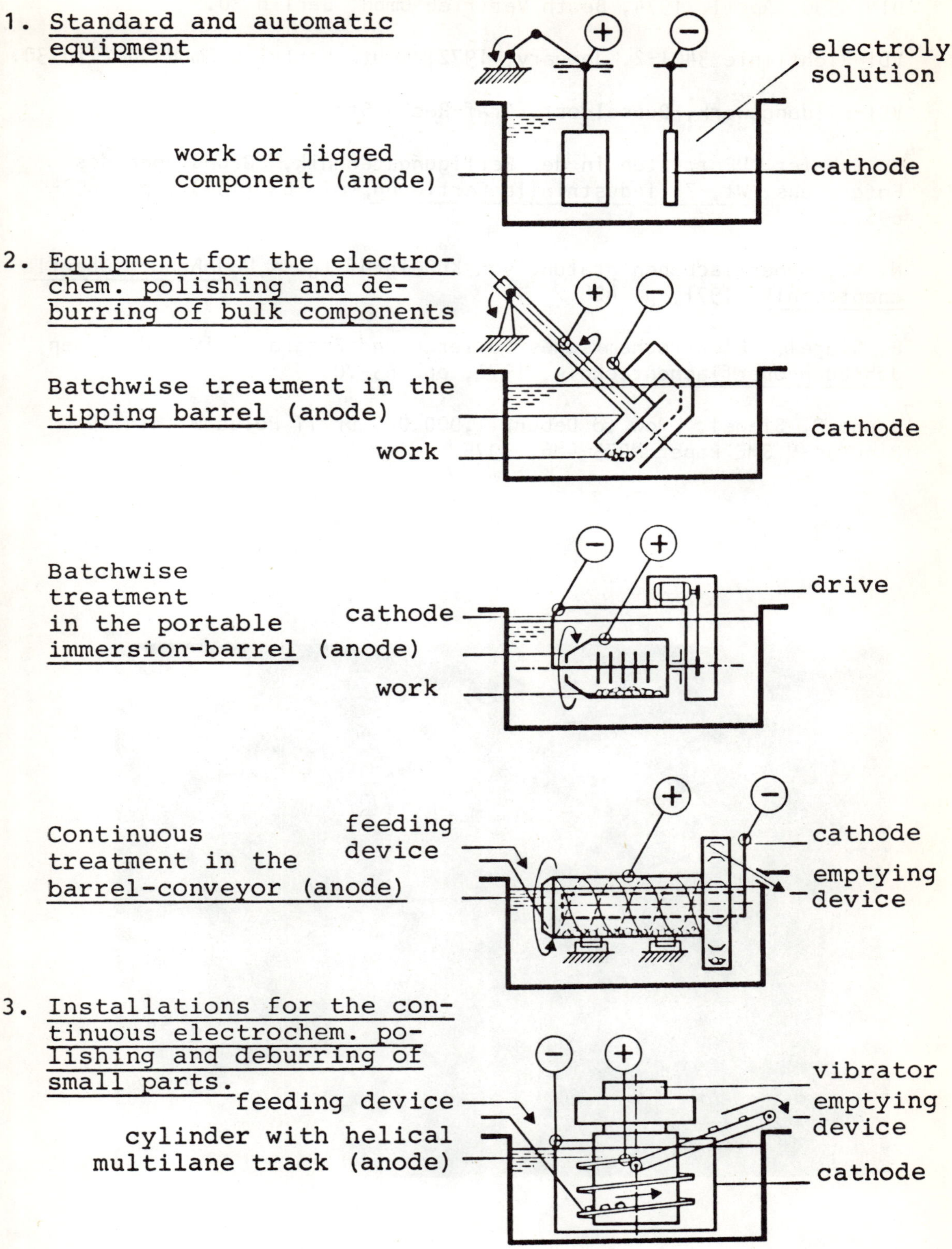

FIGURE 18-14: SCHEMATIC DIAGRAM OF ELECTROPOLISHING UNITS

References

[1]DIN 8590, April, 1974, Beuth Vertrieb GmbH, Berlin 30.

[2]VDI-Richtlinie 3401-2, January, 1972, Beuth Vertrieb GmbH, Berlin 30.

[3]VDI-Bildungswerk, Dusseldorf, Graf-Recke-Strasse.

[4]F. Schafer, "Entgraten in der Fertigungstechnik," Grundlagen des Entgratens, Wt,-Z. Industrielle Fertigung, Vol. 63 (1973), pp. 692-696.

[5]N. N.: "Oberflachenentgratung von Kugelgrafitguss," Jahrbuch Oberflachentechnik, 1971, p. 428.

[6]B. Siegel, "Elektrochemisches Polieren und Entgraten in Badanlagen," Jahrbuch Oberflachentecknik, 1971, pp. 63-70.

[7]Bruno R. Siegel, "How to Deburr 1,000,000 Drill Holes Within 15 Minutes," SME Paper MR75-486, 1975.

FIGURE 18-15: AUTOMATIC TIPPING BARREL EQUIPMENT FOR BATCH BULK ELECTROPOLISHING OF SMALL COMPONENTS (TOTAL MAXIMUM SURFACE PER BATCH APPROXIMATELY 7.5 SQ FT)

CHAPTER 19

LITTLE KNOWN DEBURRING PROCESSES

Introduction

Ten to twenty-five years ago, several deburring and finishing processes were introduced but never developed into widely used processes. Among these were chemically assisted barrel tumbling, liquid hone deburring, chlorine gas deburring, ultrasonic, and flame deburring. Although these processes are or could be in actual use today, little information about them has been published. The following paragraphs summarize some of the known information about these processes. In addition, the essentials of a new process, electrochemical brush deburring, are described.

Liquid Hone Deburring

Liquid hone deburring is a process in which an abrasive carried by water is forced over burr laden edges (Figure 19-1). Typically a 120 or finer grit of silicon carbide or aluminum oxide is used in a 50 psi (345 kPa) stream of water. Burr removal is the result of gradual erosion as the abrasive particles pass over the burrs.

The process is similar to yet distinctly different from both abrasive jet deburring and abrasive flow deburring. This process does not work by the high velocity impact of a concentrated stream as does the abrasive jet process. In liquid hone deburring, the media is forced over the edge rather than at the edge. In most applications, tooling contains and directs the water flow. Liquid hone deburring uses water to carry the abrasive particles but the water provides little force to support the particles. In abrasive flow deburring the unique properties of the silicone rubber and other carriers provide a soft "backing" for the abrasive as it works. In addition, with this type carrier, the work is concentrated at points in which flow is constricted.

The few applications in which this has been used on metals involved the removal of minute burrs from drilled holes. In two instances, the holes were small (0.060 inch (1.52 mm) diameter or smaller). In addition, the burr was the result of reaming, boring, or lapping of an adjacent surface. In the case of the lapping burr, the hole edge had to be burr free, yet no radius larger than 0.0001 inch (2.54 μm) was allowable. This requirement was met by forcing the water and abrasive mixture out the side which had the burr. In another application, the intersection of six 0.060 inch holes had to be deburred in a stainless steel part. By using a one to three minute cycle time, it was possible to remove thin reaming burrs.

Printed circuit boards are frequently "defuzzed" by this process. In this case, the abrasive removes a thin burr left from sanding the copper-

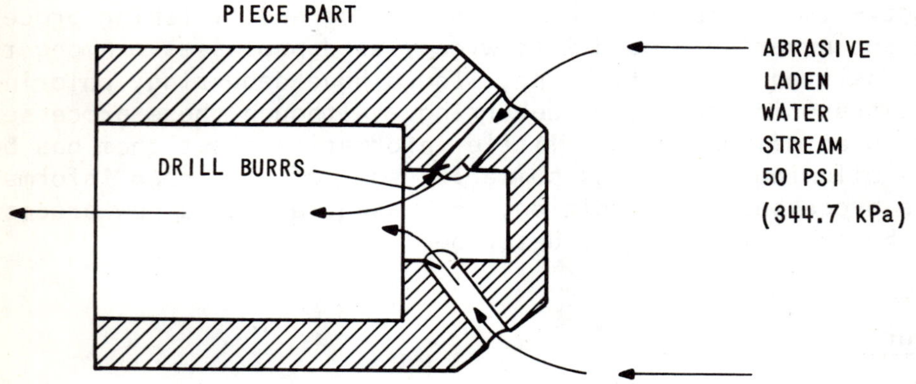

FIGURE 19-1: LIQUID HONE DEBURRING

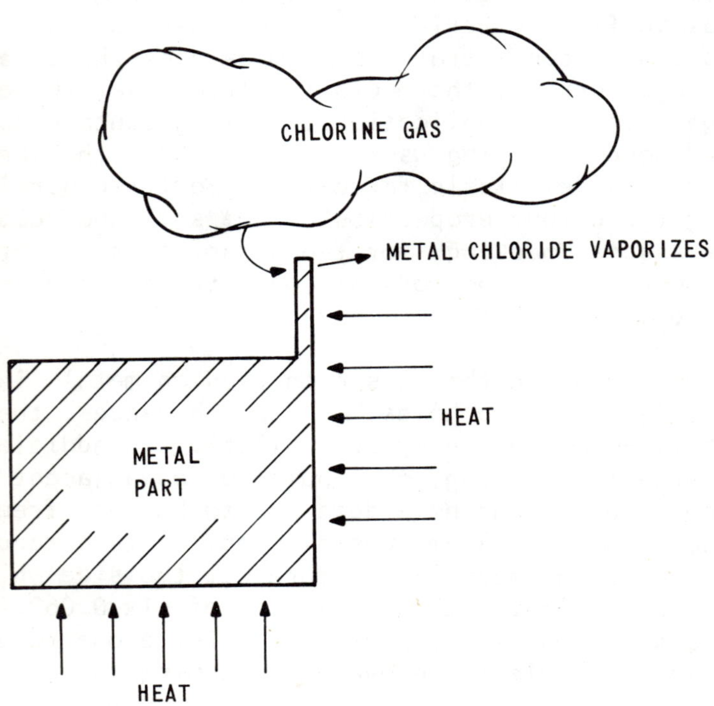

FIGURE 19-2: CHLORINE GAS DEBURRING

clad surface and as the media passes over the loose glass fibers in the holes, it knocks them out.

In general, it appears that this process will remove metal burrs of 0.0001 to 0.0005 inch (2.54 to 12.7 μm) thickness. No measurable stock loss occurs. No surface finish changes will occur unless the material is a soft aluminum or copper. Although edge radiusing up to 0.005 inch (127.0 μm) is a possibility, typical radii are in the order of 0.001 to 0.002 inch (25.4 to 50.8 μm). Because through-flow is required, this process will not do blind features.

The name of this process is not specifically connected with equipment of the same name. Commercial wet abrasive blasting can be used with suitable tooling. Small tubing pumps can also be used with a suitable holding and mixing chamber.

Chlorine Gas Deburring

In 1963, researchers at the Illinois Institute of Technology announced the results of their research on the use of chlorine as a deburring media.[1,2] Since that time, little has been heard of that process.

Chlorine reacts with many metals to produce a metal chloride which vaporizes at a relatively low temperature (Figure 19-2). Iron chloride for example vaporizes below 600° F (315° C). Thus by heating a burr to 600° F in a chlorine atmosphere, the burr essentially is vaporized.

In practice, the samples to be deburred are placed in a closed container and heated rapidly by quartz lamps or other means. Chlorine gas is then introduced which reacts with the warm metal to form the metal chlorides that immediately vaporize. Because of the relatively high surface-to-mass ratio of burrs, they heat much faster than the rest of the sample. Thus, material is removed from them much quicker than from other areas of the sample. Burrs produced in low carbon steel by drilling and by abrasive cutoff wheels were removed in 25 seconds when the temperature of the burr was 2720° F (1493° C). Low carbon steel melts at this temperature, but the part is not allowed to remain at that temperature long enough for this to happen.

Changes in workpiece size do result in this process. For example, Table 19-1 indicates how the diameter or thickness of a specimen will change in one minute as a function of temperature. It is significant that burr removal occurs in only a few seconds at high temperatures; thus Table 19-1 is not totally indicative of losses occurring in a deburring time and of stock losses. Burrs thinner than the 0.003 inch (76.2 μm) thick drilling burr described above could be removed at lower temperatures to provide the same stock loss or in a shorter amount of time at high temperatures.

Another limitation of the process is the safety which must be exercised when handling chlorine gas. Although chlorine is toxic, it is possible to construct a working unit in which it can be used. The process is limited to metals because it depends on low melting metal chlorides. The process does have some notable advantages.

- No tool is required
- No burrs result (many deburring processes, for example, produce a small burr)
- Workpiece complexity is not a factor in burr removal
- No residual stresses are induced (stressed surfaces are removed by this process)
- Surface finishes can be improved because the minute peaks heat faster than the rest of the part. (In one case a surface with a 1320 microinch (33.4 µm) finish changed to a 66 microinch (0.168 µm) finish in 90 seconds.)

TABLE 19-1: THICKNESS CHANGE RESULTING FROM CHLORINE GAS DEBURRING

Temperature (°F)	(°C)	Steel Removed (Inch)	(µm)
1000	(538)	0.0004	(10.16)
2000	(1093)	0.0040	(101.6)
2500	(1371)	0.0100	(254.0)

Flame Deburring

Oxy-acetylene torches have been used to trim flash, sprues, and gates from large cast pieces. Little information has been published to indicate where it is best used. The only obvious application is for the application just mentioned. The new miniature torches which can produce a flame only 0.003 inch wide could conceivably be used to deburr small parts particularly if only the thin loose portion of burrs had to be removed. In a related approach, Ivanov and Zhuravitsky[3] describe this application of air-arc machines for removing flash from steel and alloy castings.

Several limitations of the process are obvious:

- A heat-affected zone will be left around edges
- Fixturing is required as is some method of flame travel
- Complete removal of resolidified droplets might not occur

Chemical Vibratory and Chemical Barrel Finishing

In 1956 Sachs and Odgers[4] presented results of their study on barrel tumbling mild steel parts in a solution of oxalic acid, hydrogen peroxide, and sulphuric acid. Substantial edge radiusing occurred on these parts after seven half-hour treatments. Although this study was not directly oriented to deburring, its authors noted that material is removed preferentially from edges, corners, and projections. Surface finishes improved from 180 microinches (4.57 µm) to 20 microinches (0.51 µm) after 20 of these half-hour treatments. Thickness losses were 0.001 inch (25.4 μm) per side per hour.

In 1965, Shainskii[5] added copper sulphate to steel parts in a vibratory finishing cycle. The copper passivating film which forms when the iron dissolves is removed almost instantaneously by the vibratory media. Thus chemical and mechanical actions combine to increase metal removal rates 6 to 12 times.

In 1967, Blagodeteleva[6] extended the basic idea of chemicals plus motion to the vibratory finishing of bronze components. In addition to deburring, the process provides notable surface finish improvements. Babichev and others also reported the results of six chemical accelerators on steel burrs in the same year.[7]

In 1972, Safranec and Miller[8] indicated that a tenfold increase in metal removal rates could be obtained using chemical accelerators on zinc die-cast parts in a vibratory finishing process. Although deburring was not the main objective of the studies, Safranec has indicated that flash up to 0.02 inch (0.5 mm) thick has been removed using chemical accelerators. Surface finishes as fine as 4 microinch (0.10 μm) have been produced in a 45 minute cycle. Stock losses of 0.0003 to 0.0040 inches per hour (76.2 to 101.6 μm/hr) accompanied this finish improvement.

Although no actual data on deburring has been published for this process, it is at least logical to assume that deburring rates can be increased by at least ten times on steel, bronze, and zinc parts. Using Sachs and Odgers' data as a guide, surface finishes will improve to 20 microinch (0.51 μm). Stock losses however probably would be higher than in conventional vibratory finishing or barrel tumbling.

Ultrasonic Deburring

Ultrasonic deburring utilizes ultrasonic cavitation and a specially compounded acid slurry. The slurry is a combination of a comparatively weak etch solution and a small amount of abrasive particles. The ultrasonics provide high energy shock waves which keep the abrasives suspended and constantly bombarding the workpiece edges. The ultrasonics also increase the etching action. Apparently the combination of etchant, ultrasonics, and abrasives attack edges more than surfaces because of the microstructure changes in the area of the burr. The burrs are strain hardened material and strain hardening results in some microstructure changes.

Although first introduced in 1953, little information has been published on the process. The constituents of the slurry appear to be a closely guarded secret although a recent Soviet publication[9] does provide some data.

Wright[10] describes a number of successful applications of this process. These include deburring of brass, paliney, 416 stainless steel, 4130 steel, osmium, and other metals. Edge radiuses from 0.002 to 0.005 inch (50.8 to 126 μm) are quoted by Wright with radiusing controlled to 0.0001 inch

(2.54 μm). The surface finish on a 416 stainless steel part with 0.002 inch (5.08 μm) total tolerance was improved from 16 microinch (0.41 μm) to 8 microinch (0.02 μm) at a rate of 63 parts per hour. Other parts were deburred at a rate of 20,000 parts per day. Stock losses varied from 0.000005 to 0.001 inch (0.127 to 25.4 μm) depending upon material and time. However a more recent article[10] indicates that stock losses are always at least 0.0001 inch (2.54 μm) and typically are 0.0002 to 0.0003 inch (5.08 to 7.62 μm). Based on the applications described in the most recent articles the process is generally limited to thin burrs (in the order of 0.0005 inch (12.7 μm)). Honing burrs or small burrs on stamped electronic components are mentioned in several articles as being typical applications.

Korshunov[9] is the only author to have provided complete details of his study. Low carbon steel specimens were ultrasonically deburring in a glycerine, water, and 0.0002 inch (5 μm) diamantine (ruby waste) solution. The production rates were increased by using 6 to 8 atmospheres pressure in the working chamber.

Burrs on stamped parts were reduced from 0.0022 inch to 0.0004 inch (55 to 10 μm) in 25 minutes in this solution. A 19 kHz frequency was used on these parts with a gage pressure of 3 atmospheres. Optimum burr removal occurred with particles of 0.00012 to 0.00079 inch (3 to 20 μm) size. The addition of a two percent solution of calcinated soda increased the metal removal rate by 25 percent. Stearic acid and ethyl alcohol increased efficiency by 33 percent. A two percent solution of copper sulphate increased metal removal by 93 percent. The addition of a five percent solution of sulphuric acid and one percent of joiner's glue increased efficiency by 142 percent.

Electrochemical Brush Deburring

In 1973, Wicht and Kurze[12] described a unique variation of electrochemical deburring. In this process a rotating conductive brush is used in place of a shaped electrode. The belt driven brush spindle is connected to the negative side. As in conventional brush deburring, a variety of brush styles can be used. Although little data is given in this article, a 3000 percent increase in productivity is quoted.

Based on EC deburring and brush deburring capabilities, the advantages of this process would include speed, minimal tooling costs, stock removal concentrated at edge, edge radii capabilities from 0.001 to 0.030 inch (25.4 to 762 μm), and some localized surface improvement. Like electrochemical grinding, the percent of deburring action produced by mechanical brushing could be varied from 0 to 100 percent. Deburring time would be almost independent of burr thickness and size (load/unload time would exceed deburring time in most cases). The disadvantages would include some stray etching.

Additional Observations

One major limitation to the production use of these processes is the inability to purchase off-the-shelf equipment and supplies. In addition, in most cases it is difficult to find individuals who can provide information and assistance on these processes. Patents on some of the processes, equipment, or materials further limit the immediate production usage. Thus the use of the processes may require extensive in-plant experimentation.

Despite these limitations, these processes fill some potentially significant voids in the technology of burr and flash removal. Hopefully, within the next few years researchers and production engineers will document the actual capabilities and applications of these processes. Without some additional information on the capabilities of these processes they cannot fulfill their full potential. With an estimated U.S. expenditure of two billion dollars a year on burrs, it is important that every process near its potential.

References

[1] "New Field Opened by Chlorine Deburring," Steel, November 4, 1963, pp. 115-116.

[2] "What's New in Materials," Materials in Design Engineering, Vol. 59, January, 1964, pp. 111-113.

[3] B. G. Ivanov and Yu. I. Zhuravitsky, "Firecutting in the Trimming and Cleaning of Castings," Litelioe Prozv., No. 12, December, 1967, pp. 9-11 (in Russian).

[4] K Sachs and M. Odgers, "Smoothing of Mild Steel by Barrel Treatment in Oxalic Acid-Hydrogen Peroxide Solution," Transactions Institute of Metal Finishing, Vol. 33, 1956, pp. 245-269.

[5] M. E. Shainskii, et. al., "The Vibratory Cleaning and Polishing of Parts," Russian Engineering Journal, 1965, Vol. 45, Issue No. 9, pp. 66-70.

[6] E. S. Blagodeteleva, "Aqueous Solution for Vibrational Deburring of Sharp Edges," Soviet Patent Number 236, 183, Filed 6/6/1967.

[7] A. P. Babichev, et. al., "Accelerating the Vibratory Cleaning Process," Russian Engineering Journal, Vol. 47, 1967, No. 12, pp. 43-45.

[8] W. H. Safranek and H. R. Miller, "Vibratory Finishing With Chemical Accelerators," Plating, Vol. 59, January, 1972, pp. 38-40.

[9] B. S. Korshunov, "Ultrasonic Flash--Removal From Small Components," Russian Engineering Journal, Vol. 49, 1969, No. 12, pp. 45-46.

[10] John P. Wright, "Ultrasonic Deburring," American Machinist, Vol. 101, February 11, 1957, pp. 129-136.

[11] "Sound Waves Strip Off Burrs," Iron Age, August 5, 1965, pp. 62-63.

[12] Hans Wicht and Peter Kurze, "Deburring Still a Difficult Problem?" ZIS Mitt., Vol. 15, No. 8, August, 1973, pp. 930-935 (in German).

CHAPTER 20

AUTOMATION IN DEBURRING

Within the past ten years the automation of deburring equipment has received considerable attention. Today a wide variety of semiautomated and fully automated equipment is available. This chapter describes automatic equipment for loose abrasive processes.

In an effort to improve deburring, manufacturing innovators originally largely emphasized finding new processes. As shown in Figure 20-1, barrel tumbling eventually developed into a field of ten related processes.[1] The next approach to increasing production was to develop faster cutting compounds and media such as the microcrystalline abrasive media,[2] plastic media, and a variety of ceramic bonded preform shapes. As indicated in previous chapters, some innovators turned to chemical additives and electrochemical assists to further increase output. The latest approach to improving productivity is to provide automated equipment.

Gyrofinishing or Spindle Finishing

Gyrofinishing (now called spindle finishing in the United States) was originally reported by Squibb and Hall.[3] In the original gyrofinishing, the velocity of abrasive was very rapid. Since this caused some problems in production, low velocity abrasive was used to achieve success. Many kinds of gyrofinishing machines were made and two examples are reported here. One is a fully automatic multispindle type which is shown in Figure 20-2. Many rotating spindles are installed under a turret which revolves at a low velocity. The revolving cycle is adjusted to run a desired finishing time. Only one or two spindles are lifted from the abrasive media in exchanging the workpieces. Therefore, it is unnecessary to lift all spindles in order the change the workpieces. The other type of gyrofinishing machine is of multispindle and multitub type (Figure 20-3) and is also automated including chucking.

Reciprocating Finishing

Figure 20-4 illustrates an automatic recipro-finishing machine. The fixtures which carry workpieces are automatically loaded and unloaded and positioned in the machine and in pretreatment and posttreatment containers. Circulating fixtures have also been used.

Centrifugal Barrel Finishing

Automation has also been applied to centrifugal barrel finishing. One of the most difficult automation problems with this process involves the

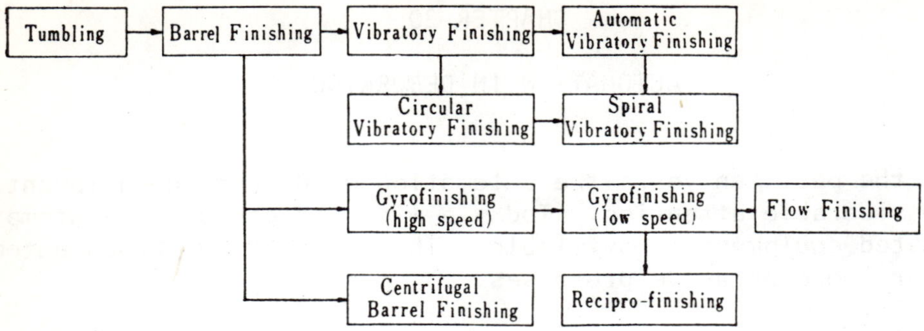

FIGURE 20-1: DEVELOPMENT OF BARREL FINISHING

tightening and removal of barrel lids. Figure 20-5 illustrates such a machine in which the lids are attached to one of the turrets. The machine contains a circular conveying system and an automatic measuring system for the initial charge of media and parts. Figure 20-6 illustrates a semi-automatic machine in which lid handling is not automated.

Bowl Type Vibratory Deburring Equipment

The first bowl type vibratory finishing equipment having automatic sieving was introduced in 1964.[5] Since that time a number of improvements have been made to such equipment. One of these is a closed type machine with a narrow outlet which gives a compressive force to the upstream part of the mass to improve finishing and sieving efficiency. A schematic diagram is shown in Figure 20-7. Note that $b \geq a > c$, to provide the compressive forces. A bowl type machine is shown in Figure 20-8.

Results of Automated Deburring

Table 20-1 and associated Figures 20-9 and 20-10 provide some background information on workpieces subjected to the processes described and the time required. As mentioned in several previous chapters, heavy burrs will not be removed by these processes without affecting dimensions.

Summary

While only automation of mass finishing processes is described in this chapter, similar automated techniques have been applied to all the other deburring processes which are in common use. Brushing, buffing, and sanding are some of the most widely automated processes which can easily be tailored to individual part requirements.[6] With the emphasis on increased

FIGURE 20-2: AUTOMATIC MULTISPINDLE GYROFINISHING MACHINE

FIGURE 20-3: AUTOMATIC MULTISPINDLE AND MULTITUB TYPE GYROFINISHING MACHINE

FIGURE 20-4: AUTOMATIC RECIPRO-FINISHING MACHINE

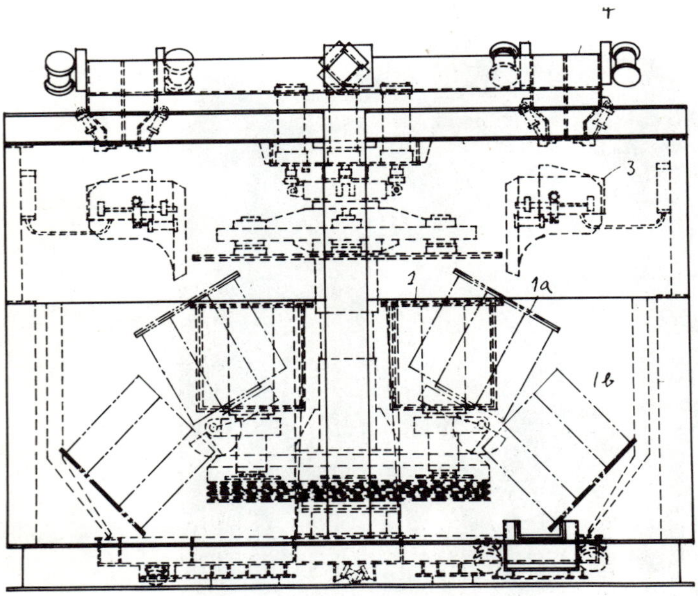

FIGURE 20-5: AUTOMATIC CENTRIFUGAL BARREL FINISHING MACHINE WITH INCLINED TUBS 1, RING-FORMED CONVEYING SYSTEMS 2, 4 FOR FINISHED AND UNFINISHED MASS RESPECTIVELY, AND MEASURING APPARATUS 3 FOR UNFINISHED MASS.

FIGURE 20-6: SEMIAUTOMATIC CENTRIFUGAL BARREL FINISHING MACHINE

FIGURE 20-7: A SCHEMATIC DRAWING OF A VIBRATORY FINISHING MACHINE WHICH AFFECTS A COMPRESSIVE FORCE TO THE MASS

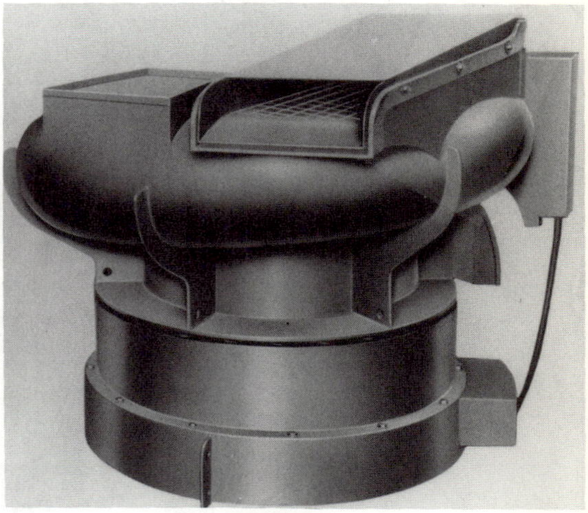

FIGURE 20-8: PHOTOGRAPH OF THE MACHINE

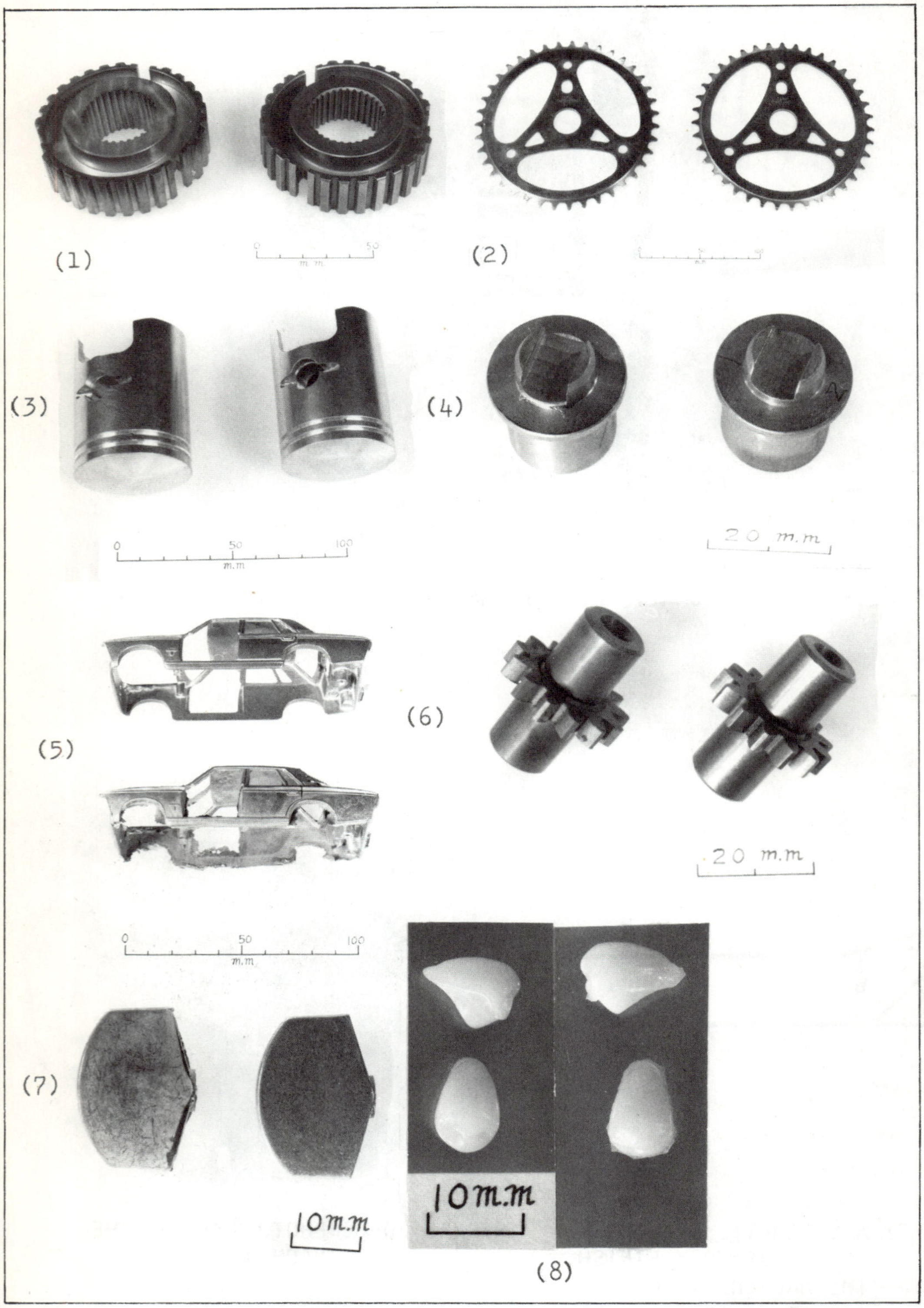

FIGURE 20-9: EXAMPLES FOR DEBURRING BY BARRELS

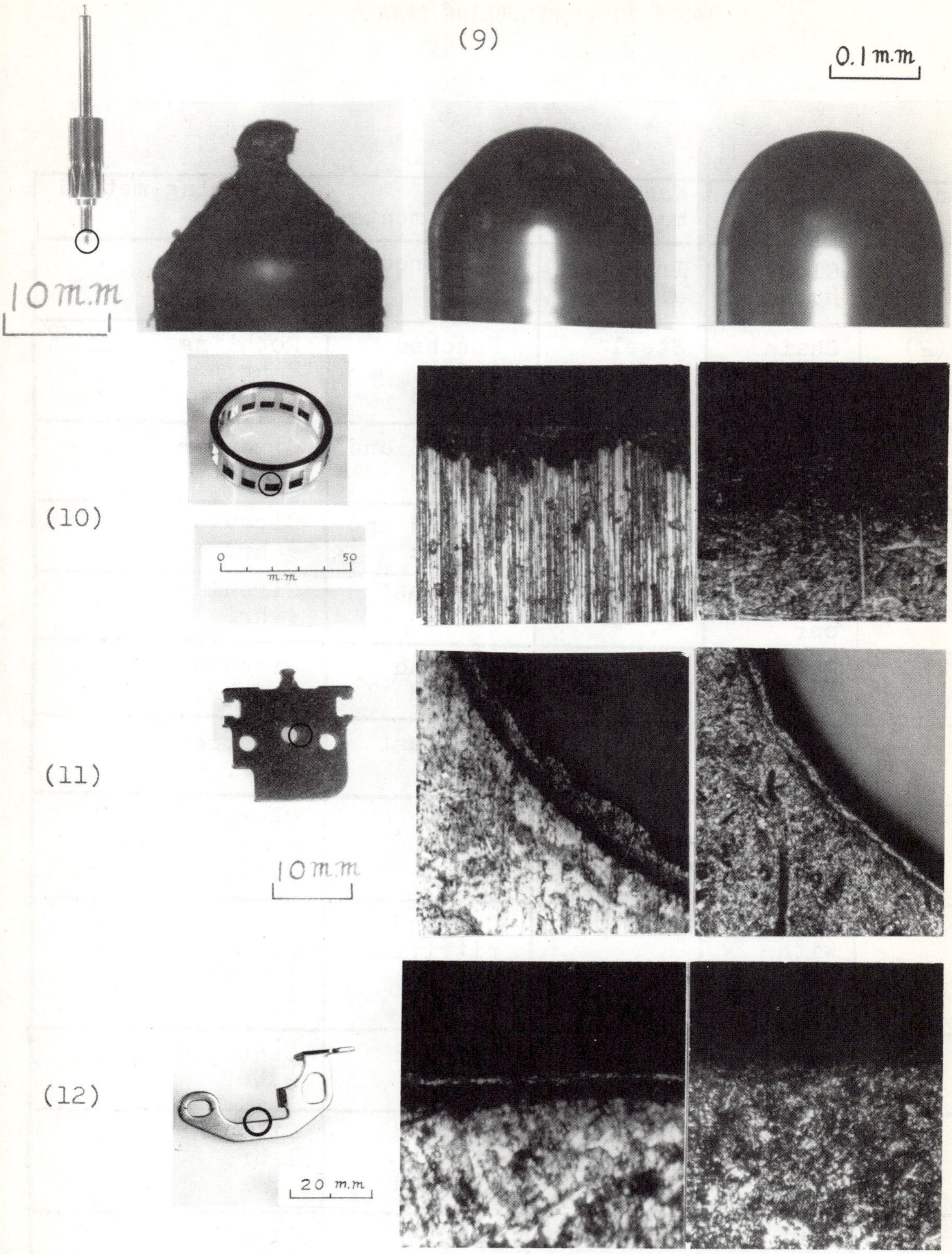

FIGURE 20-10: EXAMPLES OF DEBURRING BY BARRELS AND THEIR PHOTOMICROGRAPH (LEFT: WORKPIECES, CIRCLES INDICATE MAGNIFIED POSITIONS: CENTER: BEFORE DEBURRING; RIGHT: AFTER DEBURRING). A SCALE FOR PHOTOMICROGRAPHS IS INDICATED AT THE TOP OF THIS PAGE.

TABLE 20-1: DEBURRING DATA

Number	Name of part	Name of material	Pre-treatments	Deburring method
(1)	Clutch hub	Iron alloy	Sintered	Recipro 10 min.
(2)	Chain wheel	Steel	Punched and milled	Rotating 2 hr.
(3)	Piston	Cast aluminum	Turned and milled	Circular 5 min.
(4)	Brake piston	Steel	Turned and milled	Circular ½ hr.
(5)	Model car	Zinc alloy	Die-cast	Circular ½ hr.
(6)	Pinion	Steel	Cut and shaved	Circular 3/4 hr.
(7)	Knob	Zinc alloy	Die-cast	Rotating 3 hr.
(8)	Arti-ficial teeth	Plastics	Mold	Centrifugal 2 operations 40 min. & ½ hr.
(9)	Pivot with pinions	Steel	Cut and nitrified	Centrifugal 3 operations ½, ½ & 1 hr.
(10)	Retainer	Alminum pipe	Cut and milled	Centrifugal ¼ hr.
(11)	Part for instru-ment	Steel	Punched and pressed	Circular 1 hr.
(12)	Part for instru-ment	Steel	Punched	Circular 1 hr.

efficiency, wider use of automation is apparent for part-media separation, part load/unload, media recirculation, and part cleaning and drying.

References

[1] M. Matsunaga and Y. Hagiuda, "Researches on Barrel Finishing," Report of Inst. Ind. Sci., Tokyo University, 17-4, 1967, pp. 106-155.

[2] M. Matsunaga, "Microcrystalline Abrasive Media for Mass Finishing," Metal Finishing, 69-5, 1971, pp. 97-104.

[3] G. R. Squibb and F. T. Hall, "Gyrofinishing," Metal Ind., 19-11, 1956, p. 333.

[4] M. Matsunaga and H. Kobayashi, "Some Experiments on Centrifugal Barrel Finishing," Metal Finishing, 64-5, 1966, pp. 57-60.

[5] T. W. Black, "Vibratory Finishing Goes Automatic," Tool and Manufacturing Engineer, 58-8, 1964, pp. 72-76.

[6] M. Matsunaga, N. Kitayama, S. Veno, and H. Kobayashi, "Automation in Buffing and Barrel Finishing," SME Paper MR75-481, 1975.

CHAPTER 21

TRENDS AND NEEDS IN DEBURRING

In the past 100 years, 18 deburring processes have been added to the then-established processes of hand deburring and barrel tumbling (Figure 21-1). Five of these processes have been introduced in the past ten years. Despite the introduction of these new processes and the continuing improvement of the older processes, worldwide deburring costs amount to billions of dollars annually. That is billions of dollars to remove something that no one wanted in the first place.

Despite the large amounts spent for deburring and for solving burr related problems, little actual research has been devoted to this area which we will call the field of burr technology. The advances that have occurred since 1965 have largely been the result of evolutionary changes requested or dictated by the buyer. These changes, while substantially reducing deburring costs for some companies, have not required any indepth analysis of burrs or deburring. The following discussion is a summary of formal research and equipment trends observed in industry and a description of some needed approaches.

The Nature of Research on Deburring

As seen in Figure 21-2, deburring is actually only one facet in the attempt to minimize the costs of burrs. Burr technology involves burr prevention, control of burr properties, definition of allowable burrs, and deburring. As seen in Figure 21-3 however, the development and improvement of deburring processes is the only facet of burr technology which has received any real research emphasis. The work which has been performed in this field has primarily been reported in a qualitative sense rather than on a quantitative basis. For instance, manufacturers of tumbling media have spent considerable effort in improving the aggressiveness and life of their media. Their results however are reported as fastest cutting, longest life, etc., rather than in quantitative terms. The individual responsible for buying the most economical media from a number of competitors has no quantitative basis for making his decisions without performing his own tests. From the standpoint of defining quantitative capabilities or improvements, none of the basic approaches to minimizing the cost of burrs has received any notable emphasis (Figure 21-4).

As seen in these charts, the research in burr technology areas has been piecemeal. Some of the research has never received any notice by the industrial complex. As will be pointed out later in detail, many bits and pieces of knowledge have been generated but researchers and users have not taken an indepth systems approach to defining what is needed nor have they tried to compile the existing knowledge into an understandable package.

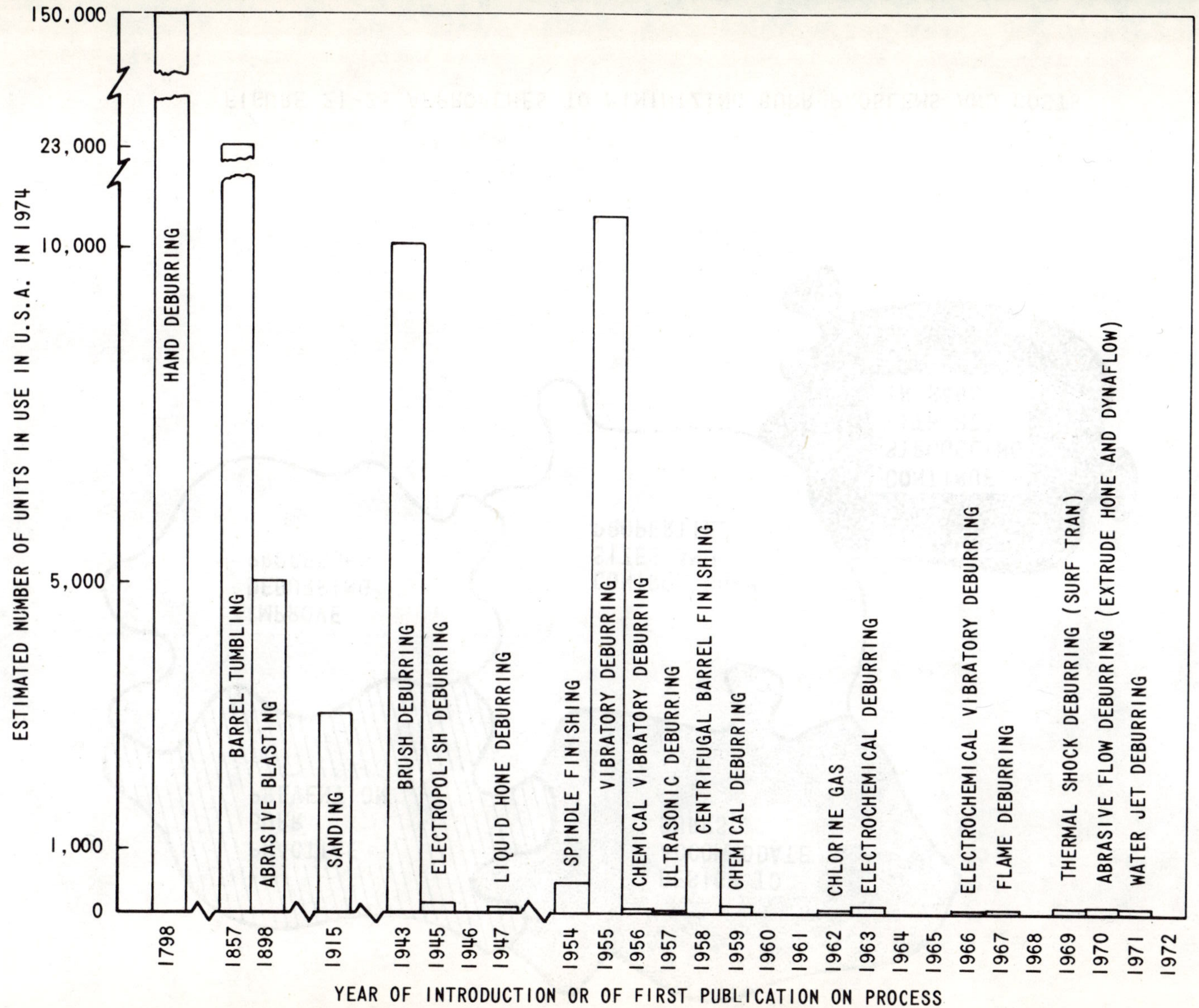

FIGURE 21-1: YEAR OF COMMERCIAL INTRODUCTION OR FIRST PUBLISHED ARTICLE ABOUT PROCESS

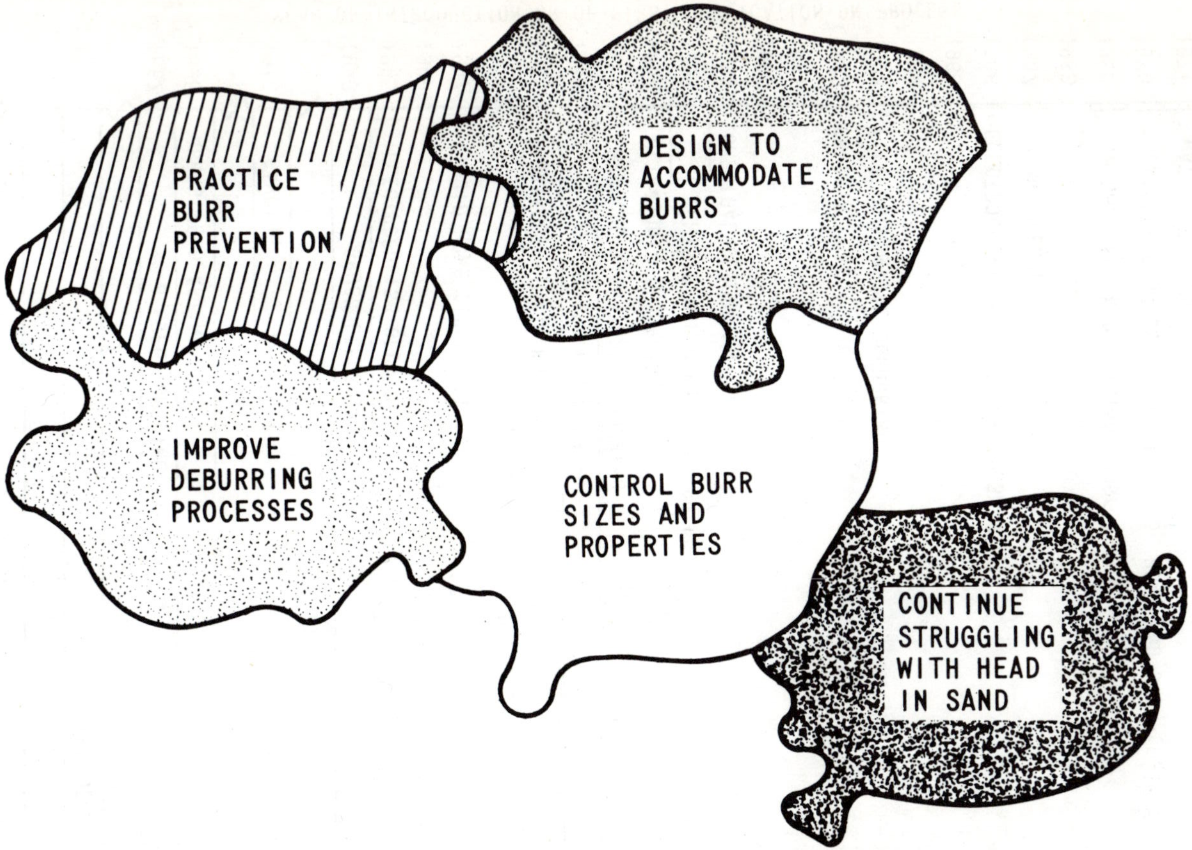

FIGURE 21-2: APPROACHES TO MINIMIZING BURR PROBLEMS AND COSTS

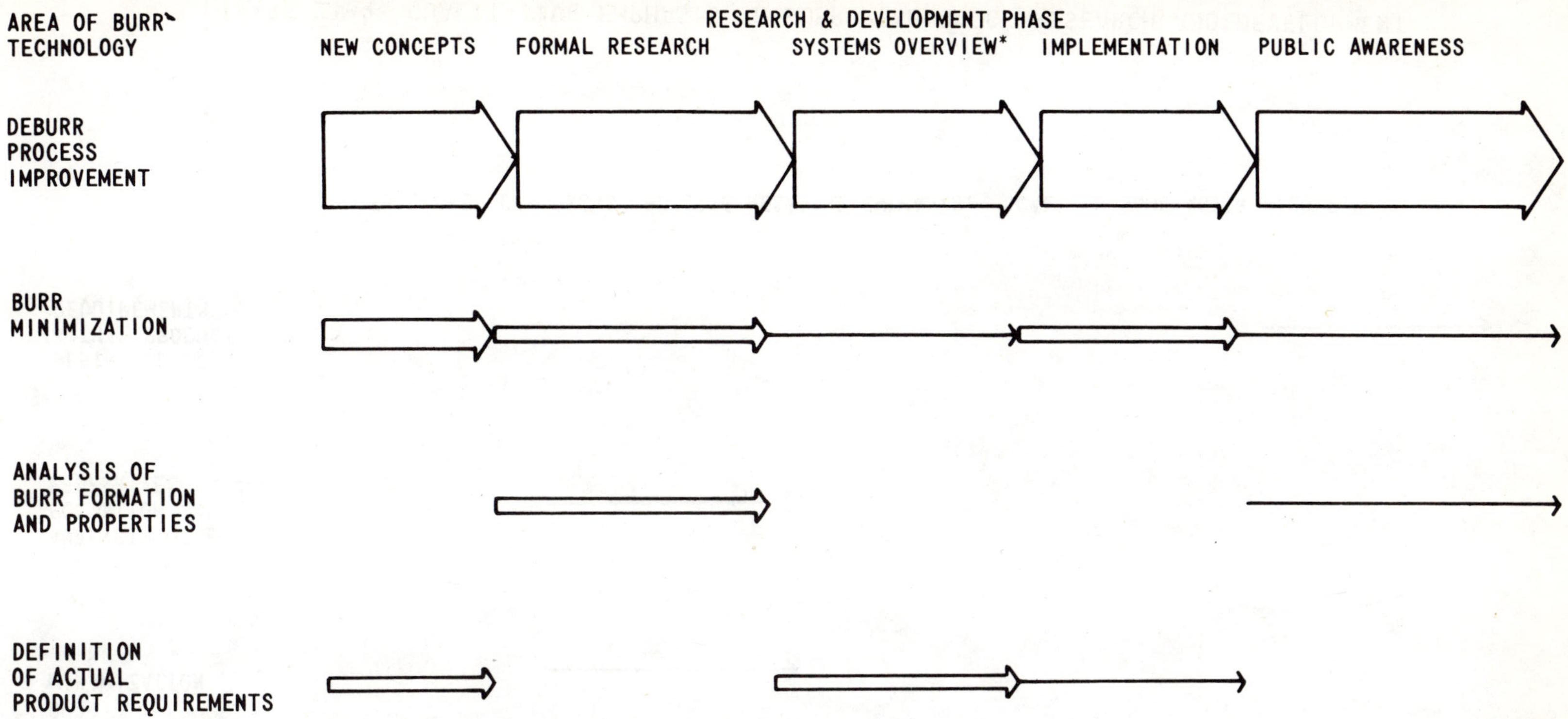

THICKNESS OF ARROWS INDICATES RELATIVE EMPHASIS.

*IN-DEPTH APPRAISAL OF WHERE THIS CONCEPT OR KNOWLEDGE PROVIDES THE GREATEST INDUSTRIAL BENEFIT.

FIGURE 21-3: QUALITATIVE EMPHASIS IN BURR TECHNOLOGY RESEARCH AND DEVELOPMENT

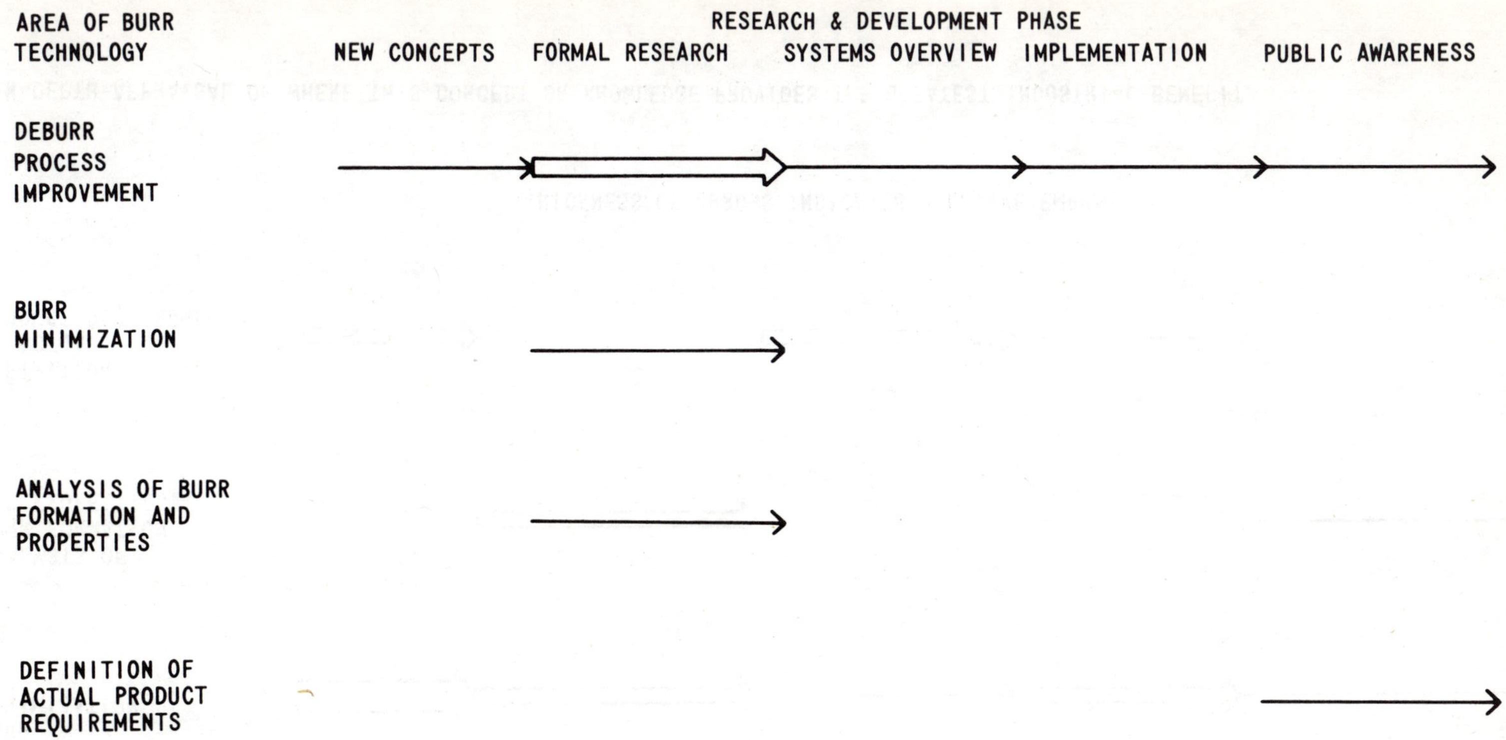

THICKNESS OF ARROWS INDICATE RELATIVE EMPHASIS. SCALE IS SAME AS IN FIGURE 3.

FIGURE 21-4: QUALITATIVE EMPHASIS IN BURR TECHNOLOGY RESEARCH AND DEVELOPMENT

The research which is reported in the following pages indicates that a trend is developing to provide a better understanding of the various facets of burr technology. The result of this work should provide better guidelines for the prevention, size minimization, and removal of burrs. In addition, in the next decade one should expect some improvement in the definition of cases in which burrs can be allowed.

Burr Formation and Properties

Stamping Operations

The majority of work performed on documenting burr properties has been for the blanking and piercing operations. Wukusik and Zeno's 1958 study[1] appears to be one of the earliest. Since that time studies or reports containing quantitative data have appeared almost yearly for these types of operations.[2-20] The majority of these reports illustrates either the effect punch-to-die clearance has on burr length or the increase in burr length which occurs as the die wears. A brief summary of these papers has been published elsewhere.[21]

While the recent work in the area of stamping burrs has provided a notable improvement in the understanding of die geometry effect on burr height, it is obvious that some of the results do not agree with those found in other tests. In addition, three aspects of our knowledge of stamping burrs are missing. Firstly, no quantitative model of burr formation in stamping operations has yet been offered. That is to say that no mathematical equations have been developed from an analytical base which describes burr formation. Until this is done, researchers will have to provide empirical results for each material of interest. Secondly, no mathematical models, analytical or empirical, have been published that will predict burr size as a function of tool geometry, material, and number of press strokes. Thirdly, no work has been reported on the burrs produced by fine edge blanking. Two notable aspects of the work which has been performed on stamping burrs are that the stamping industry has established an accepted definition of what a burr is and that it has developed convenient methods and techniques to measure burr size. Both of these are lacking in the metal cutting field.

Machining

The majority of papers written on burrs produced in machining operations involve drilling processes. Several of these papers are summarized in Chapter 4. A very brief summary of papers on other machining burrs is presented elsewhere.[21]

Burrs Produced by Processes Other Than Metal Cutting or Stamping

While little work has been reported on the specific subject of minimizing the flash produced in casting processes, the author believe this information is generally available in numerous publications on the casting processes. A special publication on this specific subject would, however, help provide a more complete understanding of the problems associated with minimizing these burr like projections.

The author has seen no data on the burrs (upset metal) produced by friction welding, flash welding, or magnetarc welding. Such data would seem as useful for minimizing input energy requirements as for minimizing the effort required to remove these burr like projections.

Equipment for Minimizing Burrs

To the author's knowledge only two inventions are being marketed for the purpose of minimizing machining burrs. The radial lip drill is one of these. With proper design it has been shown to effectively produce smaller drilling burrs than other drills. The second product is a spring loaded drill guiding mechanism[22] for use on portable drilling tools. The drill guide prevents drill walking and workpiece "cork-screwing" up the drill flutes. The spring loaded portion minimizes the thrust at drill breakthrough. By minimizing the thrust, the burr height is apparently minimized.

In assessing areas in which additional study is needed, three areas are obvious. Little effort has been applied to the non-hole producing processes. In particular, greater research effort is required in the milling processes. The grinding and turning processes generally produce relatively easy-to-remove burrs and are thus notably less significant than those of milling. The second little-studied area is that of providing analytical models of burr formation. Burrs from all processes are produced by the same basic mechanisms. Developing better mathematical models of these mechanisms will help minimize the number of empirical tests which must be performed. Information is needed on the burrs formed in plastic materials. To date, not a single study has been reported on the burrs produced in plastic parts. While most plastic features are cast, enough are machined that burrs are a significant problem.

Burr Prevention and Minimization

While the previously described work on burr formation and properties will indicate parameters which minimize burrs, little formal work has been published on burr minimization per se. More specifically, the use of low temperature melting alloys will minimize some burrs but their capabilities and limitations have not been formally explored in print. The same is true of similar approaches of burr minimization.

While some authors have written articles on techniques or devices that do not produce burrs, a closer examination reveals that a protrusion still exists at the edges. It is the author's opinion that all conventional processes must produce burrs. However, there is a distinct need to explore carefully the potential and the economics of the nontraditional processes as approaches to burr prevention.

For a field which offers so much promise for minimizing the cost of deburring, it is difficult to understand why these two subjects have received such meager attention.

Piecepart and Inspection Standards

Throughout the world, companies have wrestled with the problem of defining acceptable and unacceptable burrs; yet only one of the 1500 publications on deburring even touches on this subject.[23] No publication has delved into the implementation of these standards. It would certainly appear that a compilation and discussion of in-plant standards of defining acceptable conditions is past due.

Deburring Processes

As previously indicated, a number of improvements have been made recently in the deburring processes. Most of these advances have been reported in the trade literature but not in the technical literature. In essence, some of the advances are obviously improvements, but in almost all cases, it is still difficult to define just how significant these changes can be to specific plants.

Apparently only two groups in the world have attempted to analyze the comparative performance of the major deburring processes. F. Schafer, formerly at the Institut fur Produktionstechnik und Automatisierung in Stuttgart, Germany, has performed studies of electropolish, electrochemical, abrasive jet, and vibratory deburring. The end result of those studies will be a series of decision tables for selecting processes and parameters. Bendix (Kansas City Division) is completing a study of all deburring processes with an emphasis on obtaining quantitative comparisons of capabilities. At this time, only one report has been published by any group that even basically describes the capabilities of all 20 deburring processes and that is presented in Chapter 7. At this time there is only one other book which presents basic information on all the major burr removal processes, and it is in German.[24]

Loose Abrasive Deburring Processes

The majority of published advances have occurred in the field of loose abrasive processes such as vibratory deburring and harperizing. Matsunaga's studies[25-31] have provided the most comprehensive understanding of these processes. These include indepth empirical and analytical analyses of stock losses and part motion in vibratory finishing, barrel finishing, and centrifugal barrel finishing. Media velocity and dynamic forces on the parts are also analyzed for some cases. His observation that six-sided centrifugal barrels are three times more aggressive than round barrels has apparently not been put to use by American manufacturers. His study of media breakdown as a function of firing temperature and composition is the first of its kind.

In 1966, Robbins published a collection[32] of 120 curves illustrating the effect of machine, media, workpiece material, and time on edge radiusing in vibratory finishing. Some additional curves were published by the author in 1973.[33] A brief analysis of edge radiusing repeatability was included in this latter report as were studies of burr properties and their effects on deburring consistency. In a more recent publication,[34] the author has noted that the angle at which two surfaces meet to form an edge has a major impact on the time required to achieve a given radius.

The majority of the Soviet publications on deburring describe studies of loose abrasive processes. These studies include analyses of machine design and forces in vibratory deburring as well as studies of media and compound effects on deburring. Unlike many other studies, the authors of these describe some of the properties of the burrs which are to be removed.

Babichev[35] for example describes the relationship between metal removal, amplitude, frequency, operating time, volume of parts, media, abrasive compound, part weight, and container fill level for removing flash from plastic parts. A similar study by Babichev[36] analyzes the same effects for metal parts. The effects of chemical additives on surface finish are also described in this latter report. Shainsii's analysis of tub vibration[37] is one of the few brief but to the point analyses of machine design and action. His article also mentions the use of chemical accelerators. Bagmet[38] presents the results of a regression analysis of edge radius as a function of oscillation amplitude and frequency, treatment time and media composition. Kartyshev's study[39] of fixturing parts for vibratory finishing offers some interesting ideas for improving the efficiency of such processes. Veprintseva[40,41] presents data on effects of media filler material on vibratory finishing variables. One interesting observation of Soviet research in this area is that some researchers have utilized crushed grinding wheels in place of commercial vibratory deburring media. Less than 100 of the 1500 publications about burr related subjects were written by Soviet authors. While their work tends to be more quantitative than that of other authors, they too rely heavily on qualitative descriptions.

Although numerous equipment improvements have been made throughout the world over the past few years, the most significant have been in minimizing the cost of material handling. Flow-through systems now deburr, wash, separate parts from media, and dry and stack parts. Systems are also available which assure that delicate parts do not impinge on each other. At least one company is manufacturing hoods to muffle the noise from vibratory operations.

While considerable research has been performed on the loose abrasive processes, no one has yet compiled the data that is available into a format that can be used by manufacturing engineers to quickly select optimum deburring conditions. Similarly, no research has been reported in which the effect of burr size is related to workpiece shape, workpiece material, burr location, surface finish changes, resulting edge radius, media, abrasive compound, machine variables, and operation time. In short, no researcher has investigated all the parameters that must be considered when specifying deburring conditions. This has not been done for a single workpiece size, shape and material. Several researchers have developed data on most of the parameters, but those who have done so did not include any study of burr size effects. In discussing the state-of-the-art with suppliers of media and equipment, it is obvious that there is not even any agreement upon which deburring variables are the most significant.

Abrasive Blast Deburring

Beletskii and Zhemchugov[42] recently derived a formula for abrasive blasting that would be useful in deburring applications. This formula relates metal removal rate as a function of liquid abrasive pressure, inclination angle, nozzle diameter, abrasive concentration, and grain size. Jones and Gardos[43] performed a similar study for wet and dry blasting. The recent introduction of carbon dioxide and plastic pellets as blasting media testifies to the continuing interest in improving blasting media. As with all other processes, however, no data has been developed directly applicable to deburring needs.

Brush Deburring

Although numerous brush materials have been introduced in recent years few other advances have occurred in this field. Despite the widespread use of brushes for deburring gears only two papers have been published in 30 years which describe the burrs which can be removed by even a single brush.[44,45] One of these is contained in Chapter 14.

Electrochemical Deburring

In the past 12 years over 100 articles have been published on electrochemical deburring. As with other new processes, noticeable advances have been made in equipment used, but little documentation has been attempted. Bigony[46] describes effects of the process on hole size and resulting radius at the end of the holes. Chemform[47] has prepared a tool design manual for electrochemical deburring. Two particular equipment advances of note include complete automation of EC deburring and the use of brushes rather than machined electrodes. Wicht and Kurze[48] describe the possible applications of this electrochemical-brush deburring.

Deburring Processes For Which Little Research is Being Performed

In analyzing the work being expended to improve the processes shown in Figure 21-1, comparatively little effort can be noted for hand deburring, barrel tumbling, sanding, electropolish deburring, liquid hone deburring, spindle finishing, chemical deburring, and flame deburring. Because of the toxicity of chlorine, the chlorine gas deburring process is not being used for any known production deburring.

Deburring of Plastics

Of 1500 articles on deburring, the author has found only one which described the size of burr or flash removed from plastic components. While several companies are knowledgeable about polishing of plastics and blasting of flash, there does not appear to be a single individual in the United States, Britain, or Germany who could be called an expert in the deburring of plastics. Very few articles have ever been written on this subject and there seems to be no formal research being performed in this area.

Research Versus Production Implementation

The individuals and companies now involved in minimizing deburring costs are using at least one of the two basic approaches shown in Figure 21-5. Some have approached their problems through formal research. Many have made indepth appraisals of their production approaches and some have utilized both approaches.

Research

Much of the work currently being performed involves the formation and properties of burrs. These studies being performed by at least five independent companies or institutions all have as their basis the knowledge that control of burr properties is one of the most important approaches to minimizing deburring costs. A knowledge of how burrs form and equations for predicting burr size are essential for studies of this kind. Data are being developed on burrs produced by drilling, milling, turning, grinding, stamping, and other processes. In addition to data on burr properties, at

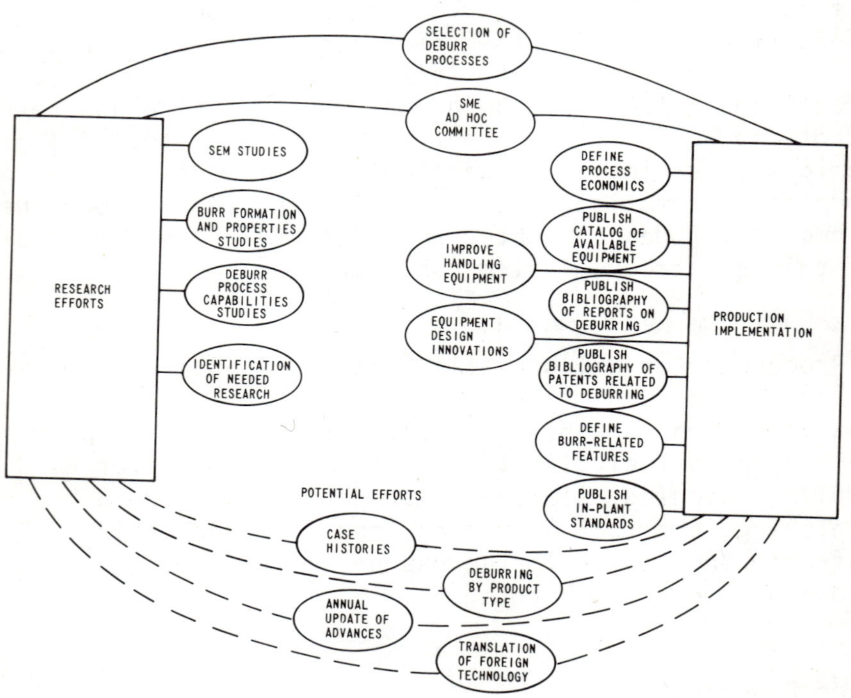

FIGURE 21-5: TRENDS IN MINIMIZING DEBURRING COSTS

least one study is also investigating effects of burrs on fatigue.

Three studies involve the use of the scanning electron microscope (SEM) to view the burr under extreme magnification. The tremendous depth of field of these instruments at magnifications up to 30,000X provide researchers with an investigative potential never before possible. In addition to verifying the physical mechanisms of burr formation, the SEM allows researchers to assess the role burrs may play in crack initiation.

One of the more promising avenues of burr formation studies concerns product geometry. As shown by Schafer in Chapter 2, burrs can be prevented or minimized by small changes in workpiece geometry. For example, Schafer observed that when a cutter exited over an edge having a 120 degree include edge angle no burr formed. A large burr formed when the edge angle was 90 degrees. This can result in major cost savings on cast parts on which some machining is required. This area of burr prevention has apparently been unnoticed by most researchers and users.

These research studies should provide industry with a better knowledge of ways to minimize burrs as well as a basic understanding of the relative economics of investing in deburring or in burr prevention or minimization. This latter field, which has never before been seriously explored, today includes researchers from Japan, West Germany, the United States, and Canada.

Research is also being performed on the quantitative capabilities of the deburring processes. Soviet, German, and U.S. researchers are trying to document the burr removal capabilities of several of the deburring processes. Although several authors have provided general information on the deburring processes, few have made any attempt to measure the properties of the burrs involved. The current trend by researchers is to analyze the process by its effect on burrs of varying thickness, length, and hardness. Some are also recording such side effects of the deburring processes as changes in surface finish, workpiece dimensions, and the characteristics of edge radiusing. While none of these studies will provide immediate cookbook solutions to all deburring problems, they will provide a rational base by which much of industry can make useful estimates. Eventually this work could form the nucleus for computerized selection of processes.

Some of the research results have already been published. Many of the studies have been summarized in chapters of this book. Soviet authors have published two volumes of research on mass finishing techniques,[3,4] in addition to at least a dozen brief but factual studies which have been translated into English. A review of Russian abstracts[49,50] implies that many additional studies have been published in Russian. Much of the English language information which is currently being obtained will be available to industry in 1976.

Although considerable research is in progress, no attempt has yet been made to identify in detail the research which needs to be performed. The newly created Society of Manufacturing Engineers Committee on Burr Technology has as one of its goals the identification of specific areas in which additional research is needed.

Production-Oriented Trends

Several efforts are being made to provide production facilities with better information and equipment for reducing deburring costs. These include providing equipment to minimize handling time and advances to minimize machine cycle time. Several approaches are being taken to provide timely, categorized information for deburring cost reductions.

The equipment advances fall into three areas:

1. Providing totally new processes and equipment,
2. Providing equipment to minimize part and media handling,
3. Providing equipment which deburrs at a faster rate.

As indicated in a previous chapter, such processes as chemical vibratory, electrochemical vibratory, and electrochemical brush deburring have recently been developed or improved enough to generate some noticeable industrial interest. Among some of the more recent innovations are a unique device for fixturing large parts in vibratory tubs,[51] a device for increasing the action in centrifugal barrel finishing,[52] the use of platinum gauze electrodes and magnetic fields for electropolish deburring,[53,54] the use of gas or liquid to support abrasive particles while an electric field moves the media past the parts,[55] and the use of magnetic fields to assist vibratory deburring.[56] The introduction of new brush materials in addition to new sanding equipment and supplies is expanding the capabilities of these older processes. The introduction of plastic coated steel balls, plastic blasting pellets, and the ice bonded vibratory finishing media further broaden the capabilities of tumbling, vibratory, and abrasive blast deburring.

The complete automation of vibratory, electrochemical, and electropolish deburring is providing noticeable savings for a number of companies. Brush deburring, power sanding, mechanized gear deburring, and abrasive blasting have been automated for some time. The use of magnetic drums for part removal and magnetic media[57] for use on nonmagnetic parts are particularly noteworthy approaches to minimizing part and media handling. Dreher and others[58,59,60] have recently presented additional detailed information on equipment trends and machines which are available.

In recent years, a number of innovations have been made in machine design (as opposed to handling equipment design). The dual shaft vibrators for faster deburring are one example. Improved tub liners, overhead drives for lower maintenance, and larger machines for faster deburring are other examples in vibratory deburring. The 160 ft^3 (4.53 m^3) vibratory finisher recently installed at Boeing[61] is one example of the trend toward larger equipment. An investigation of the number of companies producing deburring equipment (roughly 600) indicates that the field is definitely growing.

Despite the equipment innovations, much of American industry has been relatively uneducated about the equipment and knowledge which is available. This is particularly evident when individuals try to discuss and compare actual deburring capabilities of equipment and processes. Because of the growing necessity to improve productivity, to shorten process prove-in lead time, and to assure repeatable deburring, an awareness of process and equipment capability and availability is improving.

The trend in providing more meaningful production information can be attested by the increasing frequency of seminars in Russia, Germany, and the United States on the subject of deburring. The recent publication of two annotated bibliographies on burr formation, prevention, and removal[62,63] should also provide engineers and managers with immediately useful production ideas. In the near future, publications will also appear on the following topics:

- The Economics of Each Deburring Process,
- The Patents Which Have Been Issued on Deburring Process and Equipment,
- Available Equipment and Supplies for Deburring.

The recently established Society of Manufacturing Engineers Committee on Burr Technology has been established to help provide some of the information long needed by industry. The goals of this group include the following activities.

- Establish industry-wide acceptable definitions of burrs and related edge conditions.
- Provide more meaningful quantitative and qualitative information on each deburring process.
- Define appropriate measuring techniques and equipment for providing quantitative data on burrs.
- Present some typical in-plant standards used to specify allowable edge quality.
- Define at least one standard which can be used to compare the deburring efficiency of loose abrasive media.

One of the largest roadblocks to improving knowledge about burrs and deburring has been the lack of uniformity in defining a burr. Each of 132,000 manufacturing plants in the United States appears to have its own definition of a burr. To some a burr is loose material, easily removed. By this definition, however, stamping operations could never produce a burr. To some it is a sharp, ragged edge; and to others it is any material projecting past an edge. The confusion is not limited to inter-plant communication; large differences in definition can be found within a single plant. Inspection, production, and engineering departments weekly have to make decisions within their own plant about whether or not a certain metal projection is a burr. Then they have to decide what to do about it. With the current state of knowledge about how burrs form, it seems likely that some acceptable industry-wide definitions could be agreed upon within the next two years.

Defining the quantitative capabilities of each deburring process is a goal which is currently being pursued by a number of individuals. Many individuals, however, do not believe that the physical capabilities of the deburring processes can be defined quantitatively in a usable format. They believe that every part is unique or that irreconcilable differences exist among deburring machines. Despite the pessimism that exists, the capabilities of some processes are being defined. Several years may elapse before enough information is available to accurately predict the results of a process 98 percent of the time, but the data which will be available within the next two years will allow production facilities to screen out quickly the processes that are obviously unusable. Furthermore it will allow manufacturers of simple shapes to predict results accurately in 80 to 90 percent of the cases.

Defining the capabilities of a process does involve a number of variables. In addition to knowing how machine parameters influence aggressiveness, one has to know how burr size, hardness, and location affect results. One also has to note the effect of the machine parameters on edge radiusing, dimensional changes, and surface finish. Material and geometry influences must also be considered. Today's statistical approaches to testing and high-speed computers, however, can reduce test data of all these variables to meaningful production information.

Burr height has been used for many years as the measurement of burr size. As shown in previous chapters, however, burr removal time is a function of burr thickness and height. Although only a few individuals have utilized burr thickness results, it is apparent that a trend must develop in this direction if companies wish to minimize cut and try methods. The greatest roadblock to the incorporation of thickness measurements is the lack of convenient production instruments to measure both thickness and height.

One of the most time consuming and frustrating projects for many companies is developing in-plant standards describing allowable edge conditions. The SME committee hopes to publish, during the next two years, some typical standards used within companies. The purpose of this project is to minimize the continual reinvention of acceptable wording.

For twenty years, manufacturing engineers and purchasing agents have tried to determine which deburring media provided the most usable life for the least cost. Although at least two approaches now exist (one recently published[64]) for providing some indication of relative life, a national standard would provide a uniform approach. The SME committee intends to develop a workable standard to fill this need.

Within the past year, several companies have initiated in-plant manuals to define the characteristics, capabilities, and processing details of the deburring processes. This trend will continue for some time because no single publication is now available that provides such information in a format which can be used effectively by a major portion of industry. The major restriction to this approach at this time is the lack of factual information to put into such manuals. One manual author recently asked: "Where can I go to obtain unbiased information?"

Additional Areas of Interest

Several additional approaches have been proposed for providing production-oriented information on deburring. Some of these are highlighted in

Figure 21-5. While these are not yet implemented trends, they do represent a serious trend in thinking.

The case history approach has not yet been utilized as a method for communicating deburring information. This approach does have several distinct advantages over other approaches.

- Readers can identify with look-alike parts.
- It presents the rationale for choices as well as highlighting special problems and approaches which did not work.
- It results in user (as opposed to equipment vendor) input.
- It allows individual with a single unique observation to make an important contribution to industry.*

There is a distinct need for published shop-oriented "tricks of the trade." These brief examples could easily save industry hundreds of thousands of dollars a year. Those ideas could be published in the format currently used by American Machinist.

The case study and the "tricks of the trade" approach could provide a number of improvements for those utilizing hand deburring. Until a major study is made to define the best hand deburring techniques for various features and tolerance levels, these approaches offer the only hope of making significant industry-wide improvements.

Because of the wide range of part geometries requiring deburring, some individuals have suggested defining process capabilities in terms of special part geometries. As an example, individuals representing all deburring processes would be asked to compile a report on the capabilities and limitations of their process for gear deburring. Another report would be issued for drilled plates. This series of reports could also include reports on deburring of screw machine products and stamped parts. Deburring such specific types of features as intersecting holes, slots, threads, or blind features would also be appropriate report topics.

In a similar vein, a series of reports could be issued on specific deburring problems: removing the long roll-over burrs which typically are peened over in a vibratory process, removing recast material from EDM or laser machined features, and removing thick burrs from critically toleranced surfaces. Dulling edges without actually removing burrs is another problem faced by manufacturing engineers daily.

Two suggestions made at various times over the past five years are gradually being tried. The first is the use of uniform burr samples to compare the effectiveness on all deburring processes. The second is the use of an international standard for defining edge quality.

---

*The majority of technical papers are based on a large amount of experience and/or research. The case study approach allows individuals to describe notable innovations and observations in brief terms without undue concern about their abilities as writers and without worrying that they may not be experts in any field.

In comparing the deburring capabilities of processes, several individuals have suggested using a 0.5 inch (12.7 mm) cube as a specimen. Such a specimen has the advantage that all edges are exposed and that surface finish and size changes can be easily monitored. Preparing large numbers of any specimen becomes a problem when burr properties must be kept uniform. Burr properties typically vary widely on a single cut. Compounding this condition with the effects of tool wear necessitates the use of a statistical approach involving a large number of parts. One approach to minimize the number of specimens is to use a specimen with pseudo-burrs. Shim stock or thin coiled stock is spot welded or otherwise fused to the specimen in such a way that a predetermined "burr" height is provided. The thickness of the shim stock, which is closely controlled, represents burr thickness. This approach has been used for studies of the thermal energy method and vibratory deburring. It has also been used to establish inspection standards.

In an earlier chapter, Schafer described a system for defining the edge quality required on parts. The system he proposed can be used to define the edge radius required and the size and position of burrs which are allowable. Although other systems have been proposed, Schafer's system appears to be the most useful.

Schafer has also summarized a set of decision tables which are necessary to select the most economical deburring process (Figures 21-6 and 21-7).[66] While this approach has been implimented to provide computer selection of deburring processes, sufficient data is not available today to make this an accurate approach. More complete data is being developed however.

A critical annual review of equipment and process innovations is certainly in order. The majority of individuals responsible for deburring do not have the time to contact annually more than 600 manufacturers of deburring and deflashing equipment to keep abreast of the latest equipment innovations. Furthermore, it is often difficult to tell from a manufacturer's literature the significance of most innovations. An annual critical review published in one of the more popular trade journals would fill this need. Unfortunately, most magazine editors do not have sufficient knowledge of deburring to prepare a critical appraisal of such equipment innovations.

Because an increasing number of significant articles on deburring are published in languages other than English, an annual English review of these articles would be of use to equipment manufacturers and large companies. This would be of lesser benefit to small manufacturing facilities because they typically need more immediate application-oriented information and off-the-shelf equipment.

Future Needs

A considerable amount of research, both in the laboratory and in the production departments, remains to be done. In addition to the areas already mentioned, people must be identified who have some knowledge of use to industry (Table 21-1). It has been extremely difficult to identify individuals knowledgeable about any facet of deburring. At present, no good mechanism is available for soliciting case histories, tricks of the trade,

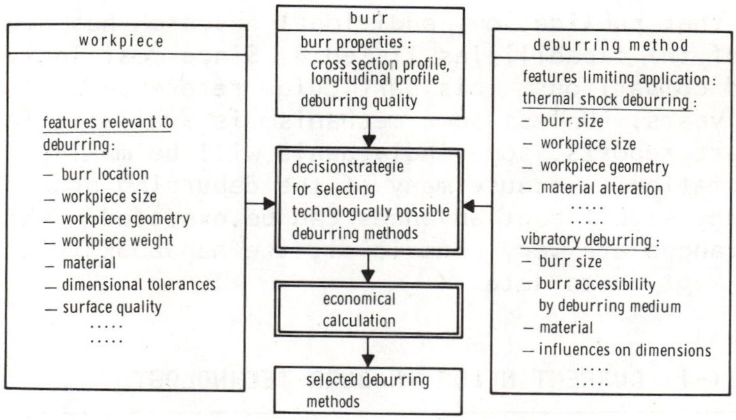

FIGURE 21-6: GENERAL DECISION NETWORK

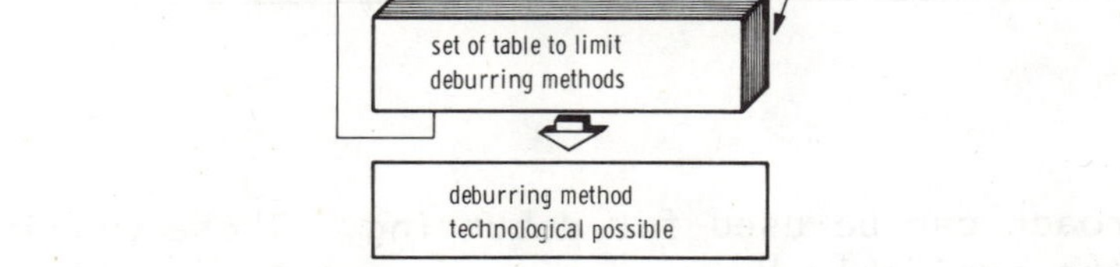

FIGURE 21-7: DECISION TABLES FOR DEBURRING

or results of research on burr prevention and removal. Because of the varied job responsibilities of these individuals and large differences in management structures, it is difficult to pinpoint the individual within a company having such knowledge.

Another problem for both the present and the future is keeping published information up-to-date. As an example, in 1974 this author coauthored a paper on the "State of the Art of General Deburring."[66] That paper which was the result of considerable empirical research combined with a review of 1500 publications on deburring contained two tables of quantitative information. Since that publication, additional research has been reported that improves some of the capabilities listed.* Since most individuals prefer to use tabled comparisons, this particular reference will probably be used for several years. Unless some mechanism is available for updating these state-of-the-art reports, some individuals will be making judgments on out-of-date information. Because many of the deburring processes are in their infancy, many significant advances can be expected. While annual reviews of these advances are very beneficial, the handbook type of information must also be kept up-to-date.

TABLE 21-1: CURRENT NEEDS IN BURR TECHNOLOGY

1.	Identify individuals who . Have knowledge of capabilities of a given mechanized process, . Have developed innovative approaches, . Have knowledge of convenient manual deburring techniques, . Are willing to help coordinate the information which is available, and . Are interested in identifying and coordinating needed research.
2.	Identify the format to be used for published deburring information
3.	Identify viable approaches that can be used to advise individuals in each plant of the current state-of-the-art.
4.	Identify approaches for obtaining better deburring information.

Conclusions

A rational approach can be used for deburring. Those willing to spend the time analyzing the available knowledge can generate significant cost savings within their plant. As indicated, a considerable amount of research is currently being performed and attempts are being made to compile the

*Particularly for Surftran's Thermal Energy Method of Deburring

available information into usable guides. One of the most significant of all trends is that, finally, individuals are beginning to quantify burr-related information.

The research reported in this book emphasizes that some companies have awakened to the fact that they can do something about the cost of burrs. The formation, properties, and removal of burrs can be treated as objectively and quantitatively as any manufacturing process. We now know what has been published on burrs and their removal. We now need only to treat burrs as the integral element of manufacturing they are. We must consider them with the same concern and with the same thoroughness we apply to other manufacturing processes. It is time we quit hiding our heads in the sand.

References

[1] C. S. Wukusik and R. S. Zeno, "Improving Punchability of Silicon Steel," Tool Engineer, Vol. 41, No. 12, December, 1958, pp. 63-70.

[2] "Tool Wear in the Cutting of Thin Gauge Steel Sheet," ASTEM Research Report No. 22, May 1, 1959 (translation and discussion of article by Otto and Werner Kienzle in Stahl und Eisen, Vol. 78, No. 12, 1958, pp. 820-828 (in German).)

[3] Otto Kienzle and Kurt Buchmann, "Beeinflussung und Prufung des Verschleisses und der Gratbildung beim Schneiden von Stahlfeinblech," Archiv fur das Eisenhuttenwesen, Vol. 34, No. 6, June, 1963, pp. 443-445 (in German).

[4] L. A. Holiga, "Punch-to-Die Clearance--Its Effect on Punched Hole Characteristics in Cold Rolled Steel," ASTME Paper 713, 1965.

[5] H. Seidenberg, "Schnittgrad und Messverfahren," Werkstatt und Betrieb, Vol. 100, No. 6, 1967, pp. 471-475 (in German).

[6] Harald Seidenberg, "Neues Verfahren zur Messung der Grathohe an Schnitteilen," Werkstatt und Betrieb, Vol. 100, No. 10, 1967, pp. 801-805 (in German).

[7] K. K. Wang, Khalil Taraman, and S. M. Wu, "An Analysis of Punching Variables by Two-Level Fractional Factorial Design," ASME Paper 69-WA/Prod-27, 1969.

[8] Harding R. Hugo, "How to Improve Metal Stamping Die Performance," Sheet Metal Industries, February, 1971, pp. 120-135.

[9] Hans Buhler and Fedor Pollmar, "Die Gratbildung beim Schneiden von Feinblech," Bander Bleche Rohre, Vol. 12, No. 3, 1971, pp. 105-111 (in German).

[10] "The Birth of a Hole," Brochure published by Dayton Progress Corporation, Dayton, Ohio, 1973.

[11]Leonard R. Allingham, "The Importance of Clearance in the Manufacture of Blanking Dies," ASTME Paper 622, 1964.

[12]H. Peter, "Der Schneidspalt des Schnittwerkzeuges," Werkstatt und Betrieb, Vol. 90, No. 1, 1957, pp. 89-92 (in German).

[13]Druckschrift F1260 uber den Obtischen Gratmesser der Optotechnischen Fabrik H. Schneider, Bad Kreuznach (in German).

[14]H. Seidenberg, "Presseneinwirkungen auf Werkzeugverschleiss und Grathohe beim Schneiden von Feinblech im gessch lossenen Schnitt," Dr.-ing. Dissertation, TH Hannover, 1965 (in German).

[15]E. Sondershaus, "Die Beurteilung der Schneidbarkeit von Feinblech durch ein Knabberverfahren," Diss. T. H. Darmstadt, 1966 (in German).

[16]H. Wentzel and H. Mehlhorn, "Erkenntnisse uber Gratbildung und Verschleiss von Schnittwerkzeugen," Fertig.-Techn. uber Betr., Vol. 14, 1964, pp. 236-240 (in German).

[17]H. Seidenberg, "Werkzeugverschleiss und Grathohe beim Schneiden von Feinblech," Stahl und Eisen, Vol. 86, 1966, pp. 1677-1688 (in German).

[18]E. Roeper, "Ursachen der Gratbildung und die Messung des Stanzgrates," Blech, Vol. 4, No. 4, 1957, p. 44 (in German).

[19]G. Oehler, "Zulassige Grathohe an Teilen aus Blech und Bandern," Blech, Vol. 15, 1968, pp. 462-463 (in German).

[20]Federico Strasser, "How Control of Burrs Aids Sheet Metal Stamping," The Iron Age, Vol. 185, No. 3, January 21, 1960, pp. 90-92.

[21]L. K. Gillespie, "An Overview of International Activities in Burr Prevention and Removal," SME Paper MRR75-02, 1975.

[22]"New Drill Guide Improves Efficiency in Drilling With Portable Tools," Cutting Tool Engineering, January/February, 1974, pp. 18-19.

[23]L. K. Gillespie, "The Burr--Manufacturing's Perennial Thorn," SME Paper MR74-992, 1974.

[24]F. Schafer, Entgraten, Krausskopf-Verlag, W. Germany, 1975 (in German).

[25]Y. Hagiuda, T. Naito, and M. Matsunaga, "Fundamental Researches on Vibratory Finishing," J. Metal Finishing Soc., Japan, Vol. 15, No. 5, 1964, pp. 180-187.

[26]M. Matsunaga, "Barrel Finishing in Japan," C.I.R.P. Annalen, Vol. 11, No. 4, 1964, pp. 221-224.

[27] M. Matsunaga and Y. Haguida, "Vibratory Finishing--Fundamental Research," Metal Finishing, Vol. 63, No. 9, September, 1965, pp. 52-57, No. 10, October, 1965, pp. 88-94, 97.

[28] M. Matsunaga and H. Kobayash, "Some Experiments on Centrifugal Barrel Finishing," Metal Finishing, Vol. 64, No. 5, 1966, pp. 57-60

[29] M. Matsunaga and Y. Hagiuda, "Researches on Barrel Finishing," Electroplating and Metal Finishing, Vol. 20, No. 3, 1967, pp. 75-80.

[30] M. Matsunaga, "Theory and Experiments on Centrifugal Barrel Finishing," The International Journal of Production Research, Vol. 5, No. 4, 1967, pp. 275-287.

[31] M. Matsunage, "Microcrystalline Abrasive Media for Mass Finishing," Metal Finishing, Vol. 69, No. 5, May, 1971, pp. 97-104.

[32] G. V. Robbins, Radius Generation in Vibratory Deburring, Bendix Corporation (Kansas City Division) Report, September, 1966 (available from NTIS under Accession No. N67-17437).

[33] L. K. Gillespie, Vibratory Deburring, Bendix Corporation (Kansas City Division) Report BDX-613-735, August, 1973 (available from NTIS).

[34] L. K. Gillespie, The Effect of Edge Angle on the Radius Produced by Vibratory Finishing, Bendix Corporation (Kansas City Division) Report BDX-613-1279, February, 1975.

[35] A. P. Babichev and V. I. Dyachenko, "Commercial Potentialities of Vibro-Tumbling," Stanki Instrum., No. 12, December, 1965, pp. 26-29 (in Russian).

[36] A. P. Babichev and V. I. Dyachenko, "Investigation of Polishing in Vibratory Drums," Russian Engineering Journal, Vol. 45, No. 6, 1965, pp. 41-44.

[37] M. E. Shainsii, et. al., "The Vibratory Cleaning and Polishing of Parts," Russian Engineering Journal, Vol. 45, No. 9, 1965, pp. 66-70.

[38] V. S. Bagmet, et. al., "Optimum Rates for Vibratory Deburring," Russian Engineering Journal, Vol 54, No. 2, 1974, pp. 72-73.

[39] B. N. Kartyshev, "Technical Methods for Bulk Vibratory Processing," Russian Engineering Journal, Vol. 49, No. 4, 1969.

[40] L. V. Veprintseva, "Service Properties of Various Fillers in Vibrational Machining of Metals," Nauch. tr. Vses. n.-i. i proekt, in-t tugoplavek. met. i tverd. splavov, No. 12, 1973, pp. 55-58 (in Russian).

[41] A. V. Veprintseva, et. al., Nauch. tr. Vses. n.-i. i proeki. in-t tugoplavek met. i tverd. splavov, No. 12, 1973, pp. 53-55 (in Russian).

[42]D. G. Beletskii and V. N. Zhemchugov, "Osnovnye Parametry Bezezhektsionnoi gidroabrazivnostruinoi obrabotki," Vestnik Machinostroeniya, No. 5, May, 1967, pp. 53-56 (in Russian).

[43]John R. Jones and Michael N. Gardos, "An Investigation of Abrasive Blasting," Lubrication Engineering, Vol. 27, November, 1971, pp. 393-400.

[44]L. K. Gillespie, Brush Deburring, Bendix Corporation (Kansas City Division) Report BDX-613-1056, March, 1974.

[45]L. K. Gillespie, "Brush Deburring of Precision Miniature Parts," SME Paper MR74-994, 1974.

[46]Frank Bigony, "Electrochemical Deburring of Aerospace Components," ASTME Paper MR69-413, 1969.

[47]Chemform Electrochemical Deburring Tool Design Manual, Brochure by Chemform Division of KMS Industries, Pompano Beach, Florida.

[48]Hans Wicht and Peter Kurze, "Entgraten--immer noch ein schwieriges Problem?" ZIS (Zentralinst Schweisstech) Mitteilungen, Vol. 15, No. 8, August, 1973, pp. 930-935 (in German).

[49]"Mechanization of the Process of Burr Removal," (Collection of Articles) Moskovskii dom Nauchnd-Teknche-Skoi Propagandy im. f.e dzerzhinskogo, 1966, (in Russian). (See Russian Engineering Journal, Vol. 52, No. 5, 1972, p. 43, also Vol. 54, No. 2, 1974, p. 71.)

[50]"Vibratory Machining of Machine and Instrument Components," (Collection of Papers) Rostov-on-Don, Institute of Agricultural Engineering, 1972, 275 pp. (in Russian). (See Referativnyi Zhurnal-Tekhnologiia Mashinostroeniia, 1973, 3B 464-469; 4B 253.)

[51]D. A. Raiskii and M. I. Zhuravlev, "Better Productivity in Vibroabrasive Machining," Russian Engineering Journal, Vol. 53, No. 11, 1973, pp. 44-45.

[52]M. E. Shainskii, "Centrifugal-Type Deburring Ring," Russian Engineering Journal, Vol. 53, No. 9, 1973, pp. 46-47.

[53]U.S. Patent 3,701,725, January 21, 1970, September 22, 1971.

[54]U.S. Patent 3,703,452, January 29, 1970, January 27, 1972.

[55]G. H. Goreunov and I. Yu. Rezentsveig, "Abrasive Machining," Russian Patent 374,154, May 13, 1969.

[56]V. A. Morozov, "Changes in the Physicomechanical Properties of Component Surfaces in Magnetic-Vibratory Machining," Vibratory Machining of Machine and Instrument Components, Rostov-on-Don, Institute of Agricultural Engineering, 1972, pp. 23-29 (in Russian). (See Referativnyi Zhurnal-Tekhnologila Mashinostroeniia, 1973, 4B 405.)

[57]U.S. Patent 3,702,758, September 10, 1970.

[58]"Ice-Bonded Vibratory Finishing Media," Product Finishing, Vol. 37, No. 4, January, 1973, pp. 120-122.

[59]M. Dreher, "Advances and Development Trends in Surface Treatment by Means of Barrels, Vibrators and Centrifugal Machines," Oberflach Entechnik (Coburg), Vol. 50, No. 11, 1973, pp. 517-518, 520, 522 (in German).

[60]"Barrel Finishing Equipment," Metal Finishing Plant and Processes, Vol. 9, No. 1, 1973, pp. 21-30.

[61]"Deburring Large Airplane Parts," American Machinist, July 1, 1975, pp. 25, 27.

[62]L. K. Gillespie, "Deburring: An Annotated Bibliography," SME Bibliography Series, 1974.

[63]L. K. Gillespie, "Deburring: An Annotated Bibliography--Vol. II," SME Report, MRR75-04.

[64]"A Comparative Media Efficiency System," Precision Metal, May, 1974, pp. 32-34.

[65]F. Schafer, "Deburring Processes in Perspective," SME Paper MR75-482, 1975.

[66]N. S. Ajlouny and L. K. Gillespie, "State of the Art of General Deburring Methods," SME Paper MR74-116, 1974.

